The Microbiome and Cancer

Nosheen Masood • Azra Yasmin
Editors

The Microbiome and Cancer

Understanding the Role of Microorganisms in Tumor Development and Treatment

Springer

Editors
Nosheen Masood
Deprtment of Biotechnology
Fatima Jinnah Women University
Rawalpindi, Pakistan

Azra Yasmin
Department of Biotechnology
Fatima Jinnah Women University
Rawalpindi, Pakistan

ISBN 978-981-97-6947-6 ISBN 978-981-97-6948-3 (eBook)
https://doi.org/10.1007/978-981-97-6948-3

This Springer imprint is published by the registered company Springer Nature Singapore Pte Ltd.
The registered company address is: 152 Beach Road, #21-01/04 Gateway East, Singapore 189721, Singapore

This book is dedicated to the radiant souls fighting cancer, who illuminate the path of courage amidst the shadows of uncertainty. To those valiant warriors battling against the tide of adversity, your resilience knows no bounds. You are the indomitable champions, the true beacons of hope in the face of tribulation. May your unwavering strength continue to inspire and uplift, lighting the way for countless others on their journey of survival.

Foreword

In the intricate tapestry of life, where every thread weaves its own story, there exists a profound interplay between the microscopic world and the realm of human health. This book explores the relationship between cancers and microorganisms.

Commencing with a compelling introduction, the foundation is laid upon which the basis of knowledge is built. From there, a venture is made into the captivating realm of the human microbiome—a vibrant ecosystem teeming with myriad life forms, each playing a pivotal role in physiological equilibrium. Delving deeper, the intricate dance between bacteria and cancer is encountered—a symbiosis fraught with both peril and possibility.

The narrative unfolds to reveal the ominous specter of viral infections, weaving a complex web that ensnares the unsuspecting host in the throes of malignancy. From hepatitis C virus in hepatocellular carcinoma to the presence of Epstein–Barr virus in cancers of various origins, a witness is borne to the intricate tapestry of causality that binds microorganisms to the genesis of cancer. Navigating through the annals of medical history, the tragic tale of Kaposi's sarcoma is uncovered, a poignant reminder of the havoc wrought by herpesviruses. The symbiotic relationship between fungal infections and cancer is confronted, unraveling the myths and realities surrounding the role of Aspergillus in lung cancer. But the exploration does not cease with bacteria, viruses, and fungi alone. It peers into the twilight realm of protists, unearthing their cryptic connection to the genesis of cancer—a revelation both profound and paradigm-shifting. Finally, attention is turned toward the horizon of cancer management and therapy, where the intricate interplay between microorganisms and modern medicine heralds a new era of innovation and progress. In traversing the vast expanse of knowledge contained within these pages, one is reminded of the indomitable spirit of human inquiry—an insatiable thirst for understanding that transcends the boundaries of time and space.

May this book serve as a guiding light, illuminating the path toward a brighter future—a future where the shadows of cancer are banished, and the promise of healing shines bright.

Department of Psychiatry
State University of New York at Buffalo, Jacobs School of Medicine and Biomedical Sciences
Buffalo, NY, USA

Zehra Agha

Preface

Cancer, one of the most complex and devastating diseases of our time, has long been studied through the lens of genetics, lifestyle, and environmental factors. However, the last few decades have witnessed an expanding horizon of research into the role of microorganisms in the onset, progression, and treatment of various cancers. This book seeks to explore this critical and often underappreciated intersection between microbiology and oncology, shedding light on how bacteria, viruses, and even protists influence cancer development and management.

In recent years, the discovery of the human microbiome has revolutionized our understanding of the delicate balance between health and disease. The microorganisms that inhabit our bodies not only participate in vital functions but also can trigger or suppress pathological conditions. The chapters of this book dive deeply into these relationships, addressing specific microorganisms implicated in cancer and the ways they interact with human cells.

The opening chapter provides an introduction to this intricate subject, setting the stage for the exploration of the human microbiome in Chap. 2. From there, the focus shifts to bacteria (Chap. 3), viruses (Chaps. 4–8), and fungi (Chap. 9), each of which has been associated with specific types of cancers. Special attention is given to well-known viral agents such as the hepatitis C virus (HCV), Epstein–Barr virus (EBV), and human papillomavirus (HPV), all of which have been shown to play major roles in the development of cancers such as hepatocellular carcinoma, nasopharyngeal carcinoma, and urogenital cancers. Protists, another group of microorganisms, are explored in Chap. 11, highlighting how these less-studied organisms might influence oncogenesis.

The final chapters focus on how these discoveries influence cancer management and treatment. As cancer therapies evolve, an understanding of the microbial component of cancer offers novel avenues for prevention, diagnosis, and treatment.

By examining the complex interplay between microorganisms and cancer, this book aims to provide a comprehensive overview of the current knowledge while inspiring further research into how the microscopic world within us shapes our health and disease outcomes. Whether you are a medical professional, researcher, or student, this book will offer a unique perspective on cancer biology and the potential that lies in targeting the microbial world for therapeutic advances.

Rawalpindi, Pakistan — Nosheen Masood
Rawalpindi, Pakistan — Azra Yasmin

Acknowledgments

The editors would like to extend their sincerest gratitude to all the contributing authors whose time and insights have been invaluable in bringing this book to fruition. Their contributions have been instrumental in shaping this work into a refined piece of art. Additionally, the editors wish to express their appreciation to all individuals involved in the publishing process, whose efforts have been essential in bringing this project to completion.

Contents

Editors and Contributors

About the Editors

Nosheen Masood is currently working as Head, Department of Biotechnology, Fatima Jinnah Women University, Rawalpindi, Pakistan. Academically, she is an assistant professor in the mentioned department. She completed her PhD in Human Molecular Genetics at the Department of Biosciences at COMSATS University, Islamabad, Pakistan. Her main research interests include molecular biology and cancer genetics, focusing on epidemiological studies, risk assessment, genomic analysis, immunohistochemistry-based diagnostics, and protein–protein interactions. She has received various prestigious awards, notably the Research Productivity Award 2014–2015 from the Pakistan Council of Science and Technology (PCST) in Category D and the Research Productivity Awards from COMSATS Institute of Information Technology for 6 consecutive years from 2010 to 2015. She has served as a reviewer for several international journals and has teaching experience in biotechnology, molecular biology, comparative genomics, environmental biology, regulation of gene expression, and molecular evolution. Dr. Masood has published several research articles in esteemed, peer-reviewed international journals and has authored or coauthored various book chapters.

Azra Yasmin is a distinguished academic at Fatima Jinnah Women University in Rawalpindi, Pakistan. She plays a pivotal role in the university's Department of Biotechnology and Department of Bioinformatics, where her expertise and dedication have made a significant impact on both the academic and research realms. Her research is primarily focused on microbiology and biotechnology. She has published numerous research papers in reputable international journals, underscoring her prominence as a scholar. Throughout her tenure at FJWU, Dr. Yasmin has been instrumental in fostering a robust academic environment. She has mentored countless students, guiding them through complex scientific inquiries and encouraging them to pursue excellence in their studies. Her innovative teaching methods and comprehensive approach to education have earned her the respect and admiration of both colleagues and students. Beyond her academic responsibilities, Dr. Yasmin actively participates in various environmental advocacy initiatives. She collaborates with governmental and nongovernmental organizations, both national and international, to promote sustainable practices and environmental conservation in Pakistan. Her contributions to microbial biotechnology and research reflect her dedication to making a positive impact on society.

Contributors

Sidra Abbas Fatima Jinnah Women University, Rawalpindi, Pakistan

Bashir Ahmad Department of Biological Sciences, International Islamic University, Islamabad, Pakistan

Neelam Batool Department of Biochemistry, Quaid-i-Azam University, Islamabad, Pakistan

Monaza Bibi Microbiology and Biotechnology Research Lab, Department of Biotechnology, Fatima Jinnah Women University, Rawalpindi, Pakistan

Nusrat Bibi Microbiology and Biotechnology Research Lab, Department of Biotechnology, Fatima Jinnah Women University, Rawalpindi, Pakistan

Hira Gul Microbiology and Biotechnology Research Lab, Department of Biotechnology, Fatima Jinnah Women University, Rawalpindi, Pakistan

Fatima Ijaz Department of Biochemistry, Quaid-i-Azam University, Islamabad, Pakistan

Muhammad Imran Department of Biological Sciences, International Islamic University, Islamabad, Pakistan

Farhat Jabeen Department of Zoology, Government College University Faisalabad, Faisalabad, Pakistan

Sara Latif Department of Biochemistry, Quaid-i-Azam University, Islamabad, Pakistan

Muhammad Suleman Malik Department of Biochemistry, Quaid-i-Azam University, Islamabad, Pakistan

Parvez Azam Malik Department of Pathology, Rai Medical College, Sargodha, Pakistan

Nosheen Masood Department of Biotechnology, Fatima Jinnah Women University, Rawalpindi, Pakistan

Ab Qayoom Naik Laboratory of Endocrinology, Department of Biosciences, Barkatullah University, Bhopal, MP, India

Hira Pervaiz Microbiology and Biotechnology Research Lab, Department of Biotechnology, Fatima Jinnah Women University, Rawalpindi, Pakistan

Maryam Pervaiz Microbiology and Biotechnology Research Lab, Department of Biotechnology, Fatima Jinnah Women University, Rawalpindi, Pakistan

Shagufta Perveen Fatima Jinnah Women University, Rawalpindi, Pakistan

Kiran Saba Department of Biochemistry, Faculty of Life Sciences, Shaheed Benazir Bhutto Women University, Peshawar, Pakistan

Aiman Sajjad Department of Zoology, Government College University Faisalabad, Faisalabad, Pakistan

Basheerulla Shaik National Institute of Technical Teachers Training and Research (NITTTR), Bhopal, MP, India

Vinoy K. Shrivastava Laboratory of Endocrinology, Department of Biosciences, Barkatullah University, Bhopal, MP, India

Fizza Ahmad Tariq Department of Biochemistry, Quaid-i-Azam University, Islamabad, Pakistan

Sumeyye Ucar Faculty of Medicine, Department of Anatomy, Erciyes University, Melikgazi/Kayseri, Turkey

Mohammad Tahir Waheed Department of Biochemistry, Quaid-i-Azam University, Islamabad, Pakistan

Azra Yasmin Department of Biotechnology, Fatima Jinnah Women University, Rawalpindi, Pakistan

Saba Zafar Department of Zoology, Government College University Faisalabad, Faisalabad, Pakistan

Tabassum Zafar Laboratory of Endocrinology, Department of Biosciences, Barkatullah University, Bhopal, MP, India

The Microbiome and Cancer: Understanding the Role of Microorganisms in Tumor Development and Treatment

1

Ab Qayoom Naik, Tabassum Zafar, Vinoy K. Shrivastava, Basheerulla Shaik, and Sumeyye Ucar

Abstract

The complex interactions between the human microbiome and cancer have become a central focus in modern biomedical research, transforming our understanding of carcinogenesis and available treatment options. This abstract explores the complex interactions between microbes and cancer, explaining how important a role they play in the growth of tumors and the development of therapeutic approaches. The microbiome of humans is a heterogeneous community of bacteria, viruses, fungi, and other microorganisms that live in different bodily niches and have a substantial impact on host physiology, immunological responses, and general health. The influence of microbiota on cancer onset, progression, and therapy response has been highlighted by recent investigations. Investigating the mechanics behind these occurrences, scientists have discovered microbial fingerprints linked to particular cancer types, emphasizing possible biomarkers for diagnosis. Furthermore, complex interactions between the immune system and the microbiome highlight the microbiota's modulatory function in immune responses against tumors. Comprehending these dynamics opens up new directions for cancer treatments, and probiotics and fecal microbiota transplantation are examples of microbial interventions that may be used in addition to traditional treatments. This abstract summarizes the state of the art and highlights the

A. Q. Naik (✉) · T. Zafar · V. K. Shrivastava
Laboratory of Endocrinology, Department of Biosciences, Barkatullah University, Bhopal, MP, India

B. Shaik
National Institute of Technical Teachers Training and Research (NITTTR), Bhopal, MP, India

S. Ucar
Faculty of Medicine, Department of Anatomy, Erciyes University, Melikgazi/Kayseri, Turkey

N. Masood, A. Yasmin (eds.), *The Microbiome and Cancer*,
https://doi.org/10.1007/978-981-97-6948-3_1

need to fully understand the microbiome's complex role in cancer in order to develop tailored, microbiome-targeted cancer treatments.

Keywords

Microbiome · Tumor development · Dysbiosis · Immune modulation · Prebiotics · Probiotics · Immunotherapy

1.1 Introduction

Microbiota is an essential component of the human environment and health. The study of various microorganisms for their significance and influence has resulted from their close relationship and presence within the human body (Altveş et al., 2020). As a result, the oral, gut, respiratory, and skin microbiota represents the different localizations of bacteria surrounding humans (Hou et al., 2022). In symbiosis with the host, the microbial populations maintain homeostasis and control immunological response (Zheng et al., 2020). On the other hand, dysbiosis of the microbiota can result in diseases such as cancer, respiratory disorders, cardiovascular diseases (CVDs), and deregulation of body systems (Carding et al., 2015; Sing et al., 2023).

The connection between cancer and microorganisms is intricate. While cancer is generally perceived as a result of host genetics and environmental influences, approximately 20% of human malignancies are associated with microorganisms (de Martel et al., 2012). Microbes located at mucosal sites can integrate into the tumor microenvironment of aerodigestive tract malignancies, and intratumoral microbes can impact cancer progression and dissemination through various mechanisms (Louis et al., 2014; Irrazábal et al., 2014). Conversely, the gut microbiota plays a role in detoxifying dietary components, mitigating inflammation, and maintaining equilibrium in host cell growth and proliferation (Jandhyala et al., 2015). The prospect of employing microbe-based approaches in cancer therapy has captured attention for over a century, ranging from early forms like Coley's toxins (considered one of the earliest cancer bacteriotherapies) to the current era's advancements in synthetic biology, including designer microbes and microbiota transplants (Felgner et al., 2016; Ikryannikova et al., 2023). Microbes and the microbiota play significant roles in carcinogenesis (Fig. 1.1), impacting a host's risk through three main avenues: manipulating the balance of host cell proliferation and death, modulating immune system function, and influencing the metabolism of host-produced factors, ingested foodstuffs, and pharmaceuticals (Bhatt et al., 2017; Belkaid & Hand, 2014).

Overview of the Human Microbiome The microbiome refers to the complex community of microorganisms, including bacteria, viruses, fungi, and archaea, that inhabit a particular environment. In the context of humans, the human microbiome encompasses the trillions of microorganisms residing in and on the body, with a significant concentration in the gut, skin, mouth, and other mucosal surfaces (Fig. 1.2). This intricate ecosystem plays a crucial role in maintaining various physiological functions and influencing human health. The diversity and composi-

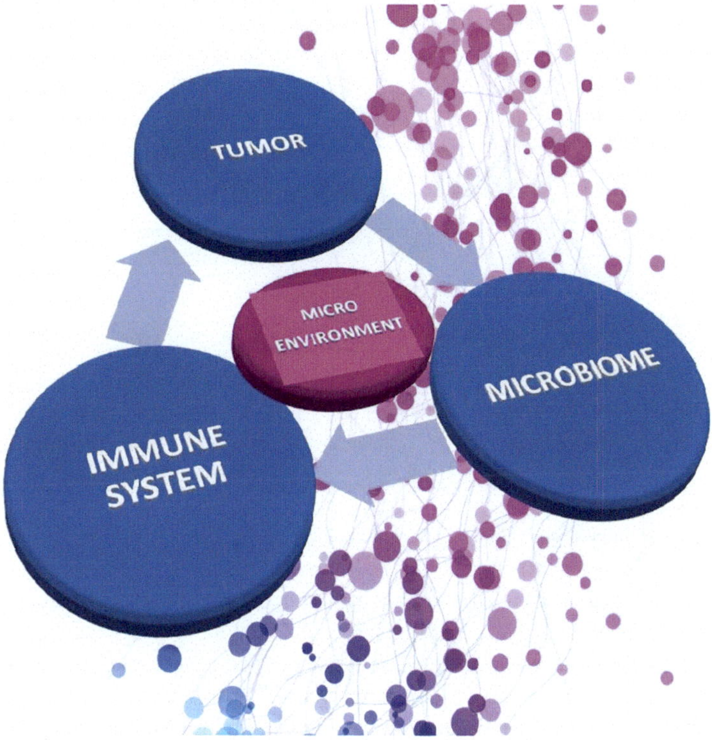

Fig. 1.1 Microbiome and its immune system modulation and tumor microenvironment

tion of the microbiome can vary significantly between individuals, influenced by factors such as genetics, diet, environment, and lifestyle (Sender et al., 2016).

Microorganisms in the human microbiome are not merely passive inhabitants; they actively participate in processes crucial for human well-being. They contribute to digestion, nutrient absorption, and the synthesis of essential vitamins. Additionally, the microbiome plays a pivotal role in regulating the immune system and protecting against pathogenic invaders. The symbiotic relationship between the host and its microbiome is dynamic, with a delicate balance that, when disrupted, can lead to various health issues (Thursby & Juge, 2017).

The human microbiome is a vast and diverse ecosystem, with the gut microbiota being one of the most extensively studied components. Comprising trillions of microorganisms, including bacteria, viruses, and fungi, the gut microbiome exerts a profound influence on human physiology. Recent advances in sequencing technologies have allowed scientists to delve deeper into understanding the composition and function of these microbial communities (Lozupone et al., 2012). The gut microbiome, for example, is involved in fermenting undigested carbohydrates, producing short-chain fatty acids, and modulating the inflammatory response. Beyond the gut, the skin microbiome contributes to the defense against pathogens and helps

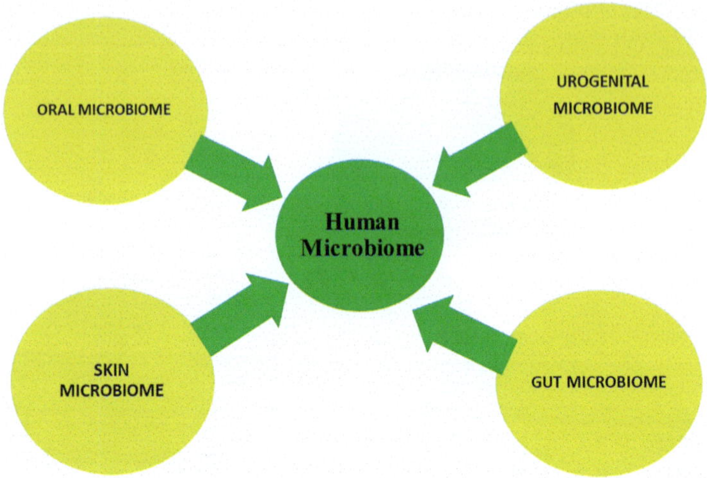

Fig. 1.2 Human microbiome, diversity, and interaction

maintain skin homeostasis. The oral microbiome, on the other hand, influences oral health and is implicated in conditions such as periodontal disease (Byrd et al., 2018). Breakthroughs in high-throughput sequencing technologies have transformed the exploration of the human microbiome, facilitating the recognition and description of varied microbial communities. These innovations have granted a deeper understanding of the microbiome's functional capabilities, enabling researchers to comprehend its involvement in processes such as nutrient metabolism, energy extraction, and modulation of the immune system (Shreiner et al., 2015).

While the microbiome is predominantly associated with health, dysbiosis—imbalances in microbial communities—has been linked to various diseases, including inflammatory bowel diseases, obesity, and even mental health disorders (Cryan et al., 2019).

Microbiome and Human Health The human microbiome, consisting of trillions of microorganisms such as bacteria, viruses, fungi, and archaea that inhabit various body surfaces, plays a crucial role in maintaining human health (Ogunrinola et al., 2020). This complex ecosystem is particularly concentrated in the gastrointestinal tract, where the gut microbiota exerts significant influence on digestion, nutrient absorption, and immune system modulation (Valdes et al., 2018). Numerous studies have highlighted the connection between the microbiome and human health, emphasizing its impact on various physiological processes. For instance, the gut microbiota contributes to the synthesis of certain vitamins, such as B vitamins and vitamin K, and aids in the breakdown of complex carbohydrates that are otherwise indigestible by human enzymes (Rowland et al., 2018; Vernocchi et al., 2020; Wan et al., 2022). A seminal study underscored the importance of microbial diversity in the gut, suggesting that alterations in the composition of the microbiome may be linked to various health conditions, including metabolic disorders, autoimmune diseases, and mental health issues (Sender et al., 2016).

Recognition of the microbiome significance in maintaining human health is increasing. Research has linked alterations in the microbiome to various health conditions, including metabolic disorders, autoimmune diseases, and mental health issues. For instance, there's a known connection between dysbiosis, a disruption in the microbial balance, and inflammatory bowel diseases such as Crohn's disease and ulcerative colitis (Bull & Plummer, 2014; Durack & Lynch, 2019). Moreover, research has shown that the microbiome plays a pivotal role in modulating the immune system, influencing its development and response to infections. The gut microbiota interacts with immune cells, contributing to the maintenance of a balanced immune system and preventing inappropriate inflammatory responses (Lazar et al., 2018; Yoo et al., 2020). This interaction is critical for protecting against pathogens and maintaining immune homeostasis. In a study, the intricate relationship between the microbiome and the immune system, emphasizing how dysbiosis, or microbial imbalance, can lead to immune dysregulation and contribute to the pathogenesis of various diseases has been extensively investigated (Campbell et al., 2023; Zhao et al., 2023).

The microbiome plays a vital role in educating and regulating the immune system. It plays a part in fostering immune tolerance, thwarting undesired immune reactions toward the body's tissues. Additionally, the microbiome has an impact on the processing of drugs and toxins, affecting both the effectiveness and safety of medications (Pryor et al., 2020; Ji et al., 2023). Furthermore, the microbiome has been implicated in the regulation of mental health and neurological function through the gut-brain axis (Sasso et al., 2023). Microbial metabolites, such as short-chain fatty acids produced by gut bacteria, can influence neural function and have been linked to conditions such as depression and anxiety (Carabotti et al., 2015; Appleton, 2018). The communication between the microbiota and the brain occurs through various pathways, such as the immune system, tryptophan metabolism, the vagus nerve, and the enteric nervous system. Factors like infection, birth delivery mode, antibiotics, nutrition, environmental stressors, and genetics can influence microbiota composition in early life, while aging and stress impact microbial diversity later in life. The gut microbiota is implicated in various conditions like autism, anxiety, obesity, schizophrenia, and Parkinson's and Alzheimer's diseases, with animal models linking neural processes to microbiome activation. Ongoing translational human studies are expected to further advance the understanding of the microbiota-gut-brain axis (Cryan et al., 2019).

Microbiome and Cancer In recent years, an emerging area of research has focused on understanding the intricate relationship between the microbiome and cancer. The human microbiome influences various aspects of cancer development, progression, and treatment response (Bhatt et al., 2017). The gut microbiota, in particular, has been implicated in modulating the efficacy and toxicity of cancer therapies such as immunotherapy. Studies have demonstrated that alterations in the gut microbiome composition can impact the host's susceptibility to certain types of cancers (Shi et al., 2023). For example, changes in the gut microbiome have been associated with an increased risk of colorectal cancer. Additionally, the microbiome can affect the metabolism of dietary components, producing metabolites that may influence cancer risk (Kandalai et al., 2023).

Understanding the intricate interplay between the microbiome and cancer holds promise for developing novel therapeutic strategies. Manipulating the microbiome, either through dietary interventions, probiotics, or fecal microbiota transplantation, is being explored as a potential avenue for enhancing cancer treatment outcomes (Soto Chervin & Gajewski, 2020). Therefore, the microbiome is a dynamic and integral component of human biology, influencing various aspects of health and disease. Investigating the link between the microbiome and cancer opens new avenues for personalized medicine and innovative approaches to cancer treatment and prevention (Sadrekarimi et al., 2022).

Microbial Diversity in Cancer Patients The human microbiome, comprising trillions of microorganisms residing in and on the human body, plays a crucial role in maintaining health and influencing various physiological processes. Recent research has increasingly focused on understanding the intricate relationship between the microbiome and tumor development. The genesis and management of tumors are intimately linked to commensal bacteria and other microorganisms that live in the human body (Sadrekarimi et al., 2022). Tumor microbiome (TM) has recently been linked to several malignancies, including lung, breast, and pancreatic cancers. The composition of TM varies among tumors, as do its effects on them. Tumor microenvironment creation, local immune modulation, tumor cell biology change, and the effectiveness of tumor treatment with drugs are all impacted by TM. Anticancer microbial drugs (GEN-001) and engineered tumor-targeting bacteria play a significant role in the therapy of cancers, and TM is anticipated to be a tumor biomarker (Chen et al., 2022).

Microbial Imbalance in Cancer Several studies have demonstrated alterations in the microbial composition of individuals with different types of cancer. In a study it was observed that the gut microbiome of melanoma patients exhibited significant differences compared to healthy individuals. The dysregulation of microbial communities in cancer patients has been associated with inflammation, immune dysfunction, and the promotion of tumor growth (Gopalakrishnan et al., 2018a, b).

Gut Microbiota and Cancer Risk Research has shown that the gut microbiota, in particular, plays a crucial role in modulating the risk of developing certain cancers. A study found that alterations in the gut microbiome can influence the host's susceptibility to colorectal cancer. The balance between beneficial and harmful bacteria in the gut may impact the local immune response and contribute to the initiation and progression of tumors (Garrett, 2015).

Microbial Signatures as Diagnostic Biomarkers Advancements in sequencing technologies have allowed for the identification of microbial signatures associated with specific cancer types (Wang et al., 2023). By analyzing the composition of the microbiome, researchers have explored the potential of using microbial biomarkers for early cancer detection. A study identified distinct microbial signatures in the oral microbiome of patients with oral squamous cell carcinoma, suggesting the potential for noninvasive diagnostic tools (Yu et al., 2023).

Role of Microbiome in Cancer Initiation The human microbiome, composed of trillions of microorganisms residing in and on the body, plays a crucial role in maintaining health and influencing various physiological processes. Emerging research has shed light on the intricate connection between the microbiome and cancer initiation.

Inflammation and Tumor Microenvironment (Inflammatory Pathways) Chronic inflammation has been implicated in the initiation and progression of various cancers. Inflammatory responses triggered by microbial dysbiosis can lead to genomic instability and DNA damage and ultimately contribute to oncogenic transformations. Key inflammatory pathways involved include NF-κB and STAT3, which are activated by microbial products, leading to the release of pro-inflammatory cytokines (Schwabe & Jobin, 2013).

Immune Cell Infiltration The microbiome modulates the tumor microenvironment by influencing immune cell infiltration. Commensal bacteria can either enhance or suppress immune responses, impacting the balance between pro-inflammatory and anti-inflammatory signals within the tumor microenvironment. This intricate interplay can influence tumor growth and the effectiveness of immunotherapies (Grivennikov et al., 2010).

Microbial Dysbiosis and Inflammation-Related Cancers Several studies have identified associations between specific microbial dysbiosis patterns and increased risk of inflammation-related cancers, such as colorectal cancer. Understanding the molecular mechanisms linking the microbiome, inflammation, and tumor microenvironment is crucial for developing targeted therapeutic strategies (Gagnière et al., 2016).

Role of the Microbiome in Cancer Initiation Continuing our exploration of the microbiome's role in cancer initiation, this page delves into the impact of microbial metabolites on cancer development. Understanding how microbial metabolites influence cellular processes can provide insights into potential therapeutic interventions and preventive strategies.

1.2 Microbial Metabolites and their Impact

Short-Chain Fatty Acids (SCFAs) Certain commensal bacteria produce SCFAs, such as butyrate, propionate, and acetate, through the fermentation of dietary fibers (Fig. 1.3). These metabolites have been shown to exhibit anti-inflammatory and antitumor properties, promoting a healthy gut environment. Conversely, alterations in SCFA production may contribute to a pro-tumorigenic milieu (Louis et al., 2014). Some microbes release toxins that can directly damage host cells and tissues, promoting an environment conducive to cancer initiation. For instance, the genotoxic effects of certain bacterial toxins can lead to DNA damage and the initiation of carcinogenesis (Arthur et al., 2012). Microbial metabolism can influence hormonal balance within the host. Dysregulation of sex hormones, such as estrogen, has been linked to the development of hormone-dependent cancers. Microbial modulation of hormone levels may contribute to the initiation of such cancers (Plottel & Blaser, 2011).

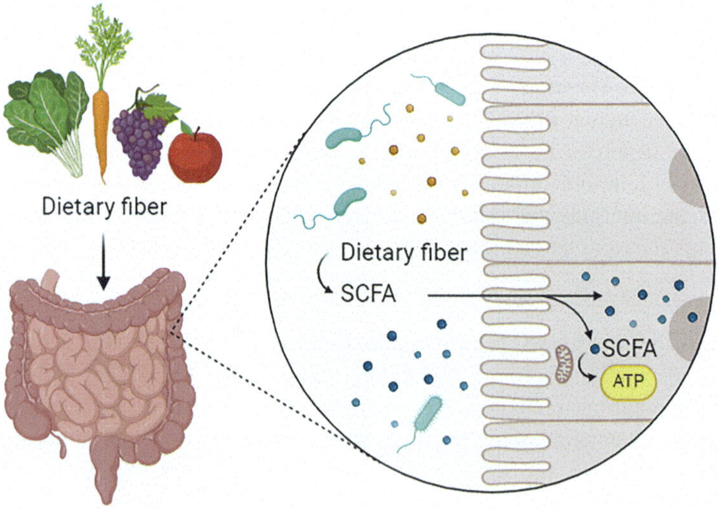

Fig. 1.3 Dietary fibers contribute short-chain fatty acid (SCFA) metabolites that exhibit anti-inflammatory and antitumor properties

Role of the Microbiome in Cancer Initiation The final page of our exploration into the role of the microbiome in cancer initiation focuses on the cross talk between inflammation, microbial metabolites, and their impact on cellular signaling pathways. A comprehensive understanding of these interactions is crucial for developing targeted interventions and personalized cancer prevention strategies.

Cross Talk between Inflammation and Metabolites (Signaling Pathways) Microbial metabolites can modulate various signaling pathways within host cells, influencing processes such as cell proliferation, apoptosis, and DNA repair. Understanding these signaling cascades is essential for unraveling the molecular mechanisms underlying cancer initiation (Viaud et al., 2013).

Epigenetic Modifications The microbiome can induce epigenetic changes in host cells, affecting gene expression patterns linked to cancer initiation. This page explores the role of microbial metabolites in epigenetic modifications and their contribution to the development of cancer (Belkaid & Hand, 2014). Insights gained from understanding the microbiome's role in cancer initiation open avenues for novel therapeutic approaches. Modulating the microbiome or targeting specific microbial metabolites may offer innovative strategies for cancer prevention and treatment (Paul et al., 2015).

Changes in the Microbiome Composition Microbial dysbiosis refers to an imbalance or disruption in the normal microbial community residing in and on the human body. This phenomenon has gained significant attention due to its potential association with various health conditions, including cancer. In this section, we explore the intricate changes that occur in the microbiome composition during microbial dysbiosis, specifically focusing on its role in cancer development. Research has shown that alterations in the gut microbiome are particularly relevant to the development of cancer. Studies have highlighted shifts in the abundance and diversity of certain bacterial species in the gut that may promote an environment conducive to carcinogenesis. These changes involve an overgrowth of potentially harmful bacteria and a decrease in beneficial commensals, leading to an imbalance that may contribute to the initiation and progression of cancer (Gagnière et al., 2016; Thomas et al., 2019).

Furthermore, dysbiosis-induced changes in the oral and vaginal microbiomes have also been implicated in the development of cancer at various sites, including the oral cavity and cervix. For instance, some studies have reported alterations in the oral and vaginal microbiomes, respectively, with an associated increased risk of cancer (Schmidt et al., 2018; Petrova & Wu, 2017). Understanding the specific alterations in the microbiome composition provides valuable insights into the potential mechanisms through which dysbiosis may contribute to cancer. This knowledge is essential for developing targeted interventions aimed at restoring microbial balance and preventing or treating cancer linked to microbial dysbiosis.

Impact on Host Immune Response The relationship between microbial dysbiosis and cancer extends beyond changes in microbial composition to the profound impact on the host's immune response. The immune system plays a pivotal role in surveilling and eliminating aberrant cells, including those undergoing malignant transformation. Dysbiosis-induced alterations in the microbiome can influence the host immune response in ways that either promote or inhibit cancer development. Several studies have reported that dysbiosis can trigger chronic inflammation, a known driver of cancer progression. Dysregulated immune responses, characterized by elevated levels of pro-inflammatory cytokines and immune cell infiltration, create a microenvironment conducive to tumor growth (Yu et al., 2017; Gopalakrishnan et al., 2018a, b). Additionally, dysbiosis-induced changes in the gut microbiome can compromise the integrity of the gut barrier, allowing the translocation of microbial products that activate inflammatory pathways systemically.

Conversely, dysbiosis may also lead to immune suppression, impairing the host's ability to mount an effective antitumor immune response. There are some research reports that have demonstrated that alterations in the gut microbiome can negatively impact the efficacy of immune checkpoint inhibitors, a common immunotherapeutic approach for cancer treatment (Viaud et al., 2013; Vetizou et al., 2015). Understanding the intricate interplay between dysbiosis and the host immune response is crucial for developing strategies to modulate the microbiome and enhance antitumor immunity. This knowledge may pave the way for novel immunotherapeutic interventions that target the microbiome to improve treatment outcomes in cancer patients. The interplay between microbes and the immune system is a complex and dynamic relationship that significantly impacts the development and progression of cancer. Microbes, particularly those residing in the human microbiome, play a pivotal role in modulating the immune response (Son & Kim, 2022).

1.3 Mechanisms of Microbial Influence on Cancer

Microbiome's role in immune system regulation: The human microbiome, comprising trillions of microbes residing in and on the body, has been recognized as a key player in immune system regulation. The gut microbiome, in particular, exerts a profound influence on the immune system through interactions with immune cells and the production of various signaling molecules (Vijay & Valdes, 2022); Kandalai et al., 2023). Commensal microbes contribute to the maintenance of immune homeostasis, promoting a balanced and responsive immune system. Research has revealed that certain microbial species within the gut microbiome can stimulate the maturation and activation of immune cells, such as T cells and dendritic cells (Zheng et al., 2020; Kogut et al., 2020). These interactions are crucial for the development of a robust immune response against pathogens and cancer cells (Ivanov & Honda,

2012; Graham & Xavier, 2023). Additionally, the microbiome helps regulate inflammation, a critical factor in cancer development, by balancing pro-inflammatory and anti-inflammatory signals (Francescone et al., 2014; Rahman et al., 2022).

1.3.1 Implications for Cancer Evasion

In the context of cancer, alterations in the microbiome composition have been associated with immune system dysregulation, creating an environment conducive to cancer evasion (Shan et al., 2022; Ge et al., 2021). Dysbiosis, an imbalance in the microbiome composition, can lead to a compromised immune response against cancer cells. This may occur through the inhibition of immune cell activation or the promotion of regulatory T cells, which suppress the immune system (Wu & Wu, 2012; Fakharian et al., 2023).

Furthermore, some microbial products or metabolites can directly influence cancer cell behavior. For instance, certain bacteria can produce molecules that modulate immune checkpoint pathways, allowing cancer cells to evade detection and destruction by the immune system (Garg et al., 2023). Understanding these microbial-mediated immune evasion strategies is essential for developing targeted therapies aimed at restoring the immune system's ability to recognize and eliminate cancer cells (Son & Kim, 2022). The intricate relationship between microbes and the immune system significantly impacts cancer development and progression. By deciphering the mechanisms through which the microbiome modulates the immune response, researchers can identify novel therapeutic strategies to enhance immune surveillance and improve outcomes for individuals with cancer (Jain et al., 2021).

1.3.2 Metabolic Interactions

Metabolic interactions play a crucial role in the intricate web of biological processes that govern the functioning of living organisms. These interactions involve the exchange and regulation of various metabolites, small molecules that are essential for energy production, growth, and maintenance of cellular functions (Judge & Dodd, 2020; Oh et al., 2023). Two prominent aspects of metabolic interactions are explored here: the impact of microbial metabolites on host metabolism and their influence on cancer cell growth and survival.

Microbial Metabolites and Their Effects on Host Metabolism The human body is home to trillions of microbes residing in the gut, collectively known as the gut microbiota. These microorganisms contribute to host metabolism through the production of a myriad of metabolites (Gebrayel et al., 2022; Horn et al., 2022). Short-chain fatty acids (SCFAs), for instance, are generated by microbial fermentation of dietary fiber and have been implicated in various physiological processes. SCFAs, including acetate, propionate, and butyrate, can influence host energy metabolism, immune function, and gut barrier integrity (Koh et al., 2016). The impact of microbial metabolites extends beyond the gut, affecting distant organs and tissues through the bloodstream.

The gut microbiota also plays a role in the metabolism of dietary polyphenols, producing bioactive compounds with potential health benefits. For example, the conversion of flavonoids by gut bacteria results in the production of metabolites with antioxidant and anti-inflammatory properties (Selma et al., 2009). Such interactions highlight the intricate relationship between microbial metabolism and host health, emphasizing the importance of maintaining a balanced and diverse gut microbiome for overall well-being.

Influence on Cancer Cell Growth and Survival Metabolic interactions between the host and microbes can have profound effects on cancer development and progression. The gut microbiota, in particular, has been implicated in influencing cancer cell growth and survival. Microbial metabolites can modulate the tumor microenvironment and impact the host's immune response to cancer. For instance, certain bacteria can produce metabolites that either promote or inhibit inflammation, affecting the tumor's ability to evade the immune system (Gopalakrishnan et al., 2018a, b).

Furthermore, microbial metabolism can influence the efficacy of cancer therapies. Recent research has highlighted the role of gut microbiota in mediating the response to immunotherapy and chemotherapy (Routy et al., 2018). Understanding the intricate interplay between microbial metabolites and cancer cell metabolism is crucial for developing targeted therapeutic strategies that exploit these interactions for improved treatment outcomes.

1.3.3 Direct Interactions with Cancer Cells

Cancer development and progression involve a complex interplay of various cellular and molecular events. Understanding the direct interactions between cancer cells and their microenvironment is crucial for unraveling the mechanisms driving tumor growth. This section explores two critical aspects: adhesion, invasion, and communication and the influence of microbial factors on tumor progression.

Adhesion, invasion, and communication: Adhesion, invasion, and communication are fundamental processes that dictate the behavior of cancer cells within the tumor microenvironment. Adhesion refers to the attachment of cancer cells to neighboring cells or the extracellular matrix (ECM), influencing their ability to migrate and invade surrounding tissues. The dysregulation of adhesion molecules, such as integrins and cadherins, has been implicated in cancer progression (Liu et al., 2018).

Invasion, a hallmark of cancer metastasis, involves the penetration of cancer cells into adjacent tissues and blood vessels. Matrix metalloproteinases (MMPs), enzymes responsible for ECM degradation, play a crucial role in facilitating cancer cell invasion (Liotta et al., 1980). Targeting these invasion-associated molecules presents a potential avenue for therapeutic interventions.

Communication between cancer cells and their microenvironment occurs through various signaling pathways. The secretion of growth factors, cytokines, and extracellular vesicles by cancer cells can stimulate angiogenesis, modulate immune responses, and promote a pro-tumorigenic environment (Hanahan & Weinberg, 2011). Unraveling the intricate signaling networks involved in cancer cell communication is vital for developing targeted therapies.

1.3.4 Microbial Factors Promoting or Inhibiting Tumor Progression

The role of microbial factors in tumor progression has emerged as a significant area of research, highlighting the intricate relationship between the host microbiome and cancer. Commensal microorganisms within the gut microbiota have been implicated in modulating immune responses and influencing tumor development (Roy & Trinchieri, 2017).

Certain microbial factors promote tumor progression by creating a pro-inflammatory microenvironment, supporting angiogenesis, and modulating the host immune system. For example, specific bacterial species have been linked to the activation of Toll-like receptors (TLRs) and the production of pro-inflammatory cytokines, contributing to chronic inflammation and promoting tumorigenesis (Rakoff-Nahoum & Medzhitov, 2009).

Conversely, some microbial factors exhibit antitumor properties. Commensal bacteria, such as *Bifidobacterium* and *Lactobacillus*, have been shown to enhance antitumor immune responses and inhibit the growth of certain cancer types (Viaud et al., 2013). Understanding the intricate balance between pro-tumorigenic and anti-tumorigenic microbial factors is essential for developing targeted therapeutic strategies. The direct interactions between cancer cells and their microenvironment involve a dynamic interplay of adhesion, invasion, and communication processes. Additionally, microbial factors within the host microbiome can either promote or inhibit tumor progression. Further research in these areas holds the promise of uncovering novel therapeutic targets for cancer treatment and advancing our understanding of the intricate mechanisms governing cancer development.

1.3.5 Microbiome as a Biomarker for Cancer

The human microbiome, consisting of trillions of microorganisms residing in and on the human body, has emerged as a potential biomarker for various health conditions, including cancer. Understanding the microbial signatures associated with different cancer types opens new avenues for early detection, prognosis, and personalized treatment strategies.

1.3.5.1 Identification of Microbial Signatures in Cancer Patients

Microbial signatures in cancer patients offer valuable insights into the complex interplay between the host and its microbiota. Various microbiome profiling techniques have been employed to unravel the intricate relationship between the microbiome and cancer.

Microbiome Profiling Techniques State-of-the-art technologies such as next-generation sequencing (NGS) and metagenomic analysis play a pivotal role in characterizing the composition and diversity of the microbiome. NGS enables the sequencing of microbial DNA, RNA, and proteins, allowing researchers to identify specific microbial taxa and their functional capabilities in cancer patients. Metagenomic approaches provide a holistic view of the entire microbial community, shedding light on potential microbial biomarkers.

These techniques have been applied to diverse sample types, including feces, saliva, and tumor tissue, to capture the full spectrum of microbial diversity associated with cancer. A study by Thomas et al. (2019) utilized metagenomic sequencing to identify distinct microbial signatures in colorectal cancer patients, highlighting the diagnostic potential of microbiome profiling in cancer research.

Potential Markers for Different Cancer Types The identification of specific microbial markers associated with different cancer types is crucial for developing targeted diagnostic and therapeutic approaches. Numerous studies have explored the microbial signatures linked to various cancers, such as colorectal, breast, and lung cancer.

For instance, in colorectal cancer, *Fusobacterium nucleatum* has been recognized as a potential biomarker due to its increased abundance in tumor tissues. In breast cancer, the abundance of specific bacterial taxa, as demonstrated by Xuan et al. (2014), may serve as diagnostic markers. Additionally, lung cancer studies, like that by Liu et al. (2018), have identified unique microbial patterns in the respiratory microbiome of cancer patients. These findings underscore the potential of microbiome-based markers for the early detection and classification of different cancer types, paving the way for personalized and precision medicine approaches.

In conclusion, the identification of microbial signatures in cancer patients through advanced profiling techniques holds promise for revolutionizing cancer diagnostics and treatment. Continued research in this field is essential to validate and refine microbial biomarkers, ultimately improving the accuracy and clinical utility of microbiome-based cancer diagnostics.

1.3.6 Diagnostic and Prognostic Applications in Healthcare

Microbiome-Based Cancer Detection Microbiome research has emerged as a promising avenue for diagnostic and prognostic applications in cancer detection.

The human microbiome, comprising trillions of microorganisms residing in and on the body, plays a crucial role in maintaining health. Recent studies have highlighted the potential of microbiome-based markers in detecting various types of cancer, offering noninvasive and sensitive methods for early diagnosis. For instance, research by Thomas G. Graeber et al. (2018) demonstrated the association between alterations in the gut microbiome and colorectal cancer. This connection opens avenues for developing microbiome-based diagnostic tools, including stool or blood tests, to detect cancer in its early stages.

Predicting Treatment Response and Outcomes Understanding and predicting treatment response is vital for personalized medicine, and recent advancements have leveraged various technologies, including genomics and machine learning, to enhance prognostic accuracy. A study by Yuan et al. (2020) showcased the potential of integrating multi-omics data for predicting treatment responses in cancer patients. Utilizing genomic, transcriptomic, and proteomic information, the study demonstrated improved accuracy in forecasting patient outcomes, aiding clinicians in making informed decisions regarding treatment plans. Moreover, liquid biopsy techniques, such as circulating tumor DNA analysis, have gained traction for their ability to monitor treatment response in real time, providing valuable insights into the effectiveness of therapeutic interventions (Diaz et al., 2016).

The integration of microbiome-based approaches and predictive technologies in cancer diagnostics holds immense potential for revolutionizing patient care. These innovative methods not only contribute to early cancer detection but also enable healthcare professionals to tailor treatment strategies based on individual patient characteristics, ultimately improving overall prognosis and outcomes.

1.4 Therapeutic Interventions

In recent years, there has been a burgeoning interest in utilizing therapeutic interventions aimed at manipulating the microbiome to promote health and prevent diseases such as cancer. Among these interventions, the use of probiotics and prebiotics has gained significant attention for their potential role in cancer prevention.

1.4.1 Probiotics and Prebiotics in Cancer Prevention

Cancer prevention strategies have expanded beyond traditional approaches to include interventions targeting the gut microbiome, which plays a crucial role in immune regulation, inflammation, and metabolism (Clemente et al., 2012). Probiotics, defined as live microorganisms that confer health benefits when administered in adequate amounts, and prebiotics, nondigestible compounds that selectively stimulate the growth and/or activity of beneficial bacteria in the gut, offer promising avenues for modulating the microbiome and reducing cancer risk (Gibson et al., 2017).

Modulating the Microbiome for Health The human microbiome, consisting of trillions of microbes residing predominantly in the gastrointestinal tract, exerts profound effects on host physiology and immune function (Thursby & Juge, 2017). Dysbiosis, characterized by alterations in the composition and function of the microbiome, has been implicated in various diseases, including cancer. Therefore, interventions aimed at restoring microbial balance hold therapeutic potential in mitigating disease risk. Probiotics contribute to microbiome modulation by enhancing the abundance of beneficial bacteria, such as *Lactobacillus* and *Bifidobacterium*, while suppressing the growth of harmful pathogens (Hill et al., 2014). Similarly, prebiotics serve as substrates for beneficial bacteria, promoting their growth and metabolic activity within the gut ecosystem (Scott et al., 2013).

Evidence Supporting the Use of Probiotics and Prebiotics Accumulating evidence from preclinical and clinical studies supports the notion that probiotics and prebiotics exert anticarcinogenic effects through diverse mechanisms, including immunomodulation, inhibition of pathogen colonization, and production of bioactive metabolites (Plaza-Díaz et al., 2019). For instance, certain probiotic strains have been shown to enhance immune surveillance against tumor development by stimulating the activity of natural killer cells and modulating cytokine production (Kau et al., 2011). Moreover, prebiotic supplementation has been associated with reduced inflammation and improved gut barrier function, thereby mitigating the risk of colorectal cancer (Gibson et al., 2017). Clinical trials investigating the efficacy of probiotics and prebiotics in cancer prevention have yielded promising results, albeit with variations in outcomes attributed to factors such as study design, microbial composition, and host characteristics (McFarland et al., 2018).

The strategic manipulation of the gut microbiome through probiotics and prebiotics represents a promising approach to cancer prevention. However, further research is warranted to elucidate the optimal formulations, dosages, and duration of intervention, as well as to identify biomarkers predictive of therapeutic response. By harnessing the symbiotic relationship between gut microbes and host physiology, these therapeutic interventions hold the potential to reshape the landscape of cancer prevention and improve public health outcomes.

1.5 Fecal Microbiota Transplantation (FMT)

Overview of fecal microbiota transplantation (FMT): Fecal microbiota transplantation (FMT) is a therapeutic procedure involving the transfer of fecal material from a healthy donor to a recipient with a disrupted or imbalanced gut microbiota. The primary aim of FMT is to restore the microbial diversity and composition within the recipient's gastrointestinal tract. This procedure gained prominence as an effective treatment for recurrent *Clostridium difficile* infection (CDI), where conventional antibiotic therapies often fail to provide long-term relief. FMT works by introducing a healthy donor's microbiota into the recipient's colon, thereby promoting the restoration of a stable and functional gut microbial community (Cammarota et al., 2014).

FMT has demonstrated success not only in treating CDI but also in addressing various gastrointestinal conditions, such as inflammatory bowel disease (IBD) and irritable bowel syndrome (IBS). The therapeutic potential of FMT extends beyond gastrointestinal disorders, as emerging research explores its impact on systemic conditions like metabolic syndrome and neurologic diseases (Moayyedi et al., 2015).

1.5.1 Potential Role in Cancer Treatment and Prevention

Recent scientific investigations have unveiled a potential role for FMT in cancer treatment and prevention, highlighting the intricate relationship between the gut microbiota and the immune system. The gut microbiome plays a crucial role in modulating systemic immune responses, and disruptions in its composition have been associated with increased cancer risk and altered responses to cancer therapies (Vetizou et al., 2015; Gopalakrishnan et al., 2018a, b).

Several studies have explored the impact of FMT on the efficacy of immunotherapy in cancer patients. By manipulating the gut microbiota through FMT, researchers aim to enhance the antitumor immune response and improve the outcomes of cancer immunotherapy. Moreover, FMT holds promise in mitigating treatment-related side effects, such as gastrointestinal complications and immunotherapy resistance (Routy et al., 2018). FMT presents a fascinating frontier in medical research, not only revolutionizing the treatment of gastrointestinal disorders but also demonstrating promising potential in the realm of cancer therapy. Further exploration of the intricate interplay between the gut microbiota and various disease processes holds the key to unlocking novel therapeutic avenues and improving patient outcomes.

1.5.2 Microbiome-Targeted Therapies

Microbiome-targeted therapies have emerged as a promising avenue in medical research, aiming to harness the intricate interplay between the human microbiome and various diseases for therapeutic purposes (Sender et al., 2016; Yadav & Chauhan, 2021). One key aspect of this field involves developing drugs to manipulate the microbiome, capitalizing on the growing understanding of the complex relationship between microbial communities and human health (Lynch & Pedersen, 2016; Berg et al., 2020; Afzaal et al., 2022). Numerous studies have highlighted the pivotal role of the microbiome in influencing host physiology, metabolism, and immune responses. With advancements in sequencing technologies and metagenomics, researchers can now explore the vast diversity of microbial species residing in the human body, paving the way for targeted interventions (Shreiner et al., 2015; Kashyap et al., 2017; Malla et al., 2019).

Developing Drugs to Manipulate the Microbiome Scientists are actively engaged in designing pharmaceuticals that can selectively modulate the composition and function of the microbiome. Probiotics, which involve the administration of beneficial microorganisms, have gained traction as a means to restore microbial balance and promote health. Additionally, prebiotics, substances that nourish beneficial microbes, are being explored to foster the growth of specific microorganisms. Furthermore, the use of antibiotics with precision to target harmful bacteria while sparing beneficial ones represents a nuanced approach to microbiome manipulation. Fecal microbiota transplantation (FMT), where the microbiome from a healthy donor is transferred to a patient, has shown remarkable success in treating certain gastrointestinal infections and is being investigated for other conditions like inflammatory bowel disease.

The development of microbiome-targeted drugs faces challenges such as the need for a deeper understanding of microbial functions, potential off-target effects, and the dynamic nature of the microbiome. Rigorous clinical trials are essential to assess the safety and efficacy of these interventions, ensuring that they provide tangible benefits without causing unintended consequences. Recent advancements in synthetic biology also open avenues for engineering microbes to deliver therapeutic payloads or produce bioactive compounds within the host, offering a novel strategy for precise microbiome manipulation.

Challenges and Opportunities in Microbiome-Based Cancer Therapies Microbiome-based cancer therapies represent a frontier where the interplay between the host immune system, tumor microenvironment, and microbial communities comes into play (Bhatt et al., 2017; Jain et al., 2021). Mounting evidence suggests that the composition of the microbiome can influence cancer development, progression, and response to treatment. Research has shown that certain bacteria within the gut microbiome can modulate the efficacy of immunotherapies, such as checkpoint inhibitors, by enhancing antitumor immune responses (Qiu et al., 2021; Aghamajidi & Maleki Vareki, 2022). Additionally, the microbiome can influence the metabolism of chemotherapeutic agents, impacting their bioavailability and efficacy (Villemin et al., 2023; Oh et al., 2021).

Microbiome-targeted therapies, particularly in the context of drug development and cancer treatment, hold immense potential but require concerted efforts to overcome challenges and harness opportunities effectively (Ting et al., 2022). The field is rapidly evolving, with ongoing research shedding light on the intricate relationships between microbial communities and human health (Aggarwal et al., 2023). As this knowledge deepens, the prospect of precision microbiome interventions to treat various diseases becomes increasingly promising, paving the way for a new era in personalized medicine (Goetz & Schork, 2018; Petrosino, 2018; Huang et al., 2023).

However, challenges abound in translating these findings into effective cancer therapies. Interindividual variability in the microbiome, coupled with the complexity of cancer biology, poses challenges in identifying universal microbial signatures or interventions. Moreover, the dynamic nature of the microbiome and its susceptibility to external factors such as diet and antibiotics necessitate a personalized approach to microbiome-based cancer therapies. Advances in metagenomic sequencing, coupled with artificial intelligence algorithms, offer tools for analyzing large datasets to discern patterns that may guide personalized treatment strategies (He et al., 2022).

References

Afzaal, M., Saeed, F., Shah, Y. A., Hussain, M., Rabail, R., Socol, C. T., Hassoun, A., Pateiro, M., Lorenzo, J. M., Rusu, A. V., & Aadil, R. M. (2022). Human gut microbiota in health and disease: Unveiling the relationship. *Frontiers in Microbiology, 13*, 999001. https://doi.org/10.3389/fmicb.2022.999001

Aggarwal, N., Kitano, S., Puah, G. R. Y., Kittelmann, S., Hwang, I. Y., & Chang, M. W. (2023). Microbiome and human health: Current understanding, engineering, and enabling technologies. *Chemical Reviews, 123*(1), 31–72. https://doi.org/10.1021/acs.chemrev.2c00431

Aghamajidi, A., & Maleki Vareki, S. (2022). The effect of the gut microbiota on systemic and anti-tumor immunity and response to systemic therapy against cancer. *Cancers, 14*(15), 3563. https://doi.org/10.3390/cancers14153563

Altveş, S., Yildiz, H. K., & Vural, H. C. (2020). Interaction of the microbiota with the human body in health and diseases. *Bioscience of Microbiota, Food and Health, 39*(2), 23–32. https://doi.org/10.12938/bmfh.19-023

Appleton, J. (2018). The gut-brain axis: Influence of microbiota on mood and mental health. *Integrative Medicine (Encinitas, Calif.), 17*(4), 28–32.

Arthur, J. C., Perez-Chanona, E., Mühlbauer, M., Tomkovich, S., Uronis, J. M., Fan, T. J., et al. (2012). Intestinal inflammation targets cancer-inducing activity of the microbiota. *Science, 338*(6103), 120–123.

Belkaid, Y., & Hand, T. W. (2014). Role of the microbiota in immunity and inflammation. *Cell, 157*(1), 121–141.

Berg, G., Rybakova, D., Fischer, D., Cernava, T., Vergès, M. C., Charles, T., Chen, X., Cocolin, L., Eversole, K., Corral, G. H., Kazou, M., Kinkel, L., Lange, L., Lima, N., Loy, A., Macklin, J. A., Maguin, E., Mauchline, T., McClure, R., … Schloter, M. (2020). Microbiome definition re-visited: Old concepts and new challenges. *Microbiome, 8*(1), 103. https://doi.org/10.1186/s40168-020-00875-0

Bhatt, A. P., Redinbo, M. R., & Bultman, S. J. (2017). The role of the microbiome in cancer development and therapy. *CA: a Cancer Journal for Clinicians, 67*(4), 326–344. https://doi.org/10.3322/caac.21398

Bull, M. J., & Plummer, N. T. (2014). Part 1: The human gut microbiome in health and disease. *Integrative Medicine (Encinitas, Calif.), 13*(6), 17–22.

Byrd, A. L., Belkaid, Y., & Segre, J. A. (2018). The human skin microbiome. *Nature Reviews Microbiology, 16*(3), 143–155.

Cammarota, G., Ianiro, G., & Gasbarrini, A. (2014). Fecal microbiota transplantation for the treatment of Clostridium difficile infection: A systematic review. *Journal of Clinical Gastroenterology, 48*(8), 693–702.

Campbell, C., Kandalgaonkar, M. R., Golonka, R. M., Yeoh, B. S., Vijay-Kumar, M., & Saha, P. (2023). Crosstalk between gut microbiota and host immunity: Impact on inflammation and immunotherapy. *Biomedicine, 11*(2), 294. https://doi.org/10.3390/biomedicines11020294

Carabotti, M., Scirocco, A., Maselli, M. A., & Severi, C. (2015). The gut-brain axis: Interactions between enteric microbiota, central and enteric nervous systems. *Annals of Gastroenterology, 28*(2), 203–209.

Carding, S., Verbeke, K., Vipond, D. T., Corfe, B. M., & Owen, L. J. (2015). Dysbiosis of the gut microbiota in disease. *Microbial Ecology in Health and Disease, 26*, 26191. https://doi.org/10.3402/mehd.v26.26191

Chen, Y., Wu, F. H., Wu, P. Q., Xing, H. Y., & Ma, T. (2022). The role of the tumor microbiome in tumor development and its treatment. *Frontiers in Immunology, 13*, 935846. https://doi.org/10.3389/fimmu.2022.935846

Clemente, J. C., Ursell, L. K., Parfrey, L. W., & Knight, R. (2012). The impact of the gut microbiota on human health: An integrative view. *Cell, 148*(6), 1258–1270.

Cryan, J. F., O'Riordan, K. J., Cowan, C. S. M., Sandhu, K. V., Bastiaanssen, T. F. S., Boehme, M., et al. (2019). The microbiota-gut-brain Axis. *Physiological Reviews, 99*(4), 1877–2013.

de Martel, C., Ferlay, J., Franceschi, S., Vignat, J., Bray, F., Forman, D., & Plummer, M. (2012). Global burden of cancers attributable to infections in 2008: A review and synthetic analysis. *Lancet Oncol, 13*(6), 607–615. https://doi.org/10.1016/S1470-2045(12)70137-7

Diaz, L. A., et al. (2016). The molecular evolution of acquired resistance to targeted EGFR blockade in colorectal cancers. *Nature, 486*(7404), 537–540.

Durack, J., & Lynch, S. V. (2019). The gut microbiome: Relationships with disease and opportunities for therapy. *The Journal of Experimental Medicine, 216*(1), 20–40. https://doi.org/10.1084/jem.20180448

Fakharian, F., Thirugnanam, S., Welsh, D. A., Kim, W. K., Rappaport, J., Bittinger, K., & Rout, N. (2023). The role of gut dysbiosis in the loss of intestinal immune cell functions and viral pathogenesis. *Microorganisms, 11*(7), 1849. https://doi.org/10.3390/microorganisms11071849

Felgner, S., Kocijancic, D., Frahm, M., & Weiss, S. (2016). Bacteria in cancer therapy: Renaissance of an old concept. *International Journal of Microbiology, 2016*, 8451728. https://doi.org/10.1155/2016/8451728

Francescone, R., Hou, V., & Grivennikov, S. I. (2014). Microbiome, inflammation, and cancer. *Cancer Journal (Sudbury, Mass), 20*(3), 181–189. https://doi.org/10.1097/PPO.0000000000000048

Gagnière, J., Raisch, J., Veziant, J., Barnich, N., Bonnet, R., Buc, E., et al. (2016). Gut microbiota imbalance and colorectal cancer. *World Journal of Gastroenterology, 22*(2), 501–518.

Garg, S., Sharma, N., Bharmjeet, & Das, A. (2023). Unraveling the intricate relationship: Influence of microbiome on the host immune system in carcinogenesis. *Cancer Reports (Hoboken, N.J.), 6*(11), e1892. https://doi.org/10.1002/cnr2.1892

Garrett, W. S. (2015). Cancer and the microbiota. *Science (New York, N.Y.), 348*(6230), 80–86. https://doi.org/10.1126/science.aaa4972

Ge, Y., Wang, X., Guo, Y., et al. (2021). Gut microbiota influence tumor development and Alter interactions with the human immune system. *Journal of Experimental & Clinical Cancer Research, 40*, 42. https://doi.org/10.1186/s13046-021-01845-6

Gebrayel, P., Nicco, C., Al Khodor, S., Bilinski, J., Caselli, E., Comelli, E. M., Egert, M., Giaroni, C., Karpinski, T. M., Loniewski, I., Mulak, A., Reygner, J., Samczuk, P., Serino, M., Sikora, M., Terranegra, A., Ufnal, M., Villeger, R., Pichon, C., … Edeas, M. (2022). Microbiota medicine: Towards clinical revolution. *Journal of Translational Medicine, 20*(1), 111. https://doi.org/10.1186/s12967-022-03296-9

Gibson, G. R., Hutkins, R., Sanders, M. E., Prescott, S. L., Reimer, R. A., Salminen, S. J., et al. (2017). Expert consensus document: The International Scientific Association for Probiotics and Prebiotics (ISAPP) consensus statement on the definition and scope of prebiotics. *Nature Reviews Gastroenterology & Hepatology, 14*(8), 491–502.

Goetz, L. H., & Schork, N. J. (2018). Personalized medicine: Motivation, challenges, and progress. *Fertility and Sterility, 109*(6), 952–963. https://doi.org/10.1016/j.fertnstert.2018.05.006

Gopalakrishnan, V., Helmink, B. A., Spencer, C. N., Reuben, A., & Wargo, J. A. (2018a). The influence of the gut microbiome on cancer, immunity, and cancer immunotherapy. *Cancer Cell, 33*(4), 570–580.

Gopalakrishnan, V., Spencer, C. N., Nezi, L., Reuben, A., Andrews, M. C., Karpinets, T. V., et al. (2018b). Gut microbiome modulates response to anti–PD-1 immunotherapy in melanoma patients. *Science, 359*(6371), 97–103.

Graeber, T. G., et al. (2018). Hypoxia-mediated selection of cells with diminished apoptotic potential in solid tumours. *Nature, 552*(7684), 126–131.

Graham, D. B., & Xavier, R. J. (2023). Conditioning of the immune system by the microbiome. *Trends in Immunology, 44*(7), 499–511. https://doi.org/10.1016/j.it.2023.05.002

Grivennikov, S. I., Greten, F. R., & Karin, M. (2010). Immunity, inflammation, and cancer. *Cell, 140*(6), 883–899.

Hanahan, D., Weinberg, RA. (2011). Hallmarks of cancer: The next generation. *Cell, 144*, 646–674. https://doi.org/10.1016/j.cell.2011.02.013

He, Q., Niu, X., Qi, R. Q., & Liu, M. (2022). Advances in microbial metagenomics and artificial intelligence analysis in forensic identification. *Frontiers in Microbiology, 13*, 1046733. https://doi.org/10.3389/fmicb.2022.1046733

Hill, C., Guarner, F., Reid, G., Gibson, G. R., Merenstein, D. J., Pot, B., et al. (2014). Expert consensus document: The International Scientific Association for Probiotics and Prebiotics consensus statement on the scope and appropriate use of the term probiotic. *Nature Reviews Gastroenterology & Hepatology, 11*(8), 506–514.

Horn, J., Mayer, D. E., Chen, S., & Mayer, E. A. (2022). Role of diet and its effects on the gut microbiome in the pathophysiology of mental disorders. *Translational Psychiatry, 12*(1), 164. https://doi.org/10.1038/s41398-022-01922-0

Hou, K., Wu, Z. X., Chen, X. Y., Wang, J. Q., Zhang, D., Xiao, C., Zhu, D., Koya, J. B., Wei, L., Li, J., & Chen, Z. S. (2022). Microbiota in health and diseases. *Signal Transduction and Targeted Therapy, 7*(1), 135. https://doi.org/10.1038/s41392-022-00974-4

Huang, G., Khan, R., Zheng, Y., Lee, P. C., Li, Q., & Khan, I. (2023). Exploring the role of gut microbiota in advancing personalized medicine. *Frontiers in Microbiology, 14*, 1274925. https://doi.org/10.3389/fmicb.2023.1274925

Ikryannikova, L. N., Gorokhovets, N. V., Belykh, D. A., Kurbatov, L. K., & Zamyatnin, A. A., Jr. (2023). Bacterial therapy of cancer: A way to the dustbin of history or to the medicine of the future? *International Journal of Molecular Sciences, 24*(11), 9726. https://doi.org/10.3390/ijms24119726

Irrazábal, T., Belcheva, A., Girardin, S. E., Martin, A., & Philpott, D. J. (2014). The multifaceted role of the intestinal microbiota in colon cancer. *Molecular Cell, 54*(2), 309–320. https://doi.org/10.1016/j.molcel.2014.03.039

Ivanov, I. I., & Honda, K. (2012). Intestinal commensal microbes as immune modulators. *Cell Host & Microbe, 12*(4), 496–508. https://doi.org/10.1016/j.chom.2012.09.009

Jain, T., Sharma, P., Are, A. C., Vickers, S. M., & Dudeja, V. (2021). New insights into the cancer-microbiome-immune axis: Decrypting a decade of discoveries. *Frontiers in Immunology, 12*, 622064. https://doi.org/10.3389/fimmu.2021.622064

Jandhyala, S. M., Talukdar, R., Subramanyam, C., Vuyyuru, H., Sasikala, M., & Nageshwar Reddy, D. (2015). Role of the normal gut microbiota. *World Journal of Gastroenterology, 21*(29), 8787–8803. https://doi.org/10.3748/wjg.v21.i29.8787

Ji, J., Jin, W., Liu, S. J., Jiao, Z., & Li, X. (2023). Probiotics, prebiotics, and postbiotics in health and disease. *MedComm, 4*(6), e420. https://doi.org/10.1002/mco2.420

Judge, A., & Dodd, M. S. (2020). Metabolism. *Essays in Biochemistry, 64*(4), 607–647. https://doi.org/10.1042/EBC20190041

Kandalai, S., Li, H., Zhang, N., Peng, H., & Zheng, Q. (2023). The human microbiome and cancer: A diagnostic and therapeutic perspective. *Cancer Biology & Therapy, 24*(1), 2240084. https://doi.org/10.1080/15384047.2023.2240084

Kashyap, P. C., Chia, N., Nelson, H., Segal, E., & Elinav, E. (2017). Microbiome at the frontier of personalized medicine. *Mayo Clinic Proceedings, 92*(12), 1855–1864. https://doi.org/10.1016/j.mayocp.2017.10.004

Kau, A. L., Ahern, P. P., Griffin, N. W., Goodman, A. L., & Gordon, J. I. (2011). Human nutrition, the gut microbiome, and immune system: Envisioning the future. *Nature, 474*(7351), 327–336.

Kogut, M. H., Lee, A., & Santin, E. (2020). Microbiome and pathogen interaction with the immune system. *Poultry Science, 99*(4), 1906–1913. https://doi.org/10.1016/j.psj.2019.12.011

Koh, A., De Vadder, F., Kovatcheva-Datchary, P., & Bäckhed, F. (2016). From dietary fiber to host physiology: Short-chain fatty acids as key bacterial metabolites. *Cell, 165*(6), 1332–1345. https://doi.org/10.1016/j.cell.2016.05.041

Lazar, V., Ditu, L. M., Pircalabioru, G. G., Gheorghe, I., Curutiu, C., Holban, A. M., Picu, A., Petcu, L., & Chifiriuc, M. C. (2018). Aspects of gut microbiota and immune system interactions in infectious diseases, immunopathology, and cancer. *Frontiers in Immunology, 9*, 1830. https://doi.org/10.3389/fimmu.2018.01830

Liu, Y. F., Zhang, S. Y., Chen, Y. Y., Shi, K., Zou, B., Liu, J., Yang, Q., Jiang, H., Wei, L., Li, C.Z., et al. (2018). ICAM-1 Deficiency in the Bone Marrow Niche Impairs Quiescence and Repopulation of Hematopoietic Stem Cells. *Stem Cell Rep, 11,* 258–273.

Liotta, L. A., Tryggvason, K., Garbisa, S., Hart, I., Foltz, CM., Shafie, S. (1980). Metastatic potential correlates with enzymatic degradation of basement membrane collagen. *Nature, 284*, 67–68.

Louis, P., Hold, G. L., & Flint, H. J. (2014). The gut microbiota, bacterial metabolites and colorectal cancer. *Nature Reviews Microbiology, 12*(10), 661–672.

Lozupone, C. A., Stombaugh, J. I., Gordon, J. I., Jansson, J. K., & Knight, R. (2012). Diversity, stability and resilience of the human gut microbiota. *Nature, 489*(7415), 220–230.

Lynch, S. V., & Pedersen, O. (2016). The human intestinal microbiome in health and disease. *The New England Journal of Medicine, 375*(24), 2369–2379. https://doi.org/10.1056/NEJMra1600266

Malla, M. A., Dubey, A., Kumar, A., Yadav, S., Hashem, A., & Abd Allah, E. F. (2019). Exploring the human microbiome: The potential future role of next-generation sequencing in disease diagnosis and treatment. *Frontiers in Immunology, 9*, 2868. https://doi.org/10.3389/fimmu.2018.02868

McFarland, L. V., Evans, C. T., & Goldstein, E. J. C. (2018). Strain-specificity and disease-specificity of probiotic efficacy: A systematic review and meta-analysis. *Frontiers in Medicine, 5*, 124.

Moayyedi, P., Surette, M. G., Kim, P. T., Libertucci, J., Wolfe, M., Onischi, C., et al. (2015). Fecal microbiota transplantation induces remission in patients with active ulcerative colitis in a randomized controlled trial. *Gastroenterology, 149*(1), 102–109.

Ogunrinola, G. A., Oyewale, J. O., Oshamika, O. O., & Olasehinde, G. I. (2020). The human microbiome and its impacts on health. *International Journal of Microbiology, 2020*, 8045646. https://doi.org/10.1155/2020/8045646

Oh, B., Boyle, F., Pavlakis, N., Clarke, S., Eade, T., Hruby, G., Lamoury, G., Carroll, S., Morgia, M., Kneebone, A., Stevens, M., Liu, W., Corless, B., Molloy, M., Kong, B., Libermann, T., Rosenthal, D., & Back, M. (2021). The gut microbiome and cancer immunotherapy: Can we use the gut microbiome as a predictive biomarker for clinical response in cancer immunotherapy? *Cancers, 13*(19), 4824. https://doi.org/10.3390/cancers13194824

Oh, S. W., Imran, M., Kim, E. H., Park, S. Y., Lee, S. G., Park, H. M., Jung, J. W., & Ryu, T. H. (2023). Approach strategies and application of metabolomics to biotechnology in plants. *Frontiers in Plant Science, 14*, 1192235. https://doi.org/10.3389/fpls.2023.1192235

Paul, B., Barnes, S., Demark-Wahnefried, W., Morrow, C., Salvador, C., Skibola, C., et al. (2015). Influences of diet and the gut microbiome on epigenetic modulation in cancer and other diseases. *Clinical Epigenetics, 7*(1), 112.

Petrosino, J. F. (2018). The microbiome in precision medicine: The way forward. *Genome Medicine, 10*(1), 12. https://doi.org/10.1186/s13073-018-0525-6

Petrova, M. I., & Wu, C. (2017). Microbiota–host interactions: A prelude to the molecular mechanisms of probiotics. *Annual Review of Microbiology, 71*, 1–18.

Plaza-Díaz, J., Ruiz-Ojeda, F. J., Vilchez-Padial, L. M., Gil, A., & Gil-Campos, M. (2019). Mechanisms of action of probiotics. *Advances in Nutrition, 10*(suppl_1), S49–S66.

Plottel, C. S., & Blaser, M. J. (2011). Microbiome and malignancy. *Cell Host & Microbe, 10*(4), 324–335.

Pryor, R., Martinez-Martinez, D., Quintaneiro, L., & Cabreiro, F. (2020). The role of the microbiome in drug response. *Annual Review of Pharmacology and Toxicology, 60*, 417–435. https://doi.org/10.1146/annurev-pharmtox-010919-023612

Qiu, Q., Lin, Y., Ma, Y., Li, X., Liang, J., Chen, Z., Liu, K., Huang, Y., Luo, H., Huang, R., & Luo, L. (2021). Exploring the emerging role of the gut microbiota and tumor microenvironment in cancer immunotherapy. *Frontiers in Immunology, 11*, 612202. https://doi.org/10.3389/fimmu.2020.612202

Rahman, M. M., Islam, M. R., Shohag, S., Ahasan, M. T., Sarkar, N., Khan, H., Hasan, A. M., Cavalu, S., & Rauf, A. (2022). Microbiome in cancer: Role in carcinogenesis and impact in therapeutic strategies. *Biomedicine & Pharmacotherapy = Biomedecine & Pharmacotherapie, 149*, 112898. https://doi.org/10.1016/j.biopha.2022.112898

Rakoff-Nahoum, S., & Medzhitov, R. (2009). Toll-like receptors and cancer. Nature reviews. *Cancer, 9*(1), 57–63. https://doi.org/10.1038/nrc2541

Routy, B., Le Chatelier, E., Derosa, L., Duong, C. P. M., Alou, M. T., Daillère, R., Fluckiger, A., Messaoudene, M., Rauber, C., Roberti, M. P., Fidelle, M., Flament, C., Poirier-Colame, V., Opolon, P., Klein, C., Iribarren, K., Mondragón, L., Jacquelot, N., Qu, B., … Zitvogel, L. (2018). Gut microbiome influences efficacy of PD-1-based immunotherapy against epithelial tumors. *Science (New York, N.Y.)*, 359(6371), 91–97. https://doi.org/10.1126/science.aan3706

Rowland, I., Gibson, G., Heinken, A., Scott, K., Swann, J., Thiele, I., & Tuohy, K. (2018). Gut microbiota functions: Metabolism of nutrients and other food components. *European Journal of Nutrition, 57*(1), 1–24. https://doi.org/10.1007/s00394-017-1445-8

Roy, S., & Trinchieri, G. (2017). Microbiota: A key orchestrator of cancer therapy. Nature reviews. *Cancer, 17*(5), 271–285. https://doi.org/10.1038/nrc.2017.13

Sadrekarimi, H., Gardanova, Z. R., Bakhshesh, M., Ebrahimzadeh, F., Yaseri, A. F., Thangavelu, L., Hasanpoor, Z., Zadeh, F. A., & Kahrizi, M. S. (2022). Emerging role of human microbiome in cancer development and response to therapy: Special focus on intestinal microflora. *Journal of Translational Medicine, 20*(1), 301. https://doi.org/10.1186/s12967-022-03492-7

Sasso, J. M., Ammar, R. M., Tenchov, R., Lemmel, S., Kelber, O., Grieswelle, M., & Zhou, Q. A. (2023). Gut microbiome-brain Alliance: A landscape view into mental and gastrointestinal health and disorders. *ACS Chemical Neuroscience, 14*(10), 1717–1763. https://doi.org/10.1021/acschemneuro.3c00127

Schmidt, B. L., Kuczynski, J., Bhattacharya, A., Huey, B., Corby, P. M., Queiroz, E. L., et al. (2018). Changes in abundance of oral microbiota associated with oral cancer. *PLoS One, 13*(6), e0199634.

Schwabe, R. F., & Jobin, C. (2013). The microbiome and cancer. *Nature Reviews Cancer, 13*(11), 800–812.

Scott, K. P., Gratz, S. W., Sheridan, P. O., Flint, H. J., & Duncan, S. H. (2013). The influence of diet on the gut microbiota. *Pharmacological Research, 69*(1), 52–60.

Selma, M. V., Espín, J. C., & Tomás-Barberán, F. A. (2009). Interaction between phenolics and gut microbiota: Role in human health. *Journal of Agricultural and Food Chemistry, 57*(15), 6485–6501. https://doi.org/10.1021/jf902107d

Sender, R., Fuchs, S., & Milo, R. (2016). Revised estimates for the number of human and bacteria cells in the body. *PLoS Biology, 14*(8), e1002533.

Shan, Y., Lee, M., & Chang, E. B. (2022). The gut microbiome and inflammatory bowel diseases. *Annual Review of Medicine, 73*, 455–468. https://doi.org/10.1146/annurev-med-042320-021020

Shi, Z., Li, H., Song, W., Zhou, Z., Li, Z., & Zhang, M. (2023). Emerging roles of the gut microbiota in cancer immunotherapy. *Frontiers in Immunology, 14*, 1139821. https://doi.org/10.3389/fimmu.2023.1139821

Shreiner, A. B., Kao, J. Y., & Young, V. B. (2015). The gut microbiome in health and in disease. *Current Opinion in Gastroenterology, 31*(1), 69–75.

Sing, T., Kaur, G., & Kaur, A. (2023). Dysbiosis—An etiological factor for cardiovascular diseases and the therapeutic benefits of gut microflora. *Advanced Gut and Microbiome Research, 2023*. https://doi.org/10.1155/2023/7451554

Son, Y. M., & Kim, J. (2022). The microbiome-immune axis therapeutic effects in cancer treatments. *Journal of Microbiology and Biotechnology, 32*(9), 1086–1097. https://doi.org/10.4014/jmb.2208.08002

Soto Chervin, C., & Gajewski, T. F. (2020). Microbiome-based interventions: Therapeutic strategies in cancer immunotherapy. *Immuno-oncology Technology, 8*, 12–20. https://doi.org/10.1016/j.iotech.2020.11.001

Thomas, A. M., Manghi, P., Asnicar, F., Pasolli, E., Armanini, F., Zolfo, M., Beghini, F., Manara, S., Karcher, N., Pozzi, C., Gandini, S., Serrano, D., Tarallo, S., Francavilla, A., Gallo, G., Trompetto, M., Ferrero, G., Mizutani, S., Shiroma, H., … Segata, N. (2019). Metagenomic analysis of colorectal cancer datasets identifies cross-cohort microbial diagnostic signatures and a link with choline degradation. *Nature Medicine*, 25(4), 667–678. https://doi.org/10.1038/s41591-019-0405-7

Thursby, E., & Juge, N. (2017). Introduction to the human gut microbiota. *Biochemical Journal, 474*(11), 1823–1836.

Ting, N. L., Lau, H. C., & Yu, J. (2022). Cancer pharmacomicrobiomics: Targeting microbiota to optimise cancer therapy outcomes. *Gut, 71*(7), 1412–1425. https://doi.org/10.1136/gutjnl-2021-326264

Valdes, A. M., Walter, J., Segal, E., & Spector, T. D. (2018). Role of the gut microbiota in nutrition and health. *BMJ (Clinical Research ed.), 361*, k2179. https://doi.org/10.1136/bmj.k2179

Vernocchi, P., Del Chierico, F., & Putignani, L. (2020). Gut microbiota metabolism and interaction with food components. *International Journal of Molecular Sciences, 21*(10), 3688. https://doi.org/10.3390/ijms21103688

Vetizou, M., Pitt, J. M., Daillère, R., Lepage, P., Waldschmitt, N., Flament, C., et al. (2015). Anticancer immunotherapy by CTLA-4 blockade relies on the gut microbiota. *Science, 350*(6264), 1079–1084.

Viaud, S., Saccheri, F., Mignot, G., Yamazaki, T., Daillère, R., Hannani, D., et al. (2013). The intestinal microbiota modulates the anticancer immune effects of cyclophosphamide. *Science, 342*(6161), 971–976.

Vijay, A., & Valdes, A. M. (2022). Role of the gut microbiome in chronic diseases: A narrative review. *European Journal of Clinical Nutrition, 76*(4), 489–501. https://doi.org/10.1038/s41430-021-00991-6

Villemin, C., Six, A., Neville, B. A., Lawley, T. D., Robinson, M. J., & Bakdash, G. (2023). The heightened importance of the microbiome in cancer immunotherapy. *Trends in Immunology, 44*(1), 44–59. https://doi.org/10.1016/j.it.2022.11.002

Wan, Z., Zheng, J., Zhu, Z., Sang, L., Zhu, J., Luo, S., Zhao, Y., Wang, R., Zhang, Y., Hao, K., Chen, L., Du, J., Kan, J., & He, H. (2022). Intermediate role of gut microbiota in vitamin B nutrition and its influences on human health. *Frontiers in Nutrition, 9*, 1031502. https://doi.org/10.3389/fnut.2022.1031502

Wang, C., Ma, A., McNutt, M. E., Hoyd, R., Wheeler, C. E., Robinson, L. A., Chan, C. H. F., Zakharia, Y., Dodd, R. D., Ulrich, C. M., Hardikar, S., Churchman, M. L., Tarhini, A. A., Singer, E. A., Ikeguchi, A. P., McCarter, M. D., Denko, N., Tinoco, G., Husain, M., … Ma, Q. (2023). A bioinformatics tool for identifying intratumoral microbes from the ORIEN dataset. *bioRxiv*: the preprint server for biology, 2023.05.24.541982. https://doi.org/10.1101/2023.05.24.541982

Wu, H. J., & Wu, E. (2012). The role of gut microbiota in immune homeostasis and autoimmunity. *Gut Microbes, 3*(1), 4–14. https://doi.org/10.4161/gmic.19320

Xuan, C., Shamonki, J. M., Chung, A., Dinome, M. L., Chung, M., Sieling, P. A., & Lee, D. J. (2014). Microbial dysbiosis is associated with human breast cancer. *PLoS One, 9*(1), e83744. https://doi.org/10.1371/journal.pone.0083744

Yadav, M., & Chauhan, N. S. (2021). Microbiome therapeutics: Exploring the present scenario and challenges. *Gastroenterology Report, 10*, goab046. https://doi.org/10.1093/gastro/goab046

Yoo, J. Y., Groer, M., Dutra, S. V. O., Sarkar, A., & McSkimming, D. I. (2020). Gut microbiota and immune system interactions. *Microorganisms, 8*(10), 1587. https://doi.org/10.3390/microorganisms8101587

Yu, T., Guo, F., Yu, Y., Sun, T., Ma, D., Han, J., et al. (2017). Fusobacterium nucleatum promotes chemoresistance to colorectal cancer by modulating autophagy. *Cell, 170*(3), 548–563.

Yu, S., Chen, J., Zhao, Y., Yan, F., Fan, Y., Xia, X., Shan, G., Zhang, P., & Chen, X. (2023). Oral-microbiome-derived signatures enable non-invasive diagnosis of laryngeal cancers. *Journal of Translational Medicine, 21*(1), 438. https://doi.org/10.1186/s12967-023-04285-2

Yuan, Y., et al. (2020). Comprehensive characterization of molecular differences in cancer between male and female patients. *Cancer Cell, 37*(5), 663–676.

Zhao, M., Chu, J., Feng, S., Guo, C., Xue, B., He, K., & Li, L. (2023). Immunological mechanisms of inflammatory diseases caused by gut microbiota dysbiosis: A review. *Biomedicine & Pharmacotherapy = Biomedecine & Pharmacotherapie, 164*, 114985. https://doi.org/10.1016/j.biopha.2023.114985

Zheng, D., Liwinski, T., & Elinav, E. (2020). Interaction between microbiota and immunity in health and disease. *Cell Research, 30*(6), 492–506. https://doi.org/10.1038/s41422-020-0332-7

Microbiome and Its Impact on Human Health: Microbiome in Various Body Organs and Its Association with Human Health and Disease

2

Muhammad Imran and Bashir Ahmad

Abstract

The human microbiome is a dynamic area of research that continues to reveal new insights into human health and the intricate relations between microorganisms and their human host. It has a pivotal role in human health by contributing to various physiological processes and influencing immune system function. Microbiomes in a variety of body organs and its association with human health have been discussed. Moreover, the importance of the digestive (oral cavity, stomach, and intestine), respiratory (nasal cavity, pharynx, larynx, trachea, and lungs), skin, urinary, and reproductive systems on human health has been discussed. The human microbiomes play a crucial role in the pathogenesis and progression of various diseases. This chapter will provide a comprehensive overview of the human microbiome and its impact on human health. We provide an overview of the potential therapeutic applications of microbiome manipulation through probiotics and prebiotics, particularly in the context of obesity, gastrointestinal diseases, and cancer treatment.

Keywords

Microbiome · Human health · Cancers · Disease pathogenesis

2.1 Human Body Residues of Different Microbiome

The human body residues of different microbiomes, which are supraorganisms with diverse types of prokaryotic and eukaryotic microbial communities, identified as a microbiome. All human bodies are colonized with microbiomes such as Eukarya,

M. Imran · B. Ahmad (✉)
Department of Biological Sciences, International Islamic University, Islamabad, Pakistan
e-mail: bashir.ahmad@iiu.edu.pk

N. Masood, A. Yasmin (eds.), *The Microbiome and Cancer*,
https://doi.org/10.1007/978-981-97-6948-3_2

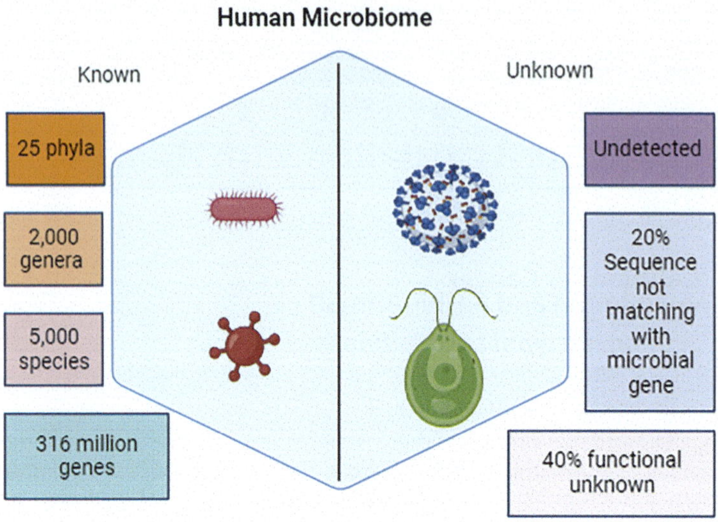

Fig. 2.1 The known and unknown human microbiome

Archaea, bacteria, and fungi (Baquero & Nombela, 2012). Metagenomics has a large number of microorganism genomes. Approximately 85% of genomic data is still mappable. Based on recent large-scale metagenomic research, human microbiome diversity consists of 25 phyla, 2000 genera, 5000 species, and up to 316 million genes shown in Fig. 2.1. However, the large diversity of human microbiomes is still unexplored (Thomas & Segata, 2019).

2.2 Human Microbiome Different Role

The human microbiome's key roles are as follows:

Diversity: The human body harbors trillions of microorganisms (known as microbiota) that help in maintaining the human body ecosystem.

Gut microbiome: The gut microbiome consists of bacteria, which help in digestion, nutrient absorption, and the synthesis of vitamins and other compounds.

Immune system: The microbiome plays a crucial role in the immune system. It helps the immune system to discriminate between pathogens and beneficial microorganisms and helps in maintaining immune balance.

Metabolism: The microbiome breaks down complex molecules that the human body cannot digest on its own, leading to the production of metabolites that can affect energy metabolism and overall health.

Disease and health: Imbalances in the microbiome, known as dysbiosis, are linked to various health conditions, including gastrointestinal disorders, autoimmune diseases, obesity, allergies, and even mental health disorders.

Research and therapeutic potential: Further research is still needed to understand the role of the human microbiome in disease and health. There is an ongoing

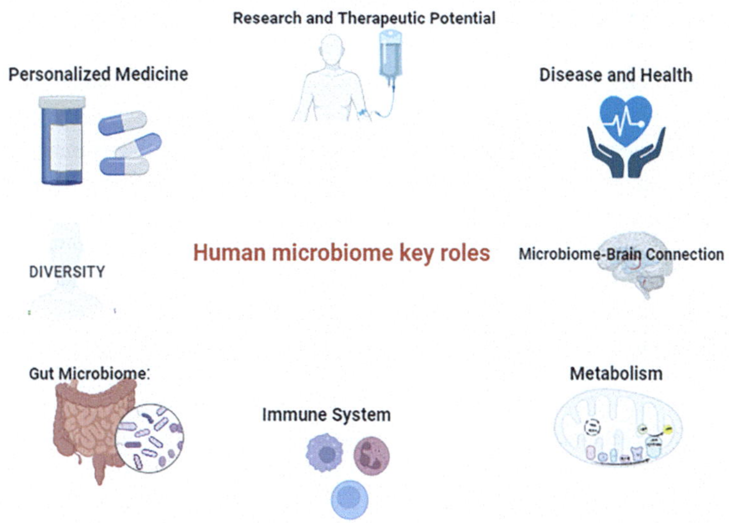

Fig. 2.2 Human microbiome's different roles in diverse areas

exploration of prospective therapeutic uses for modulating the microbiome including probiotics, prebiotics, and fecal microbiota transplantation (FMT).

Personalized medicine: Personalized medicine is a growing field that customizes medical treatments based on the unique characteristics of each individual microbiome composition. The microbiome composition is dependent on various factors, including genetics map, lifestyle, diet, and environmental exposures, leading to the concept of personalized medicine, for interventions individual's unique microbiome profile is a challenge.

Microbiome-brain connection: Emerging and recent studies have highlighted a connection between the microbiome of the digestive system and the central nervous system. Alterations in microbiome composition could impact on brain function and behavior of humans, potentially contributing to different conditions such as anxiety, depression, and neurodevelopmental disorders, as illustrated in Fig. 2.2.

Overall, the human microbiome is a fascinating and dynamic area of research that continues to reveal new insights into human health and the intricate relations between microorganisms and their human host.

2.3 Microbiota Variation across Human Organs and its Impact on Human Health

The human microbiome includes a diverse community of microorganisms that reside in the digestive tract (oral cavity, gastric, intestines), respiratory tract (nasal cavity, airway, and lungs), reproductive organs, and urinary tract. The microbiome has a pivotal role in human health by contributing to various physiological processes and influencing immune system function.

2.4 Digestive System (Oral Cavity, Gastric, and Intestine)

Oral Cavity The human oral cavity harbors different microbiomes—bacteria, fungi, viruses, and protozoa. In the mouth, it is divided into two regions: hard (teeth) and soft (oral mucosa tissue) surface. The oral cavity serves as the primary gateway for microbial entry into the human body. Within this domain, pathogenic organisms have the potential to spread in various parts of the body, triggering infections. The composition of microorganisms within the oral microbiome significantly impacts human health and disease. Some microbiome metabolizes nitrate and reduces nitrate to nitrite, followed by its transformation into nitric oxide. Nitric oxide acts as an antimicrobial agent.

Streptococcus salivarius (K12) contributes in defense mechanism by producing a bacteriocin (Gram-negative species inhibitor), an unfavorable environment that prevents periodontitis disease (Wade, 2013; Wescombe et al., 2009).

Dental caries known as tooth decay disease—it is the most common oral disease—dental caries pathogens are *Streptococcus mutans* and *Streptococcus* sobrinus, *Propionibacterium*, *Actinomyces*, *Veillonella*, *Atopobium*, *Scardovia*, *Lactobacillus acidophilus*, and *Bifidobacterium*. Dental caries manifests when acid-producing bacteria interact with food-fermentable carbohydrates residing in the oral cavity. Supragingival biofilm generates a low-pH environment, causing tooth structure demineralization and ultimately resulting in cavities formation.

Certain microorganisms cause gingivitis without proper oral hygiene. In cases of chronic bacterial infections, the microflora changes, and the accumulation of subgingival plaque impacts both (gingiva and the supportive connective tissue) (Loesche, 2007; Parahitiyawa et al., 2010; Fejerskov, 2004; Filoche et al., 2010; Williams et al., 2008; Zarco et al., 2012; Horz & Conrads, 2007).

The significant role of the oral microbiome in the health system cannot be neglected. Disruption of the oral microbiome has the potential to contribute to various chronic diseases such as osteoporosis, rheumatoid arthritis (RA), and endocarditis. The progression and development of noncommunicable diseases (NCDs) in the oral microbiome role were noted such as neuropsychiatric diseases (NPDs), obesity, diabetes, and cancer; the oral microbiome could be used to check disease status.

Comprehensive research on oral microbes is required to understand a comprehensive role and system-level functions. Future investigations could potentially unveil strategies for restoring oral microbes to a healthy state in the future (Li et al., 2000; Vieira et al., 2017; Jensen et al., 2021; Craig et al., 2018; Do et al., 2020; Knight et al., 2017; Baker et al., 2017; Shafquat et al., 2014; Cunha et al., 2019; Yang et al., 2021; Lin et al., 2019; Xue et al., 2020; Willis & Gabaldón, 2020).

Gastric In the past, there was a belief that the stomach is a sterilized organ because it was not hospitalized for bacteria. These factors include acid environment, bile acid reflux, mucus thickness, and the lactic bacteria present in the oral cavity that convert food into nitrogen oxides to counter microbial nitrogen oxides. However, simple and reliable diagnostic tests are required to explore gastric microbes—in 1982, Barry Marshall and Robin Warren discovered *Helicobacter pylori*, leading to the subsequent rejection of the notion that the stomach was a sterile environment (Yang et al., 2013; Wu et al., 2014).

Indeed, *H. pylori* is the most prevalent microorganism found in the stomach, which can change the stomach environment and change the habitat of resident microorganisms. The gastric microbiome, the community of microorganisms, is a risk factor in the development of gastric cancer.

Additionally, it has been observed that the elimination of *H. pylori* leads to an increased diversity of stomach microbial organisms. However, the interplay between *H. pylori* and commensal bacteria is unknown. A healthy gastric microbiome modulates the gastric microbiome which could prevent severe disease progression (Mitchell et al., 2017; Yu et al., 2017; Liu et al., 2019; Li et al., 2017).

Intestines The intestine is the most densely and diverse colonized organ for the microbiome and has a 1:1 bacterial-host cell ratio. Most of the bacteria live in the intestines, while a few bacteria live in the stomach and small intestines. About 90% of bacteria in the intestine are *Firmicutes* and *Bacteroides* (Arumugam et al., 2011). The composition of the gut microbiome is categorized into three distinct stages of life: from birth to the weaning phase, from weaning to eating a solid diet, and during advanced or old age.

After birth, the first gut-colonized microbe is mostly facultative anaerobes; the microbiome creates an anaerobic condition that helps *Bifidobacterium* and *Bacteroides* species to grow within 2 weeks. Naturally born infants exposed with mother's vaginal and fecal microbiota during the birthing process, while in the Caesarean section (CS) procedure, delivered babies are exposed to skin microbiota and environmental microbiome. A study demonstrated that infants who were naturally delivered were exposed to different types of *Bifidobacterium* species as compared with CT babies after 3 days (Huurre et al., 2008; Biasucci et al., 2010; Karlsson et al., 2011; Eggesbø et al., 2011).

Moreover, infants who were breastfed until the weaning age exhibited a more stable and less diverse bacterial community, particularly bifidobacteria, in comparison with children who were fed formula milk.

When solid food started, *Firmicutes* species increased. In old age, the microbiota *Bifidobacterium* number declined, while *Enterobacteriaceae* increased. In older age (>65 years), *Bacteroidetes* species increase, while *Firmicutes* becomes less abundant (Woodmansey, 2007; Harmsen et al., 2000; Fallani et al., 2011; Roger & McCartney, 2010; Claesson et al., 2011; O'Toole & Claesson, 2010; Favier et al., 2002; Klaassens et al., 2009; Fallani et al., 2010).

The composition of the gut microbiota is dependent on influenced different factors, including the distinct environments across various body locations, intestinal flow rates, and mild acid-neutral pH, which helps in obligate anaerobes' survival. The inner epithelial cell lining and mucin layer harbor minimal colonization in a healthy state, whereas the diffuse mucin layer colonizes special types of bacteria such as *Akkermansia muciniphila*; in the liquid area, *Ruminococcus* species is found in the gut lumen. In the small intestine, there are a relatively short duration of meal (3–5 h) and the presence of bile (acts as an antimicrobial agent). Molecular analysis revealed that facultative anaerobes (Proteobacteria, Bacteroides, enterococci, streptococci, and lactobacilli species were found in the jejunal and ileal part (Zoetendal et al., 2012; Booijink et al., 2010; Islam et al., 2011; Flint et al., 2012).

Different opportunistic pathogens cause diseases such as *Clostridium difficile* which is normal gut microbiota. Colonic epithelial barrier integrity and cytoskeleton damage, cellular death, and aberrant inflammatory response are the main features of the disease.

C. difficile infection (CDI)-associated symptoms are sepsis, diarrhea, pseudomembranous colitis, and potentially fatal outcomes in certain instances and even death in some cases. Gut microbiome-associated diseases are inflammatory bowel disease (IBD) (idiopathic, and chronic diseases have two common forms: Crohn's disease and ulcerative colitis (Pear et al., 1994; Song et al., 2008; Bartlett, 1981), dysbiosis of gut microbiota-associated disorders are celiac disease, irritable bowel syndrome (IBS), and colorectal cancer (CRC).

Several studies revealed that the fecal sample of IBS patients exhibited a significantly reduced lactobacillus, Bifidobacteria and Collinsella, firmicutes(lactobacilli and faecelibactrium); while an increased number of Proteobacteria (*Enterobacteriaceae* spp.), Firmicutes species (*Ruminococcus* spp. *and Veillonella, Streptococci*) and a ratio of Firmicutes to Bacteroidetes spp. was increased as compare associated to control (healthy) (Bhattarai et al., 2017; Salonen et al., 2010; Carroll et al., 2010).

The human gut microbiota communities are dependent on different stages of life cycles—these different factors are delivery mode, milk feeding method, antibiotic treatments, weaning period, anatomical regions, age, and diet use. The gut environment is dependent on different factors such as anatomical region physiology, host secretions, substrate availability, digestive flow rates, oxygen tension, and pH.

2.5 Role of Microbiome in Human Metabolism

The host metabolism is regulated mostly by commensal metabolites secreted by the microbiome, mostly the intestinal microbiome. Gut microbiome provides additional nutrition and helps in host metabolisms, generates various enzymes, degradation processes (protein and glycosaminoglycans), short-chain fatty acids (SFA), metabolites (primary and secondary) vitamins, bile acids, and amino acids production (Canfora et al., 2015; Zhang et al., 2019; Chen et al., 2019; Wu et al., 2021a, b).

2.6 Respiratory System (Nasal Cavity, Pharynx, Larynx, Trachea, and Lungs)

Nasal Cavity The nasal cavity is directly exposed to the external environment, through air inhalation—the environment comprises pathogenic microorganisms, pollutants in the air and environment, and aeroallergens. Different types of pathogenic and harmless bacteria reside in the nasal cavity which is dependent on different factors such as temperature, humidity, and respiratory tract position. The anterior part of the nares has reduced microbiome biodiversity as compared to the meatus (middle part) and sphenoethmoidal recesses. In anterior nares lined with keratinized squamous epithelium and sebaceous glands, glands produce sebum, which affects bacterial diversity. However, a study demonstrated that no significant differences among bacterial diversity in different parts such as the middle meatus, anterior nares, and inferior turbinate in healthy individuals were detected.

The anterior nares contain keratinized squamous epithelium and sebaceous glands that produce sebum. Nevertheless, a study indicated that there were no significant differences in bacterial diversity among various sections like the middle meatus, anterior nares, and inferior turbinate in healthy individuals.

However, more studies are required to obtain exit information (Yan et al., 2013; Biswas et al., 2015). The nasal microbiome in an unhealthy state is not well characterized—further research should be explored. *Staphylococcus aureus* is known as a pathogen organism that causes nasal diseases such as chronic rhinosinusitis (CRS) or is associated with nasal polyps (Zhang et al., 2015).

Pharynx, Larynx, and Trachea The pharynx, larynx, and trachea are sterile parts of the body, which makes it difficult to culture bacteria from the tract sample. However, microbiomes from the vicinity could enter the upper tract (pharynx and larynx) by the lower tract (trachea) via the oral cavity or nasal routes. The upper respiratory tract has a greater abundance of microbiomes compared with the lower region. In asthma, *Proteobacteria* have lower abundance as compared to patients vs. healthy individuals. The most common diseases are chronic obstructive pulmonary disease (COPD) and asthma (Sethi et al., 2002; Kraft, 2000; Huang et al., 2011; Abreu et al., 2012).

Lungs In many literatures, the lungs are normally a sterile organ of the body. However, during the respiration process, the lungs are exposed to environmental microbes (Dickson et al., 2016). Changes in the composition of the lung microbiome are observed in cases of lung disease. The composition of the microbiome was altered in lung disease. Pathological conditions are increased airway wall permeability and mucus production, creating favorable conditions for bacterial growth (Schmidt et al., 2014; Worlitzsch et al., 2002). Respiratory exacerbations represent acute types of respiratory dysbiosis that are associated with the respiratory ecosystem and inflammatory state of the host dysregulation. These events prompt negative effects on the host. This condition only reverses via restoration of homeostasis (Dickson et al., 2016).

Skin The skin is the most exposed and interacts with the environmental organ in the human body. The diversity of the skin microbiota remains stable but varies among individuals' skin physiology. The microbial community is divided into three main groups based on skin area: oily, dry, and moist areas. The feet are distinct unstable microbial due to regular contact with the ground (Adamczyk et al., 2020; Byrd et al., 2018). Age also significantly impacts the skin microbiome. For example, in puberty, sebaceous glands produce sebum which favors the growth of lipophilic microorganisms (*Propionibacterium*, *Corynebacterium*, and *Malassezia* spp.), while in children, the most abundant microbes are *Streptococcaceae*, *Bacteroidetes*, and *Proteobacteria* (gammaproteobacterial and beta proteobacteria). Age is associated with diseases, for instance, in children the most common disease is atopic dermatitis, while the more prominent disease is tinea versicolor in adults. *S. aureus* caused atopic dermatitis (eczema) (Seebacher et al., 2008; Havlickova et al., 2008; Kyriakis et al., 2006; Jo et al., 2017, 2016; Oh et al., 2012; Leyden et al., 1974).

Urinary System The urinary bladder was considered sterile while the pathogen microbe commonly harbored in blader, and commensal microbes were identified in the urinary tract. Different studies demonstrated that the abundance and diversity of the microbiome in the urine were lower as compared to the gut by 10^6–10^7 times (Pearce et al., 2014; Zasloff, 2007).

Anatomical proximity and gender-specific physiology exert an influence on microbial communities and their abundance. For instance, in females, the reproductive and urinary tracts are closely situated. The main source of the microbial community in the urinary tract is the vagina; urogenital microbiota in both vaginal and urine samples was common which was reported by two different studies (Komesu et al., 2020; Thomas-White et al., 2018).

In the urinary tract, the most common bacterial infection is UTI associated with *Escherichia coli*, *Enterococcus*, and *Staphylococcus*. Furthermore, in women, *Gardnerella vaginalis* causes bacterial vaginosis, and it initiates the activation of *E. coli*-induced urinary tract infection (UTI) from inactive intracellular reservoirs within the bladder (Gilbert et al., 2017; Khasriya et al., 2013).

Reproductive System The human vagina has low pH, which inhibits the growth of numerous pathogenic microorganisms. The vaginal epithelium is regulated by hormones that help in maintaining the cervical mucus (mucins) layer (mucus consists of lipids, water, protein, and glycoproteins)—it is hypothesized that the mucus protects the vaginal epithelium and provides nutrient for the vaginal microbiome—

vaginal epithelium produces glycogen; it also provides nutrients to the vaginal microbiome.

The vaginal physiology is dependent on hormonal fluctuations. For instance, during menopause, there is a reduction in mucus and glycogen levels, which modified the microenvironment for the vaginal microbiome (Muhleisen & Herbst-Kralovetz, 2016; Kurman, 2013; Han et al., 2017). The human microbiome, a diverse community of microorganisms residing in various body sites, plays a critical role in maintaining health, and disruptions in its balance are linked to numerous diseases. For instance, in the oral cavity, dysbiosis is associated with dental caries and periodontitis. On the skin, imbalances contribute to conditions like atopic dermatitis. In the nose, chronic rhinosinusitis has been linked to microbiome disruptions, while in the lungs, dysbiosis is associated with asthma and chronic obstructive pulmonary disease (COPD). The stomach can develop conditions like gastric cancer due to microbial imbalance, and in the intestines, diseases such as inflammatory bowel disease (IBD), irritable bowel syndrome (IBS), celiac disease, and colorectal cancer are tied to dysbiosis. Similarly, microbiome imbalances in the bladder are linked to urinary tract infections (UTIs), as illustrated in Fig. 2.3.

To investigate the continuous association between microbiomes and health, continued research is required. There is a lack of knowledge for establishing causal mechanisms and pathways. In the future, various methods could be employed to better comprehend how the microbiome can be manipulated for therapeutic benefits (Table 2.1).

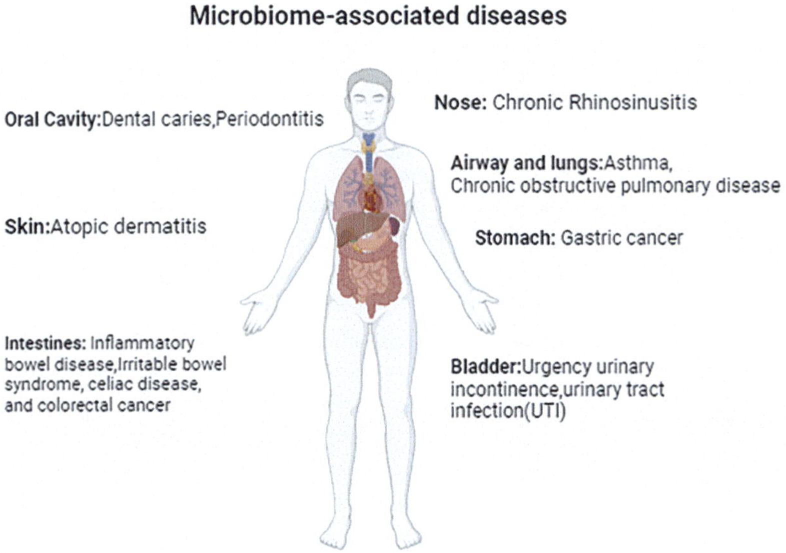

Fig. 2.3 Microbiome-associated disease on different body sites

Table 2.1 Prevalent microbiome present on different body sites

Origin	Bacterial genera	References
Oral cavity	**Bacterial genera:** *Granulicatella*, *Actinomyces*, *Prevotella*, *Corynebacterium*, *staphylococcus*, *Eikenella*, *Haemophilus*, *Streptococcus*, *Capnocytophaga*, *Gemella*, *Rothia*, *Veillonella*, *Fusobacterium*, *Prophyromonas*, *Neisseria*, *Treponema*, *Leptotrichia*, *lactobacillus*, *Peptostreptococcus*, *Eubacterium*, and *Propionibacterium* **Fungi genera:** *Candida*, *Fusarium*, *Cladosporium*, *Aspergillus*, *Cryptococcus*, and *Saccharomyces* **Viruses' genera:** Mumps, rabies, and human papillomaviruses **Protozoa genera:** *Trichomonas tenax* and *Entamoeba gingivalis*	Aas et al. (2005), Wade (2013), Kumaraswamy and Vidhya (2011), Wantland et al. (1958)
Stomach	**Bacterial genera:** *Proteobacteria*, *Bacteroidetes* (recently renamed to *Bacteroidota*), *Firmicutes* (recently renamed to *Bacillota*), *Fusobacteria*, *Lactobacillus*, *Actinobacteria*, *Veillonella*, *Clostridium*, and *H. pylori*	Zilberstein et al. (2007), Sung et al. (2016), Nardone and Compare (2015), Liu et al. (2019), Delgado et al. (2013), Bik et al. (2006), Oren and Garrity (2021)
Intestine	**Bacterial genera:** *Firmicutes* (*Bacillus*, *Enterococcus*, *Lactobacillus*, *Clostridium*, *Ruminococcus*, and *Staphylococcus aureus*), *Actinobacteria* (*Bifidobacterium*), *Bacteroides* (*Bacteroides* and *Prevotella*), *Verrucomjcobia*, *Proteobacteria* (*Helicobacter*, *Enterobacter*, *Salmonella*, *Shigella*, and *Escherichia coli*) **Viruses' genera:** *Methanosphaera stadtmanae* and *Methanobrevibacter smithii*	Arumugam et al. (2011), Greenblum et al. (2012)
Nasal cavity	**Bacterial genera**: *Actinobacteria*, *Firmicutes*, *Proteobacteria*, *staphylococcus* (*S. aureus*, *S. epidermidis*), *Propionibacterium* (*acnes*), *Corynebacterium*, or *Moraxella*	Ramakrishnan et al. (2013), Zhou et al. (2014), Bassis et al. (2014)
Pharynx, larynx, and trachea	**Bacterial genera:** *Streptococcus pneumoniae*, *Moraxella catarrhalis*, *Hemophilus influenza*, *Proteobacteria*	Bisgaard et al. (2007)

(continued)

Table 2.1 (continued)

Origin	Bacterial genera	References
Lungs	**Bacterial genera:** *Streptococcus*, *S. pneumoniae*, *Haemophilus*, *Neisseria*, *Corynebacteria*, *P. aeruginosa*, *Prevotella*, *Veillonella*, *Staphylococcus aureus*, and *Burkholderia cepacian* **Viruses' genera:** *Rhinovirus*, *Adenovirus*, measles, influenza, and Epstein-Barr **Fungi genera:** *Candida albicans*, *Aspergillus* spp., *Candida immitis*, and *Candida neoformans*	Soret et al. (2020), Hanada et al. (2018), Bassis et al. (2015), Dickson et al. (2016), Freestone et al. (2012), Kanangat et al. (1999), Lyte and Ernst (1992), Kaza et al. (2011)
Skin	**Bacterial genera:** *Propionibacterium*, *Staphylococcus*, and *Corynebacterium* **Fungi genera:** *Malassezia*, *Aspergillus*, *Cryptococcus*, *Rhodotorula*, *Epicoccum*	Findley et al. (2013), Oh et al. (2014), Grice and Segre (2011), Grice et al. (2009), Costello et al. (2009)
Urinary system	**Bacterial genera:** *Actinobacteria*, *Bacteroidetes*, *Pseudomonas*, *Escherichia*, *Streptococcus*, *Proteobacteria*, *Enterococcus*, *Citrobacter*, *Corynebacterium*, *Lactobacillus*, *Gardnerella vaginalis*, *Atopobium*, *Prevotella*, *Shigella*, and *Sneathia*	Modena et al. (2017), Mansour et al. (2020), Fouts et al. (2012)
Reproductive system	*Gardnerella*, *Prevotella*, *Atopobium*, *Sneathia*, *Megasphaera*, *Peptoniphilus*, and *lactobacillus*	Verstraelen et al. (2004), Zhou et al. (2004), Zhou et al. (2007), Ravel et al. (2011)

2.7 Microbiome for Therapeutic Applications

Within this section, we explore diverse strategies for the deliberate manipulation of the human microbiome and the potential of microbiome engineering in addressing diseases and promoting human well-being.

2.8 Probiotic Supplementation

Different studies demonstrated that reduced levels of specific microbial genera within the microbiome are significantly linked to the onset and progression of diseases. Consequently, supplements of beneficial bacteria (probiotics) can restore equilibrium to the microbiome. These strategies have been previously examined and have shown success in some cases. For example, Joung et al. conducted research to check the effect of *L. plantarum* K50 or *L. rhamnosus* GG in obese mice orally administered with a high-fat diet for 12 weeks (Joung et al., 2021).

The results showed that mice that were treated had reduced body weight, serum triglyceride, and high-density lipoprotein (HDL) cholesterol levels. *Firmicutes*/*Bacteroidetes* ratio was high as compared to control mice (nontreated obese mice). In another experiment, mice subjected to a high-fat diet exhibited a reversal in their resistance to leptin, a hormone responsible for regulating appetite. This dietary change also resulted in heightened diversity within their fecal microbiome, leading to a decrease in the number of *Proteobacteria* phylum (Borgeraas et al., 2018; Cheng & Liu, 2020).

Various clinical trials and experimental studies strongly indicate the potential of *Saccharomyces boulardii* as a biotherapeutic agent that can be used for the prevention and treatment of different gastrointestinal diseases. The use of probiotic S. boulardii shows effectiveness and positive outcomes in acute and chronic gastrointestinal conditions, as validated by clinical trials. Caution should be taken in patients with risk factors for adverse events (Kelesidis & Pothoulakis, 2012).

2.9 Prebiotic Supplementation

Prebiotics are nondigestible compounds that act as substrates for normal flora in the gut, promoting their growth and activity. Here are some common prebiotics and their uses:

Fructo-oligosaccharides (FOS): FOS are naturally occurring carbohydrates found in foods like fruits, vegetables, and grains. They stimulate the growth of *Bifidobacteria* and *Lactobacilli* in the gut. FOS helps in digestion, enhances mineral absorption, and supports the immune system.

Galacto-oligosaccharides (GOS): GOS are carbohydrates derived from lactose. They are found in human milk and certain legumes. GOS increases the growth of *Bifidobacteria* and normal flora to maintain gut health, reducing the risk of infections and supporting the immune system.

Xylo-oligosaccharides (XOS): XOS is derived from the hemicellulose fraction of plant fibers and stimulates the growth of *Bifidobacteria*, contributing to improved gut balance and potential immune system benefits.

Inulin: Inulin is a type of soluble fiber found in many plants, such as chicory root, garlic, and onions. It serves as a substrate for *Bifidobacteria*, promoting their growth and overall gut health. Inulin is used to improve bowel regularity and enhance calcium absorption.

Lactulose: Lactulose is a synthetic disaccharide (galactose and fructose). It has a prebiotic effect by stimulating the growth of beneficial bacteria. Lactulose is used for constipation and hepatic encephalopathy treatment.

Arabinoxylo-oligosaccharides (AXOS): AXOS are derived from wheat bran and other plant sources, promoting the growth of normal flora. AXOS helps in gut health and metabolic benefits.

Resistant starch (RS): RS is a type of starch that inhibits digestion in the small intestine—in the colon, it acts as a substrate for microbial fermentation. It can be found in foods like green bananas, legumes, and certain whole grains. Resistant starch promotes the growth of normal flora which are helpful for gut health.

These prebiotics can be incorporated into the diet through natural food sources or as supplements. They play a role in the composition of the gut microbiota and impact digestion, metabolism, and overall health.

An alternative approach such as probiotics is used with prebiotics. Probiotics are nondigestible substances that use the host microbiome to help in health. Prebiotics help in bacterial species growth (already present in the microbiome). The most commonly used known prebiotics are galacto-oligosaccharides (GOS), fructo-oligosaccharides (FOS), xylo-oligosaccharides (XOS), and lactulose. These substrates remodel the microbiome transitioning and change the disease state to a healthy state shown in Fig. 2.4. FOS, GOS, and XOS promote the proliferation of *Bifidobacterium* species in the human gut (Sanders et al., 2019: Macfarlane & Macfarlane, 2006: Liu et al., 2017).

The manipulation of the gut microbiota through prebiotics was used for cancer treatment. A study conducted by Han et al. demonstrated the potential effect of colorectal cancer by administering a gel to a mouse. This intervention led to enhanced antitumor efficacy, particularly when utilized combined with an immune checkpoint inhibitor protein-1 (α-PD-1) (anti-programmed cell death).

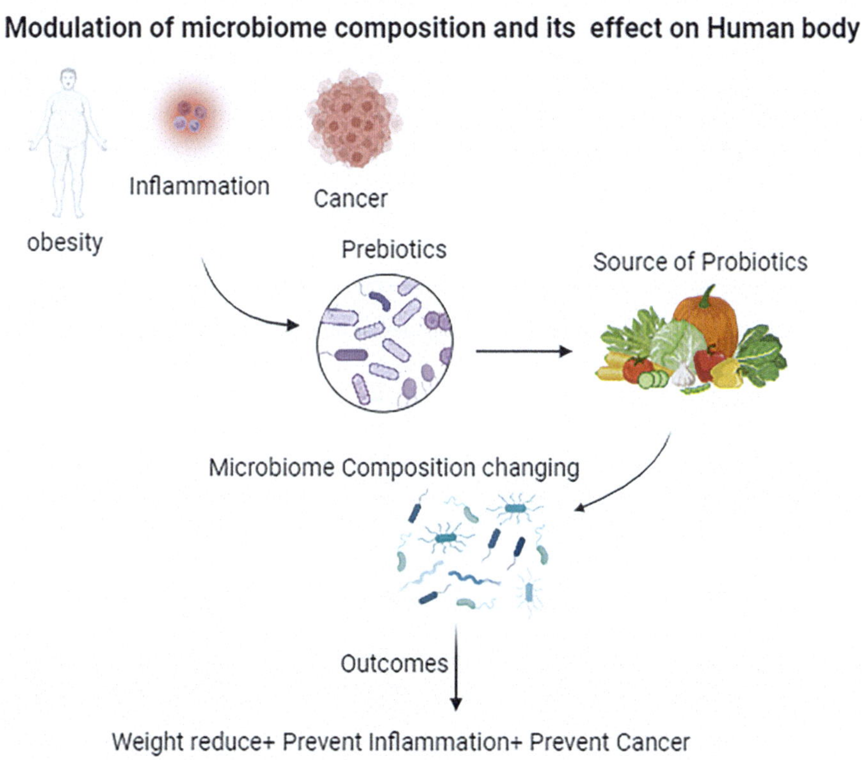

Fig. 2.4 Modulation of microbiome composition via probiotics and prebiotics. Administration of probiotics and prebiotics can change microbiome composition to maintain health

This intervention led to enhanced antitumor efficacy, particularly when utilized alongside an immune checkpoint inhibitor, namely, anti-programmed cell death protein-1 (α-PD-1).

Previous research highlighted that individuals who exhibit positive responses to immune checkpoint inhibitors had a higher abundance of beneficial bacteria in their gut microbiome, including *Akkermansia*, *Bifidobacterium*, *Faecalibacterium*, and *Ruminococcaceae*, in contrast to those who do not respond. The oral administration of the inulin gel to mice showed increased Akkermansia, Lactobacillus, and Roseburia species, T cell response within the mice with α-PD-1 (antitumor activity) (Han et al., 2021).

Conversely, those who do not respond well tend to lack this abundance. In a study by Han et al., it was demonstrated that administering inulin gel orally to mice resulted in heightened levels of *Akkermansia*, *Roseburia*, and Lactobacillus species. This, in turn, enhanced the T cell response among the mice treated with α-PD-1, leading to antitumor activity (Routy et al., 2018: Gopalakrishnan et al., 2018: Han et al., 2021) (Table 2.2).

Table 2.2 Prebiotics and probiotics are used to treat different diseases

Supplement types	Supplementation name	Outcomes	References
Probiotic	*Lactobacillus* and *Bifidobacterium* strains, *Akkermansia muciniphila*	Obesity, cardiometabolic diseases, and diabetes	Cani and de Vos (2017)
	A. muciniphila	Obesity, hypertension, IBD, and type 2 diabetes	
	L. fermentum KBL374 & KBL375	Inflammatory cytokine levels decreased, and anti-inflammatory interleukin (IL)-10 levels increased	Jang et al. (2019)
	F. prausnitzii	Colitis	Martín et al. (2014)
	Caudovirales bacteriophages	Intestinal inflammation	Norman et al. (2015)
	S. boulardii	IBD in remission	Kelesidis and Pothoulakis (2012)
	Commensal bacteria	Skin epithelial barrier function improvement	Myles et al. (2020)
	L. crispatus CTV-05	To treat UTI	Stapleton et al. (2011)
Prebiotic	FOS	Crohn's disease	Lindsay et al. (2006)
	GOS	IBS syndromes	Silk et al. (2009)
	Inulin gel	Antitumor effect	Han et al. (2021)
	Inulin-enriched FOS	Prevent colorectal cancer	Klinder et al. (2004)
	Lactose	Improvement in lactose digestion	Labayen et al. (2001)

References

Aas, J. A., Paster, B. J., Stokes, L. N., Olsen, I., & Dewhirst, F. E. (2005). Defining the normal bacterial flora of the oral cavity. *Journal of Clinical Microbiology, 43*(11), 5721–5732.

Abreu, N. A., Nagalingam, N. A., Song, Y., Roediger, F. C., Pletcher, S. D., Goldberg, A. N., & Lynch, S. V. (2012). Sinus microbiome diversity depletion and Corynebacterium tuberculostearicum enrichment mediates rhinosinusitis. *Science Translational Medicine, 4*(151), 151ra124.

Adamczyk, K., Garncarczyk, A., Antończak, P., & Wcisło-Dziadecka, D. (2020). The foot microbiome. *Journal of Cosmetic Dermatology, 19*(5), 1039–1043.

Arumugam, M., Raes, J., Pelletier, E., Le Paslier, D., Yamada, T., Mende, D. R., et al. (2011). Enterotypes of the human gut microbiome. *Nature, 473*(7346), 174–180.

Baker, J. L., Bor, B., Agnello, M., Shi, W., & He, X. (2017). Ecology of the oral microbiome: Beyond bacteria. *Trends in Microbiology, 25*(5), 362–374.

Baquero, F., & Nombela, C. (2012). The microbiome is a human organ. *Clinical Microbiology and Infection, 18*, 2–4.

Bartlett, J. G. (1981). Antimicrobial agents implicated in Clostridium difficile toxin-associated diarrhea of colitis. *The Johns Hopkins Medical Journal, 149*(1), 6–9.

Bassis, C. M., Tang, A. L., Young, V. B., & Pynnonen, M. A. (2014). The nasal cavity microbiota of healthy adults. *Microbiome, 2*(1), 1–5.

Bassis, C. M., Erb-Downward, J. R., Dickson, R. P., Freeman, C. M., Schmidt, T. M., Young, V. B., et al. (2015). Analysis of the upper respiratory tract microbiotas as the source of the lung and gastric microbiotas in healthy individuals. *MBio, 6*(2), e00037–e00015.

Bhattarai, Y., Muniz Pedrogo, D. A., & Kashyap, P. C. (2017). Irritable bowel syndrome: A gut microbiota-related disorder? *American Journal of Physiology-Gastrointestinal and Liver Physiology, 312*(1), G52–G62.

Biasucci, G., Rubini, M., Riboni, S., Morelli, L., Bessi, E., & Retetangos, C. (2010). Mode of delivery affects the bacterial community in the newborn gut. *Early Human Development, 86*(1), 13–15.

Bik, E. M., Eckburg, P. B., Gill, S. R., Nelson, K. E., Purdom, E. A., Francois, F., et al. (2006). Molecular analysis of the bacterial microbiota in the human stomach. *Proceedings of the National Academy of Sciences, 103*(3), 732–737.

Bisgaard, H., Hermansen, M. N., Buchvald, F., Loland, L., Halkjaer, L. B., Bønnelykke, K., et al. (2007). Childhood asthma after bacterial colonization of the airway in neonates. *New England Journal of Medicine, 357*(15), 1487–1495.

Biswas, K., Hoggard, M., Jain, R., Taylor, M. W., & Douglas, R. G. (2015). The nasal microbiota in health and disease: Variation within and between subjects. *Frontiers in Microbiology, 6*, 134.

Booijink, C. C., El-Aidy, S., Rajilić-Stojanović, M., Heilig, H. G., Troost, F. J., Smidt, H., et al. (2010). High temporal and inter-individual variation detected in the human ileal microbiota. *Environmental Microbiology, 12*(12), 3213–3227.

Borgeraas, H., Johnson, L. K., Skattebu, J., Hertel, J. K., & Hjelmesaeth, J. (2018). Effects of probiotics on body weight, body mass index, fat mass and fat percentage in subjects with overweight or obesity: A systematic review and meta-analysis of randomized controlled trials. *Obesity Reviews, 19*(2), 219–232.

Byrd, A. L., Belkaid, Y., & Segre, J. A. (2018). The human skin microbiome. *Nature Reviews Microbiology, 16*(3), 143–155.

Canfora, E. E., Jocken, J. W., & Blaak, E. E. (2015). Short-chain fatty acids in control of body weight and insulin sensitivity. *Nature Reviews Endocrinology, 11*(10), 577–591.

Cani, P. D., & de Vos, W. M. (2017). Next-generation beneficial microbes: The case of Akkermansia muciniphila. *Frontiers in Microbiology, 8*, 1765.

Carroll, I. M., Chang, Y. H., Park, J., Sartor, R. B., & Ringel, Y. (2010). Luminal and mucosal-associated intestinal microbiota in patients with diarrhea-predominant irritable bowel syndrome. *Gut Pathogens, 2*(1), 1–9.

Chen, S., Henderson, A., Petriello, M. C., Romano, K. A., Gearing, M., Miao, J., et al. (2019). Trimethylamine N-oxide binds and activates PERK to promote metabolic dysfunction. *Cell Metabolism, 30*(6), 1141–1151.

Cheng, Y. C., & Liu, J. R. (2020). Effect of Lactobacillus rhamnosus GG on energy metabolism, leptin resistance, and gut microbiota in mice with diet-induced obesity. *Nutrients, 12*(9), 2557.

Claesson, M. J., Cusack, S., O'Sullivan, O., Greene-Diniz, R., de Weerd, H., Flannery, E., et al. (2011). Composition, variability, and temporal stability of the intestinal microbiota of the elderly. *Proceedings of the National Academy of Sciences, 108*(supplement_1), 4586–4591.

Costello, E. K., Lauber, C. L., Hamady, M., Fierer, N., Gordon, J. I., & Knight, R. (2009). Bacterial community variation in human body habitats across space and time. *Science, 326*(5960), 1694–1697.

Craig, S. J., Blankenberg, D., Parodi, A. C. L., Paul, I. M., Birch, L. L., Savage, J. S., et al. (2018). Child weight gain trajectories linked to oral microbiota composition. *Scientific Reports, 8*(1), 14030.

Cunha, F. A., Cota, L. O., Cortelli, S. C., Miranda, T. B., Neves, F. S., Cortelli, J. R., & Costa, F. O. (2019). Periodontal condition and levels of bacteria associated with periodontitis in individuals with bipolar affective disorders: A case-control study. *Journal of Periodontal Research, 54*(1), 63–72.

Delgado, S., Cabrera-Rubio, R., Mira, A., Suárez, A., & Mayo, B. (2013). Microbiological survey of the human gastric ecosystem using culturing and pyrosequencing methods. *Microbial Ecology, 65*, 763–772.

Dickson, R. P., Erb-Downward, J. R., Martinez, F. J., & Huffnagle, G. B. (2016). The microbiome and the respiratory tract. *Annual Review of Physiology, 78*, 481–504.

Do, L. G., Ha, D. H., Bell, L. K., Devenish, G., Golley, R. K., Leary, S. D., et al. (2020). Study of mothers' and infants' life events affecting oral health (SMILE) birth cohort study: Cohort profile. *BMJ Open, 10*(10), e041185.

Eggesbø, M., Moen, B., Peddada, S., Baird, D., Rugtveit, J., Midtvedt, T., et al. (2011). Development of gut microbiota in infants not exposed to medical interventions. *APMIS, 119*(1), 17–35.

Fallani, M., Young, D., Scott, J., Norin, E., Amarri, S., Adam, R., et al. (2010). Intestinal microbiota of 6-week-old infants across Europe: Geographic influence beyond delivery mode, breast-feeding, and antibiotics. *Journal of Pediatric Gastroenterology and Nutrition, 51*(1), 77–84.

Fallani, M., Amarri, S., Uusijarvi, A., Adam, R., Khanna, S., Aguilera, M., et al. (2011). Determinants of the human infant intestinal microbiota after the introduction of first complementary foods in infant samples from five European centres. *Microbiology, 157*(5), 1385–1392.

Favier, C. F., Vaughan, E. E., De Vos, W. M., & Akkermans, A. D. (2002). Molecular monitoring of succession of bacterial communities in human neonates. *Applied and Environmental Microbiology, 68*(1), 219–226.

Fejerskov, O. (2004). Changing paradigms in concepts on dental caries: Consequences for oral health care. *Caries Research, 38*(3), 182–191.

Filoche, S., Wong, L., & Sissons, C. H. (2010). Oral biofilms: Emerging concepts in microbial ecology. *Journal of Dental Research, 89*(1), 8–18.

Findley, K., Oh, J., Yang, J., Conlan, S., Deming, C., Meyer, J. A., et al. (2013). Topographic diversity of fungal and bacterial communities in human skin. *Nature, 498*(7454), 367–370.

Flint, H. J., Scott, K. P., Louis, P., & Duncan, S. H. (2012). The role of the gut microbiota in nutrition and health. *Nature Reviews Gastroenterology & Hepatology, 9*(10), 577–589.

Fouts, D. E., Pieper, R., Szpakowski, S., Pohl, H., Knoblach, S., Suh, M. J., et al. (2012). Integrated next-generation sequencing of 16S rDNA and metaproteomics differentiate the healthy urine microbiome from asymptomatic bacteriuria in neuropathic bladder associated with spinal cord injury. *Journal of Translational Medicine, 10*(1), 1–17.

Freestone, P. P., Hirst, R. A., Sandrini, S. M., Sharaff, F., Fry, H., Hyman, S., & O'Callaghan, C. (2012). Pseudomonas aeruginosa-catecholamine inotrope interactions: A contributory factor in the development of ventilator-associated pneumonia? *Chest, 142*(5), 1200–1210.

Gilbert, N. M., O'Brien, V. P., & Lewis, A. L. (2017). Transient microbiota exposures activate dormant Escherichia coli infection in the bladder and drive severe outcomes of recurrent disease. *PLoS Pathogens, 13*(3), e1006238.

Gopalakrishnan, V., Spencer, C. N., Nezi, L., Reuben, A., Andrews, M. C., Karpinets, T., et al. (2018). Gut microbiome modulates response to anti–PD-1 immunotherapy in melanoma patients. *Science, 359*(6371), 97–103.

Greenblum, S., Turnbaugh, P. J., & Borenstein, E. (2012). Metagenomic systems biology of the human gut microbiome reveals topological shifts associated with obesity and inflammatory bowel disease. *Proceedings of the National Academy of Sciences, 109*(2), 594–599.

Grice, E. A., & Segre, J. A. (2011). The skin microbiome. *Nature Reviews Microbiology, 9*(4), 244–253.

Grice, E. A., Kong, H. H., Conlan, S., Deming, C. B., Davis, J., Young, A. C., et al. (2009). Topographical and temporal diversity of the human skin microbiome. *Science, 324*(5931), 1190–1192.

Han, L., Taub, R., & Jensen, J. T. (2017). Cervical mucus and contraception: What we know and what we don't. *Contraception, 96*(5), 310–321.

Han, K., Nam, J., Xu, J., Sun, X., Huang, X., Animasahun, O., et al. (2021). Generation of systemic antitumour immunity via the in situ modulation of the gut microbiome by an orally administered inulin gel. *Nature Biomedical Engineering, 5*(11), 1377–1388.

Hanada, S., Pirzadeh, M., Carver, K. Y., & Deng, J. C. (2018). Respiratory viral infection-induced microbiome alterations and secondary bacterial pneumonia. *Frontiers in Immunology, 9*, 2640.

Harmsen, H. J., Wildeboer-Veloo, A. C., Raangs, G. C., Wagendorp, A. A., Klijn, N., Bindels, J. G., & Welling, G. W. (2000). Analysis of intestinal flora development in breast-fed and formula-fed infants by using molecular identification and detection methods. *Journal of Pediatric Gastroenterology and Nutrition, 30*(1), 61–67.

Havlickova, B., Czaika, V. A., & Friedrich, M. (2008). Epidemiological trends in skin mycoses worldwide. *Mycoses, 51*, 2–15.

Horz, H. P., & Conrads, G. (2007). Diagnosis and anti-infective therapy of periodontitis. *Expert Review of Anti-Infective Therapy, 5*(4), 703–715.

Huang, Y. J., Nelson, C. E., Brodie, E. L., DeSantis, T. Z., Baek, M. S., Liu, J., et al. (2011). Airway microbiota and bronchial hyperresponsiveness in patients with suboptimally controlled asthma. *Journal of Allergy and Clinical Immunology, 127*(2), 372–381.

Huurre, A., Kalliomäki, M., Rautava, S., Rinne, M., Salminen, S., & Isolauri, E. (2008). Mode of delivery–effects on gut microbiota and humoral immunity. *Neonatology, 93*(4), 236–240.

Islam, K. S., Fukiya, S., Hagio, M., Fujii, N., Ishizuka, S., Ooka, T., et al. (2011). Bile acid is a host factor that regulates the composition of the cecal microbiota in rats. *Gastroenterology, 141*(5), 1773–1781.

Jang, Y. J., Kim, W. K., Han, D. H., Lee, K., & Ko, G. (2019). Lactobacillus fermentum species ameliorate dextran sulfate sodium-induced colitis by regulating the immune response and altering gut microbiota. *Gut Microbes, 10*(6), 696–711.

Jensen, E. D., Selway, C. A., Allen, G., Bednarz, J., Weyrich, L. S., Gue, S., et al. (2021). Early markers of periodontal disease and altered oral microbiota are associated with glycemic control in children with type 1 diabetes. *Pediatric Diabetes, 22*(3), 474–481.

Jo, J. H., Deming, C., Kennedy, E. A., Conlan, S., Polley, E. C., Ng, W. I., et al. (2016). Diverse human skin fungal communities in children converge in adulthood. *Journal of Investigative Dermatology, 136*(12), 2356–2363.

Jo, J. H., Kennedy, E. A., & Kong, H. H. (2017). Topographical and physiological differences of the skin mycobiome in health and disease. *Virulence, 8*(3), 324–333.

Joung, H., Chu, J., Kim, B. K., Choi, I. S., Kim, W., & Park, T. S. (2021). Probiotics ameliorate chronic low-grade inflammation and fat accumulation with gut microbiota composition change in diet-induced obese mice models. *Applied Microbiology and Biotechnology, 105*, 1203–1213.

Kanangat, S., Meduri, G. U., Tolley, E. A., Patterson, D. R., Meduri, C. U., Pak, C., et al. (1999). Effects of cytokines and endotoxin on the intracellular growth of bacteria. *Infection and Immunity, 67*(6), 2834–2840.

Karlsson, C. L., Molin, G., Cilio, C. M., & Ahrné, S. (2011). The pioneer gut microbiota in human neonates vaginally born at term—A pilot study. *Pediatric Research, 70*(3), 282–286.

Kaza, S. K., McClean, S., & Callaghan, M. (2011). IL-8 released from human lung epithelial cells induced by cystic fibrosis pathogens Burkholderia cepacia complex affects the growth and intracellular survival of bacteria. *International Journal of Medical Microbiology, 301*(1), 26–33.

Kelesidis, T., & Pothoulakis, C. (2012). Efficacy and safety of the probiotic Saccharomyces boulardii for the prevention and therapy of gastrointestinal disorders. *Therapeutic Advances in Gastroenterology, 5*(2), 111–125.

Khasriya, R., Sathiananthamoorthy, S., Ismail, S., Kelsey, M., Wilson, M., Rohn, J. L., & Malone-Lee, J. (2013). Spectrum of bacterial colonization associated with urothelial cells from patients with chronic lower urinary tract symptoms. *Journal of Clinical Microbiology, 51*(7), 2054–2062.

Klaassens, E. S., Boesten, R. J., Haarman, M., Knol, J., Schuren, F. H., Vaughan, E. E., & De Vos, W. M. (2009). Mixed-species genomic microarray analysis of fecal samples reveals differential transcriptional responses of bifidobacteria in breast-and formula-fed infants. *Applied and Environmental Microbiology, 75*(9), 2668–2676.

Klinder, A., Gietl, E., Hughes, R., Jonkers, N., Karlsson, P., McGlyn, H., et al. (2004). Gut fermentation products of insulin-derived prebiotics beneficially modulate markers of tumour progression in human colon tumour cells. *International Journal of Cancer Prevention, 1*(1), 19–32.

Knight, R., Callewaert, C., Marotz, C., Hyde, E. R., Debelius, J. W., McDonald, D., & Sogin, M. L. (2017). The microbiome and human biology. *Annual Review of Genomics and Human Genetics, 18*, 65–86.

Komesu, Y. M., Dinwiddie, D. L., Richter, H. E., Lukacz, E. S., Sung, V. W., Siddiqui, N. Y., et al. (2020). Defining the relationship between vaginal and urinary microbiomes. *American Journal of Obstetrics and Gynecology, 222*(2), 154–1e1.

Kraft, M. (2000). The role of bacterial infections in asthma. *Clinics in Chest Medicine, 21*(2), 301–313.

Kumaraswamy, K. L., & Vidhya, M. (2011). Human papilloma virus and oral infections: An update. *Journal of Cancer Research and Therapeutics, 7*(2), 120–127.

Kurman, R. J. (Ed.). (2013). *Blaustein's pathology of the female genital tract*. Springer Science & Business Media.

Kyriakis, K. P., Terzoudi, S., Palamaras, I., Pagana, G., Michailides, C., & Emmanuelides, S. (2006). Pityriasis versicolor prevalence by age and gender. *Mycoses, 49*(6), 517–518.

Labayen, I., Forga, L., Gonzalez, A., Lenoir-Wijnkoop, I., & Martinez, J. A. (2001). Relationship between lactose digestion, gastrointestinal transit time and symptoms in lactose malabsorbers after dairy consumption. *Alimentary Pharmacology & Therapeutics, 15*(4), 543–549.

Leyden, J. J., Marples, R. R., & Kligman, A. M. (1974). Staphylococcus aureus in the lesions of atopic dermatitis. *British Journal of Dermatology, 90*(5), 525–525.

Li, X., Kolltveit, K. M., Tronstad, L., & Olsen, I. (2000). Systemic diseases caused by oral infection. *Clinical Microbiology Reviews, 13*(4), 547–558.

Li, T. H., Qin, Y., Sham, P. C., Lau, K. S., Chu, K. M., & Leung, W. K. (2017). Alterations in gastric microbiota after H. Pylori eradication and in different histological stages of gastric carcinogenesis. *Scientific Reports, 7*(1), 44935.

Lin, D., Hutchison, K. E., Portillo, S., Vegara, V., Ellingson, J. M., Liu, J., et al. (2019). Association between the oral microbiome and brain resting state connectivity in smokers. *NeuroImage, 200*, 121–131.

Lindsay, J. O., Whelan, K., Stagg, A. J., Gobin, P., Al-Hassi, H. O., Rayment, N., et al. (2006). Clinical, microbiological, and immunological effects of fructo-oligosaccharide in patients with Crohn's disease. *Gut, 55*(3), 348–355.

Liu, F., Li, P., Chen, M., Luo, Y., Prabhakar, M., Zheng, H., et al. (2017). Fructooligosaccharide (FOS) and galactooligosaccharide (GOS) increase Bifidobacterium but reduce butyrate producing bacteria with adverse glycemic metabolism in healthy young population. *Scientific Reports, 7*(1), 11789.

Liu, X., Shao, L., Liu, X., Ji, F., Mei, Y., Cheng, Y., et al. (2019). Alterations of gastric mucosal microbiota across different stomach microhabitats in a cohort of 276 patients with gastric cancer. *eBioMedicine, 40*, 336–348.

Loesche, W. (2007). Dental caries and periodontitis: Contrasting two infections that have medical implications. *Infectious Disease Clinics of North America, 21*(2), 471–502.

Lyte, M., & Ernst, S. (1992). Catecholamine induced growth of gram negative bacteria. *Life Sciences, 50*(3), 203–212.

Macfarlane, S. M. G. T., & Macfarlane, G. (2006). Cummings Jt. Review article: Prebiotics in the gastrointestinal tract. *Alimentary Pharmacology & Therapeutics, 24*(5), 701–714.

Mansour, B., Monyók, Á., Makra, N., Gajdács, M., Vadnay, I., Ligeti, B., et al. (2020). Bladder cancer-related microbiota: Examining differences in urine and tissue samples. *Scientific Reports, 10*(1), 11042.

Martín, R., Chain, F., Miquel, S., Lu, J., Gratadoux, J. J., Sokol, H., et al. (2014). The commensal bacterium Faecalibacterium prausnitzii is protective in DNBS-induced chronic moderate and severe colitis models. *Inflammatory Bowel Diseases, 20*(3), 417–430.

Mitchell, D. R., Derakhshan, M. H., Wirz, A. A., Orange, C., Ballantyne, S. A., Going, J. J., & McColl, K. E. (2017). The gastric acid pocket is attenuated in H. pylori infected subjects. *Gut, 66*(9), 1555–1562.

Modena, B. D., Milam, R., Harrison, F., Cheeseman, J. A., Abecassis, M. M., Friedewald, J. J., et al. (2017). Changes in urinary microbiome populations correlate in kidney transplants with interstitial fibrosis and tubular atrophy documented in early surveillance biopsies. *American Journal of Transplantation, 17*(3), 712–723.

Muhleisen, A. L., & Herbst-Kralovetz, M. M. (2016). Menopause and the vaginal microbiome. *Maturitas, 91*, 42–50.

Myles, I. A., Castillo, C. R., Barbian, K. D., Kanakabandi, K., Virtaneva, K., Fitzmeyer, E., et al. (2020). Therapeutic responses to Roseomonas mucosa in atopic dermatitis may involve lipid-mediated TNF-related epithelial repair. *Science Translational Medicine, 12*(560), eaaz8631.

Nardone, G., & Compare, D. (2015). The human gastric microbiota: Is it time to rethink the pathogenesis of stomach diseases? *United European Gastroenterology Journal, 3*(3), 255–260.

Norman, J. M., Handley, S. A., Baldridge, M. T., Droit, L., Liu, C. Y., Keller, B. C., et al. (2015). Disease-specific alterations in the enteric virome in inflammatory bowel disease. *Cell, 160*(3), 447–460.

O'Toole, P. W., & Claesson, M. J. (2010). Gut microbiota: Changes throughout the lifespan from infancy to elderly. *International Dairy Journal, 20*(4), 281–291.

Oh, J., Conlan, S., Polley, E. C., Segre, J. A., & Kong, H. H. (2012). Shifts in human skin and nares microbiota of healthy children and adults. *Genome Medicine, 4*, 1–11.

Oh, J., Byrd, A. L., Deming, C., Conlan, S., Kong, H. H., & Segre, J. A. (2014). Biogeography and individuality shape function in the human skin metagenome. *Nature, 514*(7520), 59–64.

Oren, A., & Garrity, G. M. (2021). Valid publication of the names of forty-two phyla of prokaryotes. *International Journal of Systematic and Evolutionary Microbiology, 71*(10), 005056.

Parahitiyawa, N. B., Scully, C., Leung, W. K., Yam, W. C., Jin, L. J., & Samaranayake, L. P. (2010). Exploring the oral bacterial flora: Current status and future directions. *Oral Diseases, 16*(2), 136–145.

Pear, S. M., Williamson, T. H., Bettin, K. M., Gerding, D. N., & Galgiani, J. N. (1994). Decrease in nosocomial Clostridium difficile–associated diarrhea by restricting clindamycin use. *Annals of Internal Medicine, 120*(4), 272–277.

Pearce, M. M., Hilt, E. E., & Rosenfeld, A. B. (2014). The female urinary microbiome: a comparison of women with and without urgency urinary incontinence. *mBio, 5*(4), 1–12.

Ramakrishnan, V. R., Feazel, L. M., Gitomer, S. A., Ir, D., Robertson, C. E., & Frank, D. N. (2013). The microbiome of the middle meatus in healthy adults. *PLoS One, 8*(12), e85507.

Ravel, J., Gajer, P., Abdo, Z., Schneider, G. M., Koenig, S. S., McCulle, S. L., et al. (2011). Vaginal microbiome of reproductive-age women. *Proceedings of the National Academy of Sciences, 108*(supplement_1), 4680–4687.

Roger, L. C., & McCartney, A. L. (2010). Longitudinal investigation of the faecal microbiota of healthy full-term infants using fluorescence in situ hybridization and denaturing gradient gel electrophoresis. *Microbiology, 156*(11), 3317–3328.

Routy, B., Le Chatelier, E., Derosa, L., Duong, C. P., Alou, M. T., Daillère, R., et al. (2018). Gut microbiome influences efficacy of PD-1–based immunotherapy against epithelial tumors. *Science, 359*(6371), 91–97.

Salonen, A., de Vos, W. M., & Palva, A. (2010). Gastrointestinal microbiota in irritable bowel syndrome: Present state and perspectives. *Microbiology, 156*(11), 3205–3215.

Sanders, M. E., Merenstein, D. J., Reid, G., Gibson, G. R., & Rastall, R. A. (2019). Probiotics and prebiotics in intestinal health and disease: From biology to the clinic. *Nature Reviews Gastroenterology & Hepatology, 16*(10), 605–616.

Schmidt, A., Belaaouaj, A., Bissinger, R., Koller, G., Malleret, L., d'Orazio, C., et al. (2014). Neutrophil elastase-mediated increase in airway temperature during inflammation. *Journal of Cystic Fibrosis, 13*(6), 623–631.

Seebacher, C., Bouchara, J. P., & Mignon, B. (2008). Updates on the epidemiology of dermatophyte infections. *Mycopathologia, 166*, 335–352.

Sethi, S., Evans, N., Grant, B. J., & Murphy, T. F. (2002). New strains of bacteria and exacerbations of chronic obstructive pulmonary disease. *New England Journal of Medicine, 347*(7), 465–471.

Shafquat, A., Joice, R., Simmons, S. L., & Huttenhower, C. (2014). Functional and phylogenetic assembly of microbial communities in the human microbiome. *Trends in Microbiology, 22*(5), 261–266.

Silk, D. B. A., Davis, A., Vulevic, J., Tzortzis, G., & Gibson, G. R. (2009). Clinical trial: The effects of a trans-galactooligosaccharide prebiotic on faecal microbiota and symptoms in irritable bowel syndrome. *Alimentary Pharmacology & Therapeutics, 29*(5), 508–518.

Song, H. J., Shim, K. N., Jung, S. A., Choi, H. J., Lee, M. A., Ryu, K. H., et al. (2008). Antibiotic-associated diarrhea: Candidate organisms other than Clostridium difficile. *The Korean Journal of Internal Medicine, 23*(1), 9.

Soret, P., Vandenborght, L.-E., Francis, F., Coron, N., Enaud, R., Avalos, M., Schaeverbeke, T., et al. (2020). Respiratory mycobiome and suggestion of inter-kingdom network during acute pulmonary exacerbation in cystic fibrosis. *Scientific Reports, 10*(1), 3589.

Stapleton, A. E., Au-Yeung, M., Hooton, T. M., Fredricks, D. N., Roberts, P. L., Czaja, C. A., et al. (2011). Randomized, placebo-controlled phase 2 trial of a lactobacillus crispatus probiotic given intravaginally for prevention of recurrent urinary tract infection. *Clinical Infectious Diseases, 52*(10), 1212–1217.

Sung, J., Kim, N., Kim, J., Jo, H. J., Park, J. H., Nam, R. H., et al. (2016). Comparison of gastric microbiota between gastric juice and mucosa by next generation sequencing method. *Journal of Cancer Prevention, 21*(1), 60.

Thomas, A. M., & Segata, N. (2019). Multiple levels of the unknown in microbiome research. *BMC Biology, 17*(1), 1–4.

Thomas-White, K. J., Gao, X., Lin, H., Fok, C. S., Ghanayem, K., Mueller, E. R., et al. (2018). Urinary microbes and postoperative urinary tract infection risk in urogynecologic surgical patients. *International Urogynecology Journal, 29*, 1797–1805.

Verstraelen, H., Verhelst, R., Claeys, G., Temmerman, M., & Vaneechoutte, M. (2004). Culture-independent analysis of vaginal microflora: The unrecognized association of Atopobium vaginae with bacterial vaginosis. *American Journal of Obstetrics and Gynecology, 191*(4), 1130–1132.

Vieira, A. T., Castelo, P. M., Ribeiro, D. A., & Ferreira, C. M. (2017). Influence of oral and gut microbiota in the health of menopausal women. *Frontiers in Microbiology, 8*, 298600.

Wade, W. G. (2013). The oral microbiome in health and disease. *Pharmacological Research, 69*(1), 137–143.

Wantland, W. W., Wantland, E. M., Remo, J. W., & Winquist, D. L. (1958). Studies on human mouth protozoa. *Journal of Dental Research, 37*(5), 949–950.

Wescombe, P. A., Heng, N. C., Burton, J. P., Chilcott, C. N., & Tagg, J. R. (2009). Streptococcal bacteriocins and the case for streptococcus salivarius as model oral probiotics. *Future Microbiology, 4*(7), 819–835.

Williams, R. C., Barnett, A. H., Claffey, N., Davis, M., Gadsby, R., Kellett, M., et al. (2008). The potential impact of periodontal disease on general health: A consensus view. *Current Medical Research and Opinion, 24*(6), 1635–1643.

Willis, J. R., & Gabaldón, T. (2020). The human oral microbiome in health and disease: From sequences to ecosystems. *Microorganisms, 8*(2), 308.

Woodmansey, E. J. (2007). Intestinal bacteria and ageing. *Journal of Applied Microbiology, 102*(5), 1178–1186.

Worlitzsch, D., Tarran, R., Ulrich, M., Schwab, U., Cekici, A., Meyer, K. C., et al. (2002). Effects of reduced mucus oxygen concentration in airway Pseudomonas infections of cystic fibrosis patients. *The Journal of Clinical Investigation, 109*(3), 317–325.

Wu, W. M., Yang, Y. S., & Peng, L. H. (2014). Microbiota in the stomach: New insights. *Journal of Digestive Diseases, 15*(2), 54–61.

Wu, Q., Sun, L., Hu, X., Wang, X., Xu, F., Chen, B., et al. (2021a). Suppressing the intestinal farnesoid X receptor/sphingomyelin phosphodiesterase 3 axis decreases atherosclerosis. *The Journal of Clinical Investigation, 131*, 9.

Wu, J., Wang, K., Wang, X., Pang, Y., & Jiang, C. (2021b). The role of the gut microbiome and its metabolites in metabolic diseases. *Protein & Cell, 12*(5), 360–373.

Xue, L., Zou, X., Yang, X. Q., Peng, F., Yu, D. K., & Du, J. R. (2020). Chronic periodontitis induces microbiota-gut-brain axis disorders and cognitive impairment in mice. *Experimental Neurology, 326*, 113176.

Yan, M., Pamp, S. J., Fukuyama, J., Hwang, P. H., Cho, D. Y., Holmes, S., & Relman, D. A. (2013). Nasal microenvironments and interspecific interactions influence nasal microbiota complexity and S. Aureus carriage. *Cell Host & Microbe, 14*(6), 631–640.

Yang, I., Nell, S., & Suerbaum, S. (2013). Survival in hostile territory: The microbiota of the stomach. *FEMS Microbiology Reviews, 37*(5), 736–761.

Yang, I., Arthur, R. A., Zhao, L., Clark, J., Hu, Y., Corwin, E. J., & Lah, J. (2021). The oral microbiome and inflammation in mild cognitive impairment. *Experimental Gerontology, 147*, 111273.

Yu, G., Torres, J., Hu, N., Medrano-Guzman, R., Herrera-Goepfert, R., Humphrys, M. S., et al. (2017). Molecular characterization of the human stomach microbiota in gastric cancer patients. *Frontiers in Cellular and Infection Microbiology, 7*, 302.

Zarco, M. F., Vess, T. J., & Ginsburg, G. S. (2012). The oral microbiome in health and disease and the potential impact on personalized dental medicine. *Oral Diseases, 18*(2), 109–120.

Zasloff, M. (2007). Antimicrobial peptides, innate immunity, and the normally sterile urinary tract. *Journal of the American Society of Nephrology, 18*(11), 2810–2816.

Zhang, Z., Adappa, N. D., Doghramji, L. J., Chiu, A. G., Cohen, N. A., & Palmer, J. N. (2015, August). Different clinical factors associated with Staphylococcus aureus and Pseudomonas aeruginosa in chronic rhinosinusitis. *International Forum of Allergy & Rhinology, 5*(8), 724–733.

Zhang, X., Browman, G., Siu, W., Basen-Engquist, K. M., Hanash, S. M., Hoffman, K. L., et al. (2019). The BE GONE trial study protocol: A randomized crossover dietary intervention of dry beans targeting the gut microbiome of overweight and obese patients with a history of colorectal polyps or cancer. *BMC Cancer, 19*(1), 1–11.

Zhou, X., Bent, S. J., Schneider, M. G., Davis, C. C., Islam, M. R., & Forney, L. J. (2004). Characterization of vaginal microbial communities in adult healthy women using cultivation-independent methods. *Microbiology, 150*(8), 2565–2573.

Zhou, X., Brown, C. J., Abdo, Z., Davis, C. C., Hansmann, M. A., Joyce, P., et al. (2007). Differences in the composition of vaginal microbial communities found in healthy Caucasian and black women. *The ISME Journal, 1*(2), 121–133.

Zhou, Y., Mihindukulasuriya, K. A., Gao, H., La Rosa, P. S., Wylie, K. M., Martin, J. C., et al. (2014). Exploration of bacterial community classes in major human habitats. *Genome Biology, 15*, 1–18.

Zilberstein, B., Quintanilha, A. G., Santos, M. A., Pajecki, D., Moura, E. G., Alves, P. R. A., et al. (2007). Digestive tract microbiota in healthy volunteers. *Clinics, 62*(1), 47–56.

Zoetendal, E. G., Raes, J., Van Den Bogert, B., Arumugam, M., Booijink, C. C., Troost, F. J., et al. (2012). The human small intestinal microbiota is driven by rapid uptake and conversion of simple carbohydrates. *The ISME Journal, 6*(7), 1415–1426.

Bacteria and Cancer

3

Nusrat Bibi and Azra Yasmin

Abstract

Microbial infections are the major contributor of almost 20–30% of cancer cases, and dysbiosis of the gut microbiota is progressively considered a hallmark of cancer. According to the WHO, a number of bacterial infections are linked with common malignancies, and a variety of bacteria are classified as class I carcinogens. Despite the lack of clarity in the mechanisms involved in bacterial carcinogenesis, chronic inflammation and the creation of carcinogenic compounds have been identified as key pathways that develop association between cancer and bacteria. Surprisingly, microbes are capable of eradicating tumor cells. Even though chemotherapy, surgery, radiation, and immunotherapy are effective treatments against cancer, these treatments also have drawbacks such as toxicity and tumor resistance. This chapter focuses on the dual nature of bacteria as cancer progressor as well as cancer suppressor. The chapter highlights the role of bacteria in cancer development specifically colorectal, gastric, lung, gall bladder, and breast cancers with reference to association of bacteria with various types of cancers along with oncogenic molecular pathways together with bacterial therapeutics to improve outcomes and minimize adverse effects. Promising both on their own and in combination with chemotherapy, bacteria can elicit anticancer responses and act as carriers of medicines or genes. This chapter also highlights clinical studies involving bacteria-based cancer treatments and tumor regression induced by bacteria.

N. Bibi · A. Yasmin (✉)
Microbiology & Biotechnology Research Lab., Department of Biotechnology, Fatima Jinnah Women University, Rawalpindi, Pakistan
e-mail: azrayasmin@fjwu.edu.pk

N. Masood, A. Yasmin (eds.), *The Microbiome and Cancer*,
https://doi.org/10.1007/978-981-97-6948-3_3

Keywords

Bacteria · Microbiome · Dysbiosis · Cancer progression · Angiogenesis Antiangiogenesis · Cancer mechanisms · Oncogenic molecular pathways Cancer regression · Immune response · Bacterial metabolites · Bacteriocins Anticancer · Combined therapy · Probiotic bacteria · Bacteriobot

Cancer is considered as a group of heterogeneous diseases marked by uncontrolled growth and spread of aberrant cells, which eventually results in death (Gurunathan et al., 2018). Tumors are masses of malignant cells that evolve into cancer; these tumors produce DNA mutations that result in the acquisition of epigenetic modifications that promote oncogenesis and carcinogenesis in a number of disorders (Torres et al., 2018). A tumor is an atypical mass of tissue, which could be cancerous or benign. Benign tumors are defined as those that are localized to a certain area and do not spread to nearby cells. Malignant tumors are defined as those that have moved to other areas of the cell; this process is known as metastasis (Ashu et al., 2019).

Several studies have demonstrated the processes of angiogenesis and lymph angiogenesis, malignant cell migration, resistance to anoikis, and the epithelial-mesenchymal transition (EMT) which are among the many subsequent events that are involved in cancer metastasis (Hu et al., 2012).

Cancer is a multistage series of genetic changes that results in the transformation of normal cell into tumor cells. These changes are induced by several reasons like drinking, smoking, sunlight, exposure to physical or chemical factors, and eventually chronic microbial infection including bacteria (Al-Hilu & Al-Shujairi, 2020). Tumor suppressor genes, which typically act as "brakes" to stop cell growth and division, and proto-oncogenes, which speed up cell growth and division, are the two gene types that are mutated. Tumor suppressor gene inhibition and hyper-activation of proto-oncogenes result in cancer development (Chial, 2008). The best characterized oncogenes encode RAS, JUN, FOS, ABL, RAF, GSP, SIS, FMS, and BCL2, which are among other proteins involved in the control of transcription, growth factor signaling, and kinase activity. Mutations transform proto-oncogenes into oncogenes. RAS is a well-known oncogene, and the most often mutated isoform of RAS in malignancies is KRAS. One of the most well-studied tumor suppressor genes is the one that codes for p53 (Hobbs et al., 2016).

According to the information based on the Global Cancer Observatory (GLOBOCAN0, 19.3), million incident cases of cancer were reported globally in 2020 (Sung et al., 2021). Breast and lung cancers were the most prevalent cancers worldwide, accounting for 12.5% and 12.2% of all newly diagnosed cases in 2020. In 2020, colorectal cancer accounted for 10.7% of all new cases and was the third most frequent cancer with 1.9 million new cases. The highest death rates from lung cancer, stomach cancer, liver cancer, and esophageal cancer are found in developed nations. The highest death rates from lung cancer, breast cancer, liver cancer, cervical cancer, and stomach cancer are found in developing nations (He, 2020). The prevalence of various types of cancers around the world as updated by the World Cancer Research Fund International is depicted in Fig. 3.1.

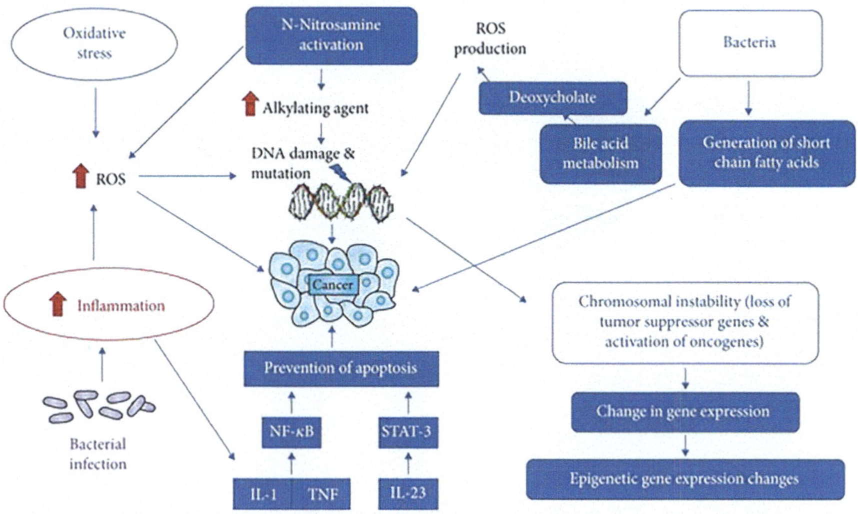

Fig. 3.1 Different molecular mechanisms of carcinogenesis induced by bacterial infection and oxidative stress (Sheweita & Alsamghan, 2020)

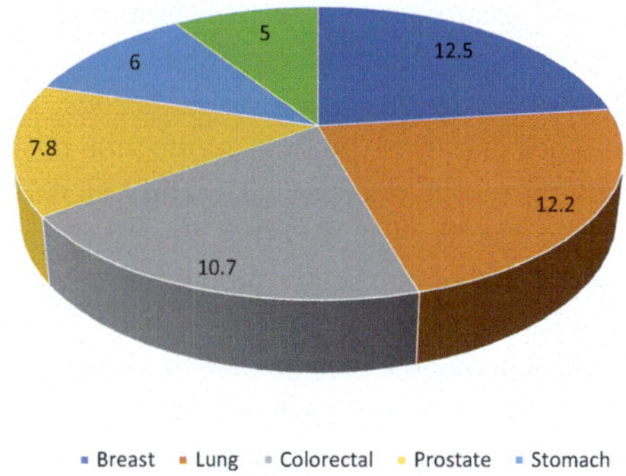

(a) Percentage of most prevalent cancers round the world

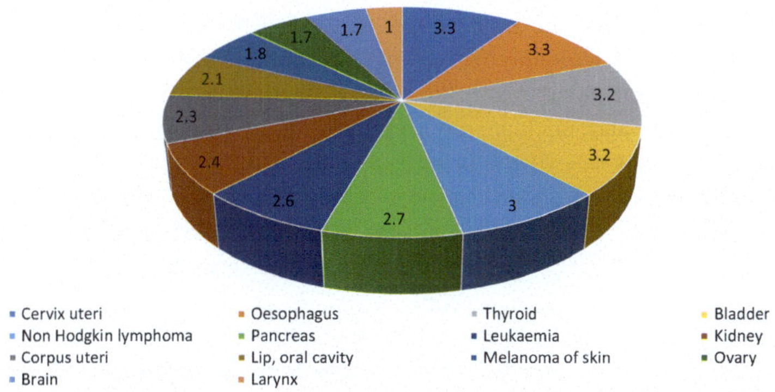

(b) Prevalence of different cancers around the world showing percentage range of 1–5

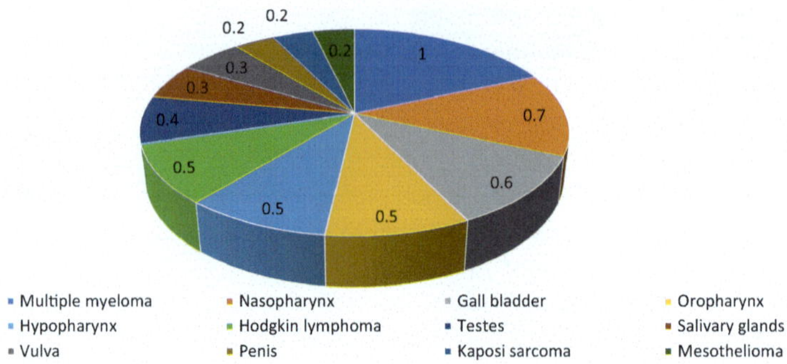

(c) Prevalence of different cancers around the world showing percentage below 1

3.1 Relationship Between Cancer and Human Microbiome

The human microbiome comprises an intricate network of bacteria that are distributed throughout many bodily regions, such as the skin, respiratory system, reproductive tract, oral cavity, and gastrointestinal tract. These microorganisms include archaea, bacteria, eukaryotes, fungi, and viruses (Sender et al., 2016). The characterization of the entire human microbiome is possible because of high-throughput DNA sequencing technologies that involved in the collection of genetic material of all the microorganisms living in our organism. Recently, it is discovered that, in addition to the gut, other human body niches that were previously thought to be sterile are home to native bacterial populations, thanks to research made possible by these cutting-edge approaches (Sender et al., 2016).

Microorganisms are tiny but potent cancer and immunity-influencing agents. Over 30 trillion microbes constitute the human microbiome, and research has revealed that these organisms are crucial to both human health and disease (Motta et al., 2021). Malnutrition, overnutrition, inflammatory and viral ailments especially those affecting the gastrointestinal system, and medication use can all disrupt the gut microbiota (Kolodziejczyk et al., 2019). Antibiotic exposure on several occasions throughout life and in the neonatal period is known to cause dysbiosis, an unfavorable change in the quantity, makeup, and function of the microbial population, and has been attributed to specific cancers (Cao et al., 2018).

Due to poverty and unclean environments, microbial infections have been recently identified as one of the primary causes of many types of cancer, especially in developing and underdeveloped nations (Goodman & Gardner, 2018). Infections triggered by 11 pathogenic species are categorized as class I carcinogens by international cancer research institutes. These include the human papillomavirus, human T-cell lymphotropic virus, human immunodeficiency virus, Epstein-Barr virus, human herpesvirus, hepatitis B and hepatitis C viruses, *Helicobacter pylori*, *Clonorchis sinensis*, *Opisthorchis viverrini*, *Schistosoma haematobium*, and human papillomavirus (De Martel et al., 2012). Because viruses have an impact on human genes, they have received the most attention (Saleh & Halawa, 2020).

3.2 Role of Bacteria as Cancer Inducer

Tumor and microbial infection are used to be widely recognized as distinct conditions with little relationship to one another. Cancer was long thought to not be infectious or communicable until it was found that bacteria are the root cause of many infectious diseases. Therefore, it was unacceptable to believe that bacterial involvement would cause cancer (Nath et al., 2010). Russell initially proposed the involvement of bacteria in cancer in 1890. Thomas Glover noted a few years later in 1926 that it was possible to continuously isolate certain bacteria from neoplastic tissue. It was discovered in 1931 that acid-fast bacteria were linked to Hodgkin's disease (Elise, 1931). Later, in 1941, George Mazet discovered that bacterial infections were consistently connected to both Hodgkin's disease and leukemia. White (1976) asserted a protective effect of antibodies against anticancer microorganisms. In 1953, highly polymorphic bacteria were isolated from cancer tissue (Diller, 1974).

Since cell changes take a long period, the actual genesis and source of tumor cells are unknown (Dong & Xing, 2018). Cell transformation may happen on its own in the presence of carcinogenic physical, chemical, and biological elements (Al-Hilu & Al-Shujairi, 2020). Although there isn't enough data to support the idea that bacteria cause tumors, microbiota may accelerate the development of cancer (Morales-Sánchez & Fuentes-Pananá, 2014). Even though there have been theories about it being a transmissible cause of cancer since the sixteenth century, research on the direct role of bacteria in the progression of cancer has just recently begun. The link between bacteria and tumors was altered when bacteria were found in tumor tissue (Zhao et al., 2022).

3.3 Mechanisms Behind the Carcinogenic Potential of Bacteria

The mechanism supporting bacteria's capacity to cause cancer has recently come to attention. These mechanisms include inflammation, immune evasion, the generation of pro-carcinogenic metabolites, DNA damage, and genomic instability. According to various investigations, the bacteria are tumor-stimulating and tumor-promoting. They can also create toxins that alter the signals that regulate cell growth (Lax, 2005). In the ideal scenario, the immune system would prevent tumors from growing. However, a persistent bacterial infection would infiltrate the system and trigger immune responses, which would then contribute significantly to the development of cancer by releasing pro-inflammatory cells' cytokines and free radicals like reactive oxygen species (ROS) (Schoppmann et al., 2002; Baik et al., 1996).

3.4 Bacteria Involved in Incidence of Various Types of Cancers and Mechanisms of Carcinogenesis

In organs lacking bacteriomes of their own, dysbiosis of the bacterium may lead to cancer. This has been found in gastric cancer (GC), lung cancer (LC), colorectal cancer (CRC), and hepatocellular cancer (HCC). *Helicobacter pylori* can directly cause stomach cancer by inducing epithelial damage that results in inflammation and following carcinogenesis (Schwabe & Jobin, 2013). Recent study data suggests that a variety of bacteria may play a role in the genesis of various cancer types. These include *Helicobacter pylori*, which has been linked to gastric and mucosa-associated lymphoid tissue cancer (Choi, 2021); *Salmonella typhi*, which has been linked to gallbladder cancer (Koshiol et al., 2016); *Bacteroides fragilis*, which has been related to colon cancer (Cheng et al., 2020); and *Chlamydia trachomatis*, which has been associated to cervical cancer (Zhu et al., 2016). This has inspired scientists to investigate the mechanisms by which these bacteria support oncogenesis in order to verify their involvement. Table 3.1 lists the major pathways of carcinogenesis along with the bacteria involved in the process.

Table 3.1 Bacteria involved in carcinogenesis along with key mechanisms of carcinogenesis

Type of cancer	Bacteria involved	Key mechanisms of carcinogenesis	References
Gallbladder cancer	*Salmonella typhi*, *Salmonella enterica*	Variations in the p53 gene's sequence; protein kinase activation; cytolethal distending toxin B (CdtB); biliary deoxycholate; derivatives of cholic acid; 5-alpha, 6-alpha-epoxide-cholesterol; upregulation of PI3K pathway; and activation of the MAPK and AKT pathways	Tsay et al. (2018); Shi et al. (2016); Scanu et al. (2015)
Gastric cancer	*Helicobacter pylori*	Essential requirement for constitutive pS-STAT3-mediated IL-11 and IL-6 signaling in *Helicobacter*-induced gastric carcinogenesis; Wnt/b-catenin pathway, Correa pathway	Balic et al. (2020); Piao et al. (2016); Correa & Piazuelo (2012)

(continued)

Table 3.1 (continued)

Type of cancer	Bacteria involved	Key mechanisms of carcinogenesis	References
Colorectal cancer	*Streptococcus bovis*, *Helicobacter pylori*, *Bacteroides fragilis*, *Enterococcus faecalis*, *Clostridium septicum*, *Fusobacterium* spp., and *Escherichia coli*	Toxin secretion by *Bacteroides fragilis*; NF-μB activation; TNF-α and IL-17A expression; β-catenin expression; induction of NF-κB, Stat3 signals, and IL-17R; induction of colibactin (clbB) and *Bacteroides fragilis* toxin (BFT) gene expression; increased colonial interleukin-17; damage to colonic epithelial DNA, resulting in tissue damage and chronic intestinal inflammation and Wnt/b-catenin and Th17 adaptive immunity activation	Cheng et al. (2020); Hwang et al. (2019); Dejea et al. (2013); Chung et al. (2018); Mármol et al. (2017); Gagnière et al. (2016); Bonnet et al. (2014); Sears & Garrett (2014); Castellarin et al. (2012)
Oral squamous cell carcinoma	*Fusobacterium nucleatum*	Invasion of oral squamous cell carcinoma	Xu et al. (2007a, b)
Lung cancer	*S. pneumoniae (SP)* *Chlamydia pneumoniae*	SP acts as an oncogenic driver to accelerate the onset and spread of lung cancer by interacting with PAFR through its surface protein PspC. The connection between PspC and PAFR promotes lung tumorigenicity by starting SP adhesion, triggering the NF-kB and PI3K/AKT signaling pathways, and inducing activation-mediated inflammatory responses; changes in apoptosis and/or cell programming-signaling; overexpression of miRNA-328; via activating γδ T cells in the lung; creation of IL-1b and IL-23 that are dependent on Myd88; production of reactive oxygen species; and increased release of cytokines, including IL-8, IL-10, and TNF	Li et al., (2023); Shen et al. (2018); Khan et al., (2016); Singh et al., (1994)

(continued)

Table 3.1 (continued)

Type of cancer	Bacteria involved	Key mechanisms of carcinogenesis	References
Bladder cancer	*Staphylococcus albus hemolytic*, *Staphylococcus aureus*, *Klebsiella* spp., *Proteus mirabilis*, and *E. coli*	Formation of N-nitrosamines; DNA methylation; reactive chemical species	Hicks et al. (1982); Calmels et al. (1988); Tricker et al. (1991); Umbenhauer et al. (1985); Kroft & Oyasu (1994)
Breast cancer	*Methylobacterium radiotolerans*, *Sphingomonas Yanoikuyae*, *Escherichia coli*, and *Staphylococcus*, *Clostridiales; Fusobacterium nucleatum*	Bioactive metabolites such as estrogens, short-chain fatty acids, amino acid metabolites, or secondary bile acids are secreted by the microbiota; dysbiosis, in vitro induction of genomic instability and double-stranded breaks in DNA fap-2 dependent attachment of the bacteria to breast cancer tissue, which inhibits tumor growth by generating the metabolite trimethylamine N-oxide (TMAO), which stimulates CD8+ T cells' antitumor immunity and slows tumor progression and metastasis Gal-GalNa	Parhi et al. (2020); Mikó et al. (2019); Urbaniak et al. (2016); Wang et al. (2016); Bleyer & Welch (2012)

3.5 Association of Bacteria with Various Types of Cancers along with Oncogenic Molecular Pathways

Extensive study has been carried out on the carcinogenic potential of some microorganisms. According to reports, toxic bacterial metabolites and long-term inflammation are major factors that encourage neoplasticity. Reactive oxygen species (ROS) are produced excessively during chronic inflammation, and this is believed to be the cause of malignant cell proliferation. There is also a possibility that various virulence factors and bacteria-dependent mechanisms are important in the development of bacteria-induced malignancies (Prasad et al., 2022). Various oncogenic molecular pathways involved in carcinogenesis are listed in Fig. 3.2.

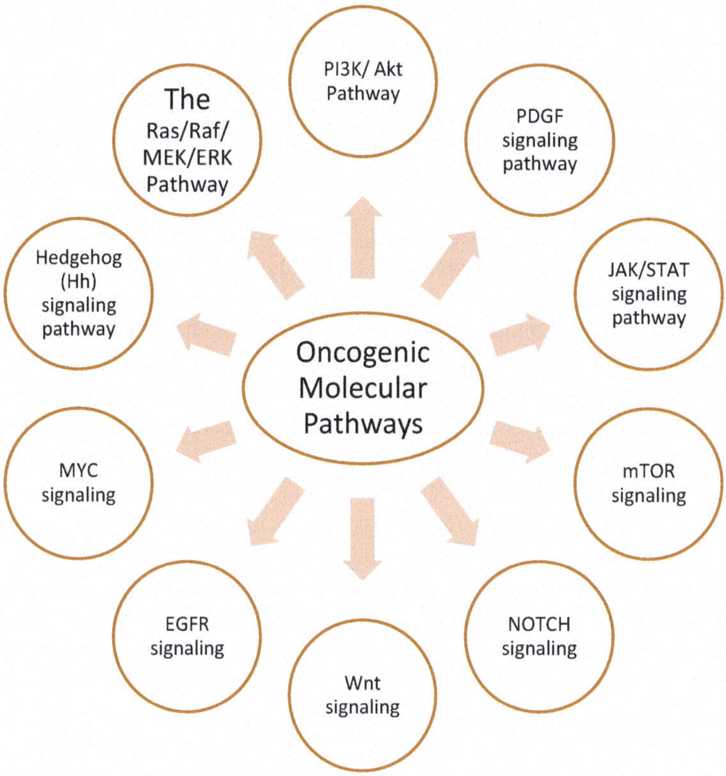

Fig. 3.2 Various oncogenic molecular pathways involved in carcinogenesis

3.6 Gastric Cancer

3.6.1 *Helicobacter pylori* and Gastric Carcinogenesis

Microbial infections are capable of modulating host transformation for progression of malignant features via production of oncogenic metabolites that are involved in inflammation responses, dysregulation of cellular metabolism, and integrity along with genetics and epigenetic manipulations. The best example of carcinogenic bacteria is *Helicobacter pylori* infection. *Helicobacter pylori* is declared as class I carcinogens by the World Health Organization (WHO) (Eyvazi et al., 2020). The bacteria release CagA, an oncogenic protein that disrupts the epithelial cell division in gastric mucosa. Besides, inducing inflammatory stress, cell cycle disruption, and cellular polarity, the bacteria are also involved in the activation of oncogenic pathways such as β-catenin pathways, PI3K/AKT/GSK3β, and STAT3 that results in aberrant expression of activation-induced cytidine deaminase (AID)—all these are oncogenic promoting events (Tabassam et al., 2009; Bronte-Tinkew et al., 2009; Matsumoto et al., 2007).

H. pylori activates a number of important pathways, which are well known for their roles in these processes. These pathways include NF-κB, AP-1, and MAPK. By activating integrin, *H. pylori* injects CagA into the host gastric epithelial cells (Kwok et al., 2007). Furthermore, multiple signaling pathways are activated by phosphorylation of CagA by Src family kinases or Abl kinase. Interaction of phosphorylated CagA activates SHP2. Erk-MAP kinase signaling is amplified by CagA-SHP2 in both Ras-dependent and Ras-independent ways (Higashi et al., 2004). CagA-SHP2 facilitates the contact between cells and the extracellular matrix and dephosphorylates focal adhesion kinase (FAK). Both signals cause a morphological alteration in cells known as the "hummingbird phenotype," which improves the ability of cells to migrate (Tsutsumi et al., 2006). Nonphosphorylated CagA hinders intracellular signaling networks by interacting with E-cadherin to bring about disruption in E-cadherin-β-catenin complex. The accumulation of nuclear β-catenin allows the transcription of oncogenes. CagA also directly activates β-catenin by interacting with MET which leads to the activation of PI3K-AKT signaling (Murata-Kamiya et al., 2007; Suzuki et al., 2009). CagA triggers the signal transducer and activator of transcription 3 (STAT3) pathway which is driven by host immune response and linked to the development of cancer and gastritis caused by *H. pylori* (Jackson et al., 2007; Bronte-Tinkew et al., 2009; Menheniott et al., 2015). According to a recent study, CagA also attaches to 25 recognized components of the host cell in order to take control of several signaling pathways that are connected to apoptosis, genetic instability, inflammation, and cell polarity (Knorr et al., 2019). In addition to CagA, the Cag secretion system makes use of outer membrane vesicles to deliver *H. pylori* peptidoglycan into the host cells which activates control cell migration, proliferation, and death regulating PI3K-AKT (Nagy et al., 2009). Besides CagA, vacuolating toxin A (VacA) is another important factor that determines *H. pylori*'s pathogenicity. VacA protein is encoded by *H. pylori* gene VacA and is associated with several cellular functions, including vacuolation, membrane-channel creation, apoptosis, pro-inflammatory response, and cancer (Yamaoka, 2010). VacA stimulates β-catenin in a PI3K-dependent way, according to Nakayama et al.'s findings (Nakayama et al., 2009) (Fig. 3.3).

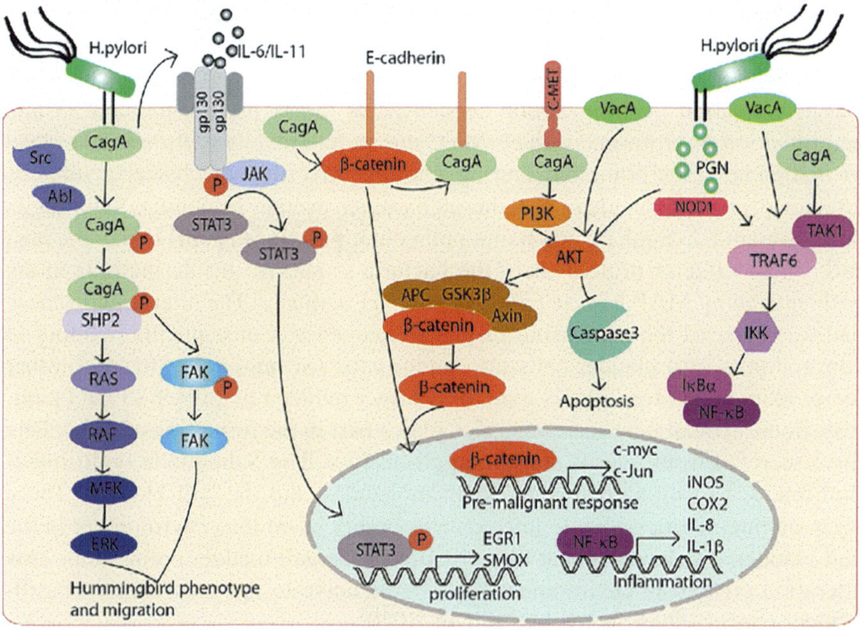

Fig. 3.3 The role of *Helicobacter pylori* in the molecular etiology of gastric cancer. Phosphorylated CagA induces the hummingbird phenotype in epithelial cells and stimulates cell migration through the MEK-ERK and FAK signaling pathways. Nonphosphorylated intracellular CagA disrupts PI3K-AKT signaling or the E-cadherin-β-catenin complexes, activating β-catenin. CagA releases IL-6/IL11 and stimulates gp130 to initiate the JAK-STAT3 pathway (Zhao et al., 2020)

3.7 Gallbladder Cancer

3.7.1 *Salmonella typhi* and Gallbladder Cancer Carcinogenesis

According to data of GLOBOCAN 2020, 115,949 new cases of GBC along with 84,695 deaths were registered round the world in the year 2020 (Sung et al., 2021). Key parameters contributing to GBC are mutations at molecular level and exposure to carcinogenic agents and bacteria. *Salmonella typhi* is the causative agent of typhoid fever. Besides producing typhoid toxin, it also produces nitroso-chemical compound which is a carcinogenic toxin. Typhoid toxins possessing carcinogenic potential also have potential to damage and alter the cell cycle in infected cells. *Salmonella typhi* is a biofilm producer, and the extracellular polymeric substances (EPS) produced in the biofilm are primary cause of chronic infection and cholecystitis, which exposes the epithelium to carcinogenic toxins made by *S. typhi* (Di Domenico et al., 2017). At the molecular level, the synthesis of cytolethal toxin is mainly driven by three genes: CdtA, CdtB, and CdtC. The CdtB gene exhibits oncogenic potential by creating double-stranded DNA break in the host cell's DNA. CdtB makes covalent link with Pertussis toxin (Plt; Plt A and Plt B) resulting in the formation of unique triplet structure, i.e., CdtB-PltA-PltB. This triplet on entering into the target cell nucleus and with the assistance of Plt A and Plt B unit, CdtB perform the host cell DNA damage thus provoking gall bladder cancer (Di Domenico et al., 2017) (Fig. 3.4).

The alteration in the genomic sequences of tumor protein p53 (TP53 and amplification of proto-oncogene c-MYC due to Salmonella typhi infection lead to malignant transformation. During chronic infection, the bacteria releases effector protein molecules i.e. Salmonella outer protein E (SopE), Salmonella outer protein E2 (SopE2), and Salmonella outer protein B (SopB inside the host cell. These effector proteins help the bacteria to survive inside the host along with enhancing MAP Kinase pathway and Akt pathway. The alteration in these pathways proves the expression of proto-oncogenes. Consequently, leading to conversion of gall bladder cells into tumor cells (Scanu et al., 2015). Another factor responsible for GBC is genotoxic effect induced by cGAS-STING pathway. Some special type secretions also play a part in the formation of GBC. Bile juice secreted from liver is stored in gall bladder. The Salmonella typhi infection releases many carcinogenic toxin molecules into the gall bladder. These toxin on interacting with bile juice creates highly hazardous environment in the gall bladder. The environment directly affects the gall bladder epithelium. This alteration creates an environment that is conducive to the progression of gallbladder cancer (Sheweita & Alsamghan, 2020).

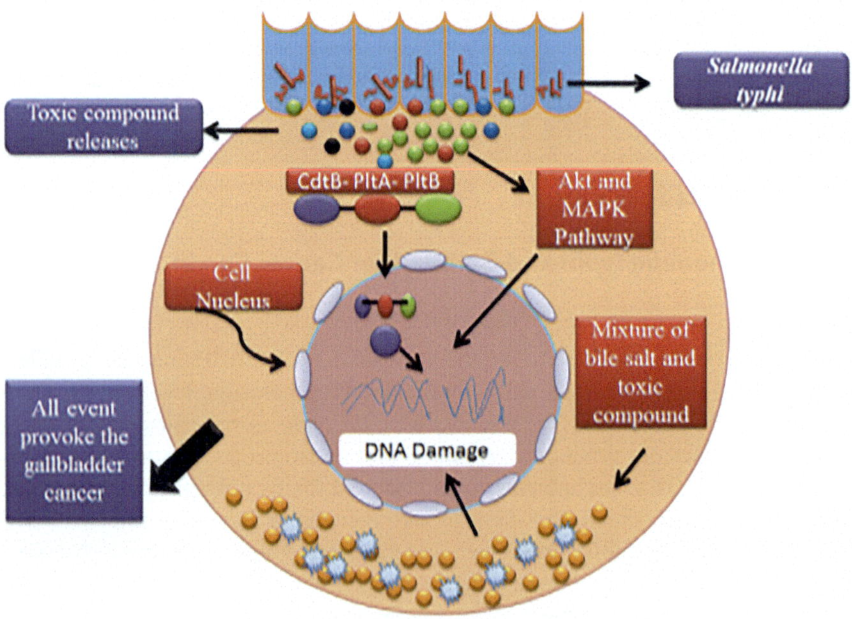

Fig. 3.4 Overview of oncogenesis mechanism promoted by *S. typhi* in the gallbladder (Upadhayay et al., 2022)

3.8 Colorectal Cancer

By far the most common gastrointestinal cancer identified in 2020 (almost two million cases), according to the World Health Organization, was colon cancer. The major risk factor responsible for colon cancer includes lifestyle and nutritional choices like obesity, consumption of processed or red meat, and alcohol. On the other hand, an increasing amount of data points to the significance of gut microbes and their dysbiosis (imbalance) in the emergence of colon cancer. Accordingly, numerous investigations have documented the pathogenic consequences of particular bacterial species, such *Helicobacter pylori*, *E. Coli*, *Fusobacterium nucleatum*, *Bacteroides fragilis*, and *Streptococcus bovis* (Gagnière et al., 2016; Ray et al., 2023).

3.8.1 *Fusobacterium nucleatum* and Colorectal Cancer Carcinogenesis

Chronic inflammation, production of toxins, or biosynthesis of genotoxins are the main processes involved in the cancer progression by microorganisms including bacteria (Candela et al., 2014). For example, *F. nucleatum* adopted several carcinogenic mechanisms—most noticeable of them are chronic infection, interaction of the cell surface molecules of these bacteria with the immune system, immune evasion, and immune suppression. Moreover, the virulence factors of the *F. nucleatum* nucleus such as FadA, Fap2, and LPS and cell wall extracts may act as effector molecules in the conversion of normal epithelial cells into oncogenic cells (Gholizadeh et al., 2017). Relationship between driver and passenger bacteria is involved in the development of colorectal cancer. The disease onset with driver bacteria, which initiate tumorigenesis, changes the intestinal environment leading to an overgrowth of passenger bacteria and the eventual progression of colorectal cancer (Nistal et al., 2015). Carcinogenesis is a result of a highly complex mechanism of genetic mutation brought on by bacterial invasion, which altered the intestinal environment and damaged DNA (Carbonero, et al., 2012). Because the production of nitric oxide and other oxygen radicals from infected and inflammatory tissues contributes to the processing of carcinoma, bacteria and other microorganisms may cause infection and inflammation in tissues such as colitis, hepatitis, and gastritis, which are cancers in humans at different sites (Ohshima & Bartsch, 1994).

The progression of CRC may also be involved in compounds produced from protein fermentation. Increased consumption of protein is linked with a rise in waste products in the colon, including amines, ammonium, sulfide, nitrate, branched-chain amino acids, and H2S. According to research, hydrogen sulfide (H2S), a by-product of protein breakdown, is genotoxic and promotes inflammation (Greer & O'Keefe). Moreover, H2S concentrations in CRC patients are higher than in healthy persons (Kanazawa et al., 1996), and their colons' capacity to detoxify is less, which increases genotoxic effects (Carbonero et al., 2012) (Fig. 3.5).

Intestinal bacteria also produce genotoxic metabolites which can induce mutations by binding to particular cell surface receptors and altering intracellular signal

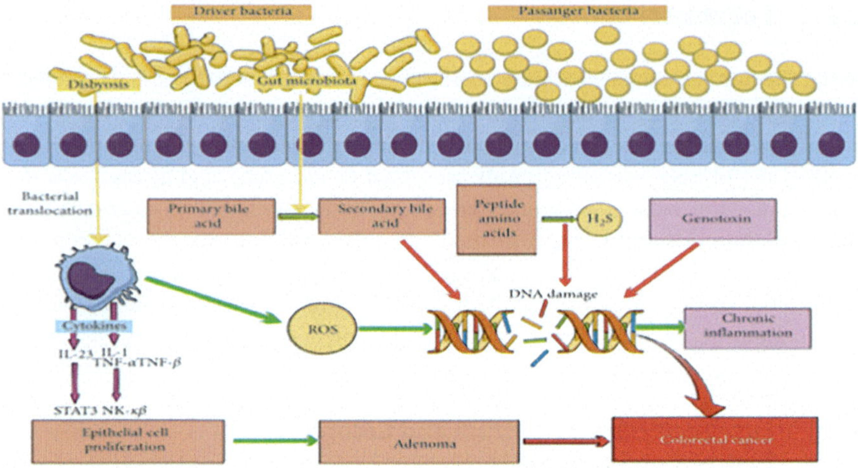

Fig. 3.5 The mechanism demonstrating the role of the driver and passenger bacteria to stimulate the colorectal cancer by changing the composition of the gut microbiota (Nistal et al., 2015)

transduction, leading to the development of colorectal cancer (CRC) (Azcárate-Peril et al., 2011). In certain cases, the toxins damage DNA directly by attacking the genome through enzymatic attack or indirectly by inducing an inflammatory response that releases free radicals. Additionally, they have been observed to have an impact on DNA repair pathways (Candela et al., 2014). For example, enterotoxigenic strains of *Bacteroides fragilis* can colonize human mucosa without causing any symptoms; nevertheless, in certain instances, they release the *B. fragilis* toxin (BFT), which can cause inflammatory diarrhea.

BFT is a zinc-dependent metalloprotease that rapidly modifies the structure and function of colon epithelial cells, causing the tumor suppressor protein E-cadherin to break (Zhu et al., 2013; Rhee et al., 2009; Sears & Garrett, 2014). E-cadherin is a calcium-dependent protein that plays essential functions in the behavior of epithelial cells, the creation of tissues, and the suppression of cancer (Roy et al., 2008). One of the earliest stages in the development of a tumor is the increase in permeability of polarized epithelial cells in the colon caused by the loss of this protein (Soler et al., 1999). Additionally, BFT triggers the transcription factor NF-κB, which releases cytokines that exacerbate mucosal inflammation (Sears, 2009).

Even though BFT is thought to be one of the primary toxins in the formation of colorectal cancer (CRC), new research has revealed that the toxins most actively transcribed in the tumor tissues of CRC patients are derived from *Shigella flexneri*, *Salmonella enterica*, and *E. coli*. This implies that enterobacterial toxins have a significant role in the development of tumors (Schwabe & Wang, 2012). In the colon, some commensal *E. coli* strains have the ability to directly manufacture colibactin toxin. The colon cells' DNA is broken into two strands by this toxin. When cellular repair processes are unable to repair these DNA damages, chromosomal abnormalities and subsequent mutations cause tumorigenesis (Cuevas-Ramos et al., 2010).

3.9 Lung Cancer

Lung cancer is one of the leading causes of death around the world with increased prevalence in developing countries (Shankar et al., 2019) with a 5-year survival rate below 18% (Zappa & Mousa, 2016). For a very long time, healthy people's lungs were even thought to be sterile, despite the fact that they come in contact with the outside air. Recent investigations have confirmed that healthy individuals have lung microbes. Earlier research has employed culture-independent molecular techniques to examine the most prevalent bacterial phyla, such as *Firmicutes*, *Proteobacteria*, and *Bacteroidetes*; notable genera include *Prevotella*, *Veillonella*, and *Streptococcus* (Segal et al., 2013; Morris et al., 2013; Bassis et al., 2015; Marsh et al., 2016). Moreover, specific bacteria, including *Enterobacteriaceae*, *Haemophilus*, *Methylobacterium*, *Ralstonia*, and *Tropheryma* species, are present in the lungs (Dickson et al., 2015).

Decline in symbiotic bacteria and an increase in pathogenic bacteria lead to dysbiosis in the microbiome. The continuous production of inflammatory mediators by pathogenic bacteria can cause chronic inflammation. This can impact cell apoptosis and increase the number of mutations. Furthermore, bacterial metabolites such as reactive oxygen and nitrogen directly damage DNA or alter cellular communication to create an environment that is conducive to cancer (Dejea et al., 2013).

3.9.1 *Chlamydia pneumonia* and Lung Cancer Carcinogenesis

Chronic chlamydiosis was one of the earliest bacterial diseases associated to lung cancer. A 1.6-fold increased risk of developing lung cancer has been linked with a prolonged pneumocyte infection caused by *Chlamydia pneumoniae* (Tsay et al., 2018). There have been several explanations put up to explain the link between lung cancer and the bacterial infection *Chlamydophila pneumoniae*. It has been proposed that the generation of ROS during inflammation, which results in DNA damage, may be one of the possible explanations of this connection (Laurila et al., 1997). Lung inflammation can accelerate the rate of cell division and raise the possibility of a mutation through a fixed rate of DNA damage, which can lead to the development of tumorigenesis (Jackson et al., 2000).

The effector molecules generated from *C. pneumoniae* have been shown to induce immunosuppression, cause chronic inflammation, inhibit tumor suppressor mechanisms, and cause cell transformation through oncogene transfer within the internal environment of host cells (Tsay et al., 2018). Furthermore, the proteins of *C. pneumoniae* generated in host cells during a chronic lung infection can migrate to various organelles, including the nucleus, endoplasmic reticulum, and mitochondria, thereby influencing a variety of significant biological activities and ultimately contributing to the development of cancer (Hess et al., 2003). Furthermore, it has been discovered that *C. pneumoniae* can effectively increase the production of IL-6, TNF-α, and IL-1β in host monocytic cells, which could potentially aid in the development of cancer (Redecke et al., 1998; Gaydos, 2000).

The host cell's mitochondria can be targeted by *C. pneumoniae* proteins, leading to cell malfunction, according to an in silico investigation. According to a study by Alshamsan et al. (Alshamsan et al., 2017), 513 of these proteins target cytoplasmic cellular activities, whereas 183 out of 1112 target the mitochondria. This can then change apoptosis, interfere with regular cell proliferation, and start the lung cancer process. These proteins have the ability to construct a host cell proteome by being integrated into intracellular organelles (Budisan et al., 2021). The chlamydial heat shock protein-60 (cHSP60), which is produced even when chlamydia is dormant, is an endotoxin-like protein that may mediate the carcinogenic activity of *C. pneumoniae* (Bulut et al., 2009). The family of proteins known as heat shock proteins is extremely conserved in all species, including prokaryotes and eukaryotes, that play important role in cellular stress responses. Regarding chlamydia, it is possible that cHSPs are similar to human HSPs in that they activate TLR4, which in turn causes the release of IL-6, TNF-α, and IL-1β, hence eliciting autoimmune reactions (Budisan et al., 2021) (Fig. 3.6).

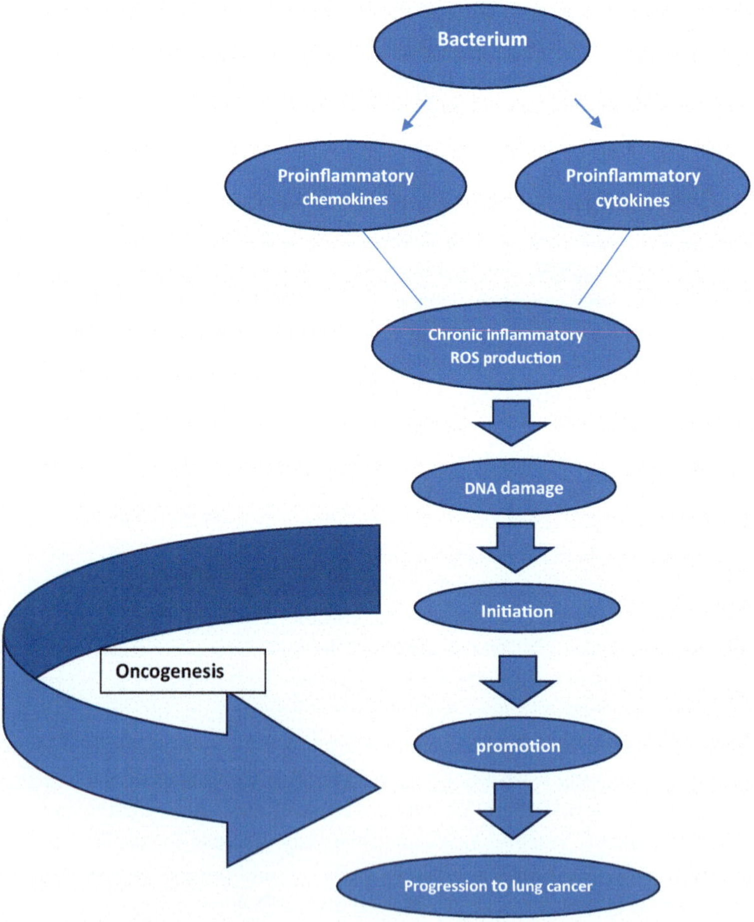

Fig. 3.6 Mechanism of lung cancer progression

3.10 Breast Cancer

Breast cancer is the second most prevailing cancer in women worldwide (Alkabban & Ferguson, 2023). It is a heterogeneous malignancy with distinct molecular subtypes. With the best clinical prognosis, the luminal A subgroup exhibits good response to endocrine therapy and is distinguished by estrogen receptor (ER) expression and function. Compared to the previous subtype, luminal B tumors exhibit lower levels of ER and have higher rates of proliferation. Roughly 15% of all invasive breast tumors belong to the human epidermal growth factor receptor 2 (HER2)-positive subpopulation, which is both progesterone receptor (PR)- and ER-negative. Comparing it to luminal-like tumors, it is more aggressive. In conclusion, hormone receptors and HER2 are not expressed in triple-negative breast tumors (TNBC), the subtype with the lowest survival rate and the most difficult treatment (Bernardo et al., 2023). Although mutation carriers make up a small percentage of breast cancer patients, the BRCA1 and BRCA2 genes have been identified as genetic risk factors for breast cancer (Armstrong et al., 2019).

3.11 The Role of Distinct Microbiota in Breast Cancer Tumorigenesis and Metastasis

Dysbiosis, or an abnormal composition of the microbiome, is a hallmark of breast cancer. The gut microbiome seems to contribute to the pathophysiology of the illness; however, the exact involvement of the breast microbiome in breast cancer is unknown. The gut microbiota secretes bioactive metabolites (short-chain fatty acids, reactivated estrogens, amino acid metabolites, or secondary bile acids) that influence breast cancer. Breast cancer behavior has been demonstrated to be modulated by such blood-borne microbial metabolites. Due to their formation in a "gland" (the microbiome in this example) and subsequent circulatory transportation to distant locations of action, these metabolites bear resemblance to human hormones. These metabolites typically constitute essential components of the tumor microenvironment (Mikó et al., 2019).

Through a variety of processes, microbiota has a complex involvement in the onset, progression, and management of BC. Based on current studies and categorization, different microbiota mostly influences BC tumorigenesis and metastasis in the following ways. First, the development of BC may be aided by bacterial DNA damage or gene alterations (Fig. 3.7). Second, the way that people with BC metabolize estrogen can be influenced by their microbiota. Third, the microbiota may create compounds that have an indirect impact on the development of cancer (Fig. 3.2). In addition, bacteria have the ability to directly impact immunological signaling pathways that are connected to tumor networks (Laborda-Illanes et al., 2020; Alpuim Costa et al., 2021).

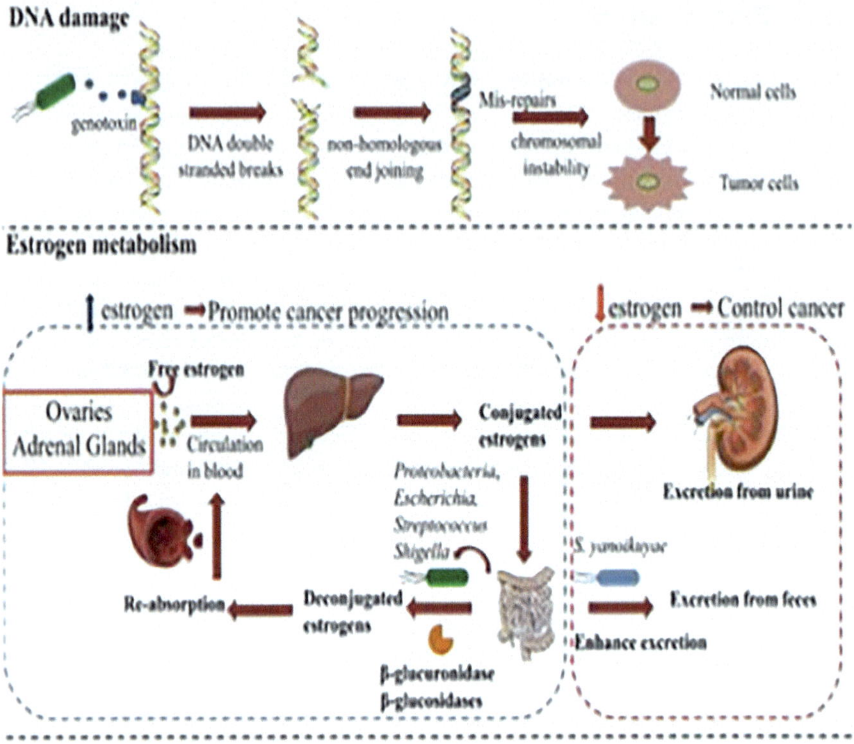

Fig. 3.7 The microbiota affects breast cancer tumorigenesis and metastasis by DNA damage and regulating the estrogen metabolism (Meng et al., 2022)

3.12 Role of Bacteria as Tumor Suppressor

Currently, tumors pose a serious risk to people's health and lives. Conventional therapy for malignant tumors involves chemotherapy and radiation. However, because of their high toxicity to normal cells, incapacity to treat deep tumor tissues, and potential to induce drug resistance in tumor cells, traditional therapies do not always produce satisfying outcomes (Fan et al., 2022). Bacteria owing to numerous unique features, including their capacity to target specific tumors, their high motility, their immunogenicity, and their ability to serve as gene or drug carriers, are now viewed as a potentially effective platform for cancer therapy. Promising outcomes have been observed in the use of various types of bacteria for solid and metastatic tumor therapy. The field of synthetic biology has allowed for the controlled expression of therapeutic proteins in bacteria. In addition, nanomaterials have been widely employed to alter bacteria for photothermal, magnetothermal, photodynamic, and targeted drug delivery, as well as to boost the antitumor activity of synergistic cancer therapies (2022).

3.13 Tumor-Targeting Features of Bacteria

It is widely recognized that both active and passive mechanisms play a critical role in the bacteria's capability to target solid tumor. Additionally, it is suggested that living bacteria could accidentally get stuck in the leaky vasculature of a solid tumor. Following the bacterial infection-induced bleeding, they gather within the tumor (Lee, 2012). However, active mechanisms of bacteria in tumor microenvironment may vary according to oxygen tolerance. These mechanisms most likely include chemotactic factors and particular signal molecules generated by dying tumor cells as well as a low oxygen microenvironment (Lee, 2012). *Bifidobacterium* and *Clostridium* which are obligate anaerobes cannot survive in oxygen rich environment and tend to migrate toward anoxic regions of tumors. Anoxic microenvironment is the key attribute of tumors since normal cells do not possess completely deoxygenated microenvironment. This scenario let the accumulation of anerobic bacteria in the regions of tumor (Dang et al., 2001).

Tumor nutrient-rich environment is also attractive to various bacteria; for instance, *Escherichia* and *Salmonella* sense the nutrient-rich region in and around the tumor and accumulate in tumors. As bacteria can activate numerous pathways to precisely target tumors, these strategies are actually neither mutually exclusive nor entirely independent (Garza-Morales et al., 2020). Moreover, like "biorobots," bacteria with motility—a fundamental feature of all bacteria—can enter tumor tissues quickly.

3.14 Mechanisms of Bacterial Action in Cancer Therapy

In cancer treatment trials, live strains of *Clostridium* and *Streptococci* were initially employed. Tumor treatment using microorganisms can be accomplished by a range of methods (Leschner et al., 2009). Many bacteria have been revealed to possess the ability to eradicate tumors via various mechanisms, including the genus of *Pseudomonas*, *Caulobacter*, *Listeria*, *Proteus*, *Bifidobacteria*, and *Salmonellae*. Some of the bacterial mechanisms include their ability to carry out RNA interference as well as prodrug cleavage, as well as their utilization of bacterial toxicity, production of enzymes, biofilm formation, ribocin production, and immunotherapeutic components (Felgner et al., 2016). The abovementioned species' potential as cancer treatments has also been investigated in animal models (Alexandroff et al., 2010; Danino et al., 2013). To conduct the studies on humans with various cancers, further work must be done.

3.15 Potential Anticancer Mechanisms of Bacteria

3.15.1 Anticancer Potential by Triggering Immune Response

Bacteria exhibited anticancer potential by activating host immune cells, i.e., T lymphocytes and cytokines, via an immune response that recognizes cancer cells as antigens and destroys them (Kucerova & Cervinkova, 2016). Outer membrane of gram-negative bacteria comprises of lipopolysaccharide (LPS) or flagellin, while that of gram-positive bacteria contains lipoprotein. These structures are regarded as pathogen-associated molecular model (PAMP) responsible for the activation of toll-like receptors (TLRs) via the surface-bound pattern recognition receptors (PRRs). The activation of TLR signaling pathway results in the upregulation of expression of certain genes encoding different pro-inflammatory cytokines such as TNF-α, IL-1, and IL-6, type I interferons, as well as anti-inflammatory molecules, cyclooxygenase-2 (COX-2) (Dunne et al., 2011; Shi et al., 2020). Different types of TLR can recognize different PAMPs.

3.15.1.1 Activation of Cytokines

Certain microbes such as *Salmonella typhimurium* activate certain pathways that enhances the production of cytokine such as IL-1β, TNF-α, and Il-18. These cytokines destroy tumors (Phan et al., 2015). When the toll-like receptor (TLR4) becomes active and certain inflammations result, LPS (lipopolysaccharides) secretes IL-1β, which kills cancer cells. Additionally, bacteria that cause BMDM to activate inflammasomes harm cancer cells by interfering with macrophages' ability to phagocytose injured cancer cells. Therefore, via interacting with the inflammasome pathway, *Salmonella typhimurium* ΔppGpp has therapeutic efficacy against cancer (Phan et al., 2015).

3.15.1.2 Activation of T Cell Lymphocytes

Escherichia coli (*E. coli*) destruct tumors by producing lymphocyte T cells. Additionally, *E. coli* promotes the development of CD4+ T lymphocytes, which aid in the anticancer activity (Stern et al., 2015). These T cell lymphocytes are capable of eradicating colon cancer. A trial conducted on severe infected mice showed cancer reduction upon the administration of T cell lymphocytes (Erdman et al., 2003).

3.15.1.3 Activation of Tumor Necrosis

Salmonella enterica serovar *Typhimurium* has been reported to initiate the production of tumor necrosis factor (TN-α) which plays vital role toward formation of drainage in tumors. The way it functions also determines how neutrophilic granulocytes are produced. The neutrophils facilitate the bacteria's movement to the tumor, the site of infection. Research has demonstrated that an increase in the host's neutrophil count determines the outcome of bacterially mediated tumor treatment; hence, an increasing estimate of necrosis can lead to the complete elimination of identified tumors (Westphal et al., 2008).

3.15.2 Anticancer Potential of Bacteria Via Discharged Substances

Bacteria release certain types of metabolites which could be toxins or enzymes that are potent to show regression toward the development of tumors. On entering the infected cells, these metabolites alter the functioning of the cell which eventually causes the cells to die.

3.15.2.1 Bacteriocins

Bacteriocins are cationic peptides synthesized by ribosome and are produced by almost all groups of bacteria. Gratia was the first who discovered the first bacteriocin in *Escherichia coli*, and this bacteriocin was named as colicin (Gratia, 1925). Bacteriocins have exhibited cytotoxicity activity against tumor cells and emerged as potential candidate to develop new anticancer therapeutic (Baindara et al., 2018). Normal body cell membranes are neutral, but cancer cells have negatively charged cell membrane owing to increased content of anionic phosphatidylserine, sialylated gangliosides, o-glycosylated mucins, and heparin sulfates. This negative charge brings strong interaction between cancer cell and bacteriocins. Colicins are the bacteriocins that are reported to have anticancer potential against human cancer cell lines like colon cancer, breast cancer, and bone cancer (Kaur & Kaur, 2015). Various types of bacterial strains listed in Table 3.2 have been reported in the literature that showed promising anticancer activities against the different cancer cell lines.

Table 3.2 Bacteriocin with anticancer activity produced by bacteria

Bacteriocin	Bacteriocin producing bacteria	Oncogenic cell lines	References
Rhamnosin	*Lacticaseibacillus rhamnosus*	CCA cell lines	Kerdkumthong et al. (2023)
Lysostaphin	*Staphylococcus simulans*	CCA cell lines	Kerdkumthong et al. (2023)
Nisin	*Lactococcus lactis*	HepG2, HeLa, A549, HDFa, and THLE-2	Thanjavur et al. (2022)
Enterocin LNS18	*E. thailandicus*	HepG2	Al-Madboly et al. (2020)
Bovicin HC5	*S. bovis* HC5	MCF-7, HepG2	Rodrigues et al. (2019)
Duramycin	*S. cinnamoneus*	AsPC-1, Caco-2, Colo320, CT116, JJN3, Lovo, MCF-7, MDA-B-231, MIA PaCa-2	Rodrigues et al. (2019)
Azurin	*P. aeruginosa*	MCF-7, UISO-Mel-2, osteosarcoma (U2OS)	Nguyen & Nguyen (2016)
Colicin E3	*E. coli*	P388, HeLa, HS913T	Kohoutova et al. (2014)
Microcin E492	*K. pneumoniae*	HeLa, Burkitt lymphoma variant (RJ2.25)	Kaur & Kaur (2015)

3.15.2.2 Phenazine1, 6-Di-Carboxylic: A Redox Active Secondary Metabolite

Phenazine are produced and secreted by bacteria as redox active secondary metabolites. *Pseudomonas aeruginosa* strains produce different phenazine metabolites like phenazine 1,6-di-carboxylic acid (PDC) and phenazine 1-carboxylic acid (PCA). From *Streptomyces* species, first PDC phenazine was isolated that proved to be an effective tool for controlling and preventing the biofilm formation in *Candida albicans* (Morales et al., 2013). PDC phenazine also depicted substantial broad-spectrum activity against various cell cancer cell lines such as MCF7 cell lines, HeLa, and HT29 and less activity toward DU145 (human prostate cancer cell lines) (Dasgupta et al., 2015).

3.15.3 Anticancer Potential of Bacteria through Biofilms

Antitumor potential of biofilm can be explored owing to its potential to deliver therapeutics to prevent the cancer progression. Chemotherapy using hydroxyurea and doxorubicin has induced and promoted the formation of biofilm in *P. aeruginosa* (Groizeleau et al., 2016). The SOS response, an activated DNA repair mechanism, causes the bacterial cells to proliferate as a biofilm on cancer cells, and the bacteria acquire phenotypes that damage or penetrate the cancer cells (Podlesek & Žgur Bertok, 2020). The proteins and DNA in the biofilm make a coat on cancer cells that block metastasis (Adnan et al., 2017). The polysaccharide discharged by *Streptococcus agalactiae* inhibited the tumor cell attachment to endothelial cells (Miyake et al., 1996). The biofilm derived iron oxide nanowires were loaded with doxorubicin to test their carrier potential—they could induce cytotoxicity in human breast tumor cells (Kumeria et al., 2016). However, further study is required on the role of microbial biofilm in metastatic diversion.

3.15.4 Potential of Bacteria as Vehicles for Gene Therapy for Cancer

Gene delivery systems involve chemical or physical approach for introduction of foreign RNA or DNA into host cells to express gene expression. Most commonly used vectors for gene delivery are viruses and bacteria. Vector carrying killing gene is introduced into malignant cell to kill the tumor cells. Various bacteria such as *Escherichia*, *Salmonella*, *Clostridium*, *Listeria*, and *Bifidobacterium* are employed as gene delivery vehicles. It is a very effective strategy to treat different types of cancers by delivering a gene that directly produces therapeutic protein to the cancer sit (Divyashree et al., 2022).

3.15.5 Potential Role of Bacteria in Antiangiogenesis Therapy

Angiogenesis is the formation of new blood vessels, and growth and metastasis of solid tumors depend on angiogenesis. Blocking the tumor angiogenesis could be an effective approach for treatment of solid tumors. In an investigation, the researchers used *Bifidobacterium adolescentis* as a vector to stop angiogenesis which exhibited quite promising results (Li et al., 2003). Antiangiogenic protein (endostatin) was efficiently delivered by *Bifidobacterium longum* to murine liver tumors, and it exhibited excellent anticancer activity (Xu et al., 2007a, b; Fu et al., 2005). Moreover, on delivering antiangiogenic bacterial vaccines against vesicular endothelial growth factor receptor 2 (VEGFR-2) in animal models of malignant melanoma, lung cancer and colorectal carcinoma exhibited efficacious results (Niethammer et al., 2002). Bacteria also showed good invasion capability to the tumor regions unaccessible by systemically administered agents by utilizing their unique metabolic features, which enhances the chances of tumor eradication by inhibiting angiogenesis (Gardlik et al., 2011). Furthermore, the anticancer effect can be enhanced by coadministering tumor necrosis factor-linked apoptosis inducing ligand (TRAIL) and endostatin (Hu et al., 2009).

3.15.6 Potential Role of Bacteria in Combined Therapy

3.15.6.1 Combination of Bacteriotherapy with Chemotherapy or Radiotherapy

Effectiveness of chemotherapy can be boosted along with reduced chemotherapy by combining it with bacteriotherapy. Tumors can be sensitized with bacteria and chemotherapy in order to explore its potential for gene or drug delivery (Sedighi et al., 2019). Results of many clinical and preclinical study administration of probiotics during or after radiotherapy resulted in reversal of damaged normal tissue (Giralt et al., 2008; Blanarova et al., 2009). Combination of radiation therapy and the bacterial strain *Clostridium novyi*-NT against mice bearing HCT116 tumors resulted in tumor shrinkage. Additionally, administering X-rays 5–15 Gy with injections of *Salmonella* 2 × 105 CFU into mice bearing B16F10 or Cloudman S91 melanomas suppressed the tumor growth and enhanced the survival rate as well (Sedighi et al., 2019). Using rat model experiments to reduce the negative gastrointestinal effects of chemotherapy, probiotic formula VSL-3 gave positive results (Bowen et al., 2007).

3.15.6.2 Anticancer Potential of Combined Treatment with Bacteria and Viruses

Therapeutic antitumor potential of virus and bacteria can be enhanced via combination strategy. Due to attachment of virus with bacteria, hide both from immune system. Lytic viruses cause lysing of tumors resulting in release of new viruses which further infect nearby cancer cells and kill them. Research conducted in mice model by using which combination of two rationally diverse microorganisms, i.e., vesicular stomatitis virus and nonpathogenic *E. coli*, demonstrated improved therapeutic outcomes (Cronin et al., 2014).

3.16 Potential of Probiotic Bacteria in Cancer Treatment

Probiotics can tolerate the acidic gastrointestinal environment and colonize mucosal surface for prolonged time. Owing to these attribute, probiotics could be utilized as vehicles to deliver various therapeutics such as cytokines, drugs, enzymes, or DNA in recent years, and such approach has been successfully applied in colorectal cancer treatment (Górska et al., 2019). Intragastric applications of the recombinant strains of *Lactobacillus lactis* resulted in expression of anti-inflammatory compounds like IL-10, antioxidants, cytokine, or human interferon-β with improved intestinal inflammation and demonstrated cytoprotective effect (Kaye et al., 1994). The catalase expressing *L. lactis* reduced the production of reactive oxygen species (ROS) like H2O2, showing decline in colonic damage and inflammation (Leblanc et al., 2008). Recently, studies are conducted to target the inhibition of inflammatory-related carcinogenesis by using different combinations of probiotic vectors expressing various therapeutic molecules, cytokines, or drugs. The main limitation of probiotic strains with recombinant vectors is the use of antibiotic resistance genes as selective markers in cloning that could be transferred to resident intestinal microbiota (Cano-Garrido et al., 2015).

3.17 Bacteria-Based Cancer Drugs with Status of Clinical Trials

Several phase I clinical trials are performed both in in vitro and in vivo studies of tumor-bearing rats and mice to explore anticancer therapeutic potential of bacteria. Variations in clinical trial result were observed because of trial designs, the types of patients recruited for the studies, and other host-related factors. The results of the clinical trials so far have been disappointing. A pool analysis in bacterial vaccine trials against cancer depicted only 3.6% of the patients benefited from the provided vaccination (Klebanoff et al., 2011). There is a dire need to address all these issues to improve vaccines to get desired outcomes against cancer. Table 3.3 represents bacteria-based cancer drugs that are in clinical trials.

Table 3.3 Bacteria-based cancer drugs that are in clinical trials

Bacterial strain	Trial phase	Cancer type	References
Clostridium novyi-NT	I	Colorectal cancer	Staedtke et al. (2016)
Listeria monocytogenes	II	Metastatic pancreatic tumors	Le et al. (2015)
	II	Cervical cancer	Basu et al. (2018)
S. typhimurium Ty21a VXM01	I	Pancreatic cancer	Schmitz-Winnenthal et al. (2018)

3.18 Bacteriorobot

An innovative bacteria-based method for tumor treatment is known as the bacteriorobot and involves the use of bacteria as microsensors to treat strong tumors (Park et al., 2013). Microrobots have been constructed utilizing the nanotechnology (Gong et al., 2012; Zhao et al., 2022). In an investigation, a microrobot therapy was developed by encapsulating *Salmonella typhimurium*, a therapeutic bacterium in alginate microbeads which is biocompatible/biodegradable. The flagellated bacteria (*Salmonella typhimurium*) are attached on microbead to make a bacteria-based microrobot in targeted tumor site (Park et al., 2013). The bacteria-encapsulated delivery system is safer than the direct inoculation of bacteria as it protects the bacteria from being attacked by the immune system (Park et al., 2015).

3.19 Conclusion

Being involved in both cancer progression and regression, bacteria act as double edge sword. Various bacteria can induce carcinogenesis via damage to regulatory proteins and DNA, releasing free radicals, using molecular mechanisms involving bacterial proteins to activate specific inflammatory proteins. There was substantial evidence in the literature correlating various bacterial species to cancers. It is important to comprehend the intricate interaction existing between humans and microbiological flora in order to understand the pathophysiology of many cancers.

Therapeutic potential of bacteria to treat solid tumors has been known for decades. However, owing to adverse uncontrollable side effects, the clinical application of bacteriotherapy never became routine. To combat this problem, attenuated bacterial species showing antitumor potential have been identified and studied. These species of bacteria have little or no side effects and are considered safe for cancer therapeutic application. Use of bacteria alone is not fully effective against cancer. Their modifications as anticancer agents, immunogenic antigens, or antioncogenes and their combine therapy with other therapeutic processes will upgrade their potential for cancer therapy. A deeper comprehension of bacteria-based cancer therapy is made possible by the expanding body of knowledge in immunology, cancer biology, and bacterial pathogenesis. The ability to review bacterial-based cancer therapy from a future viewpoint has been made possible by genetic manipulation and new molecular technologies. Researchers continue to face challenges in the field of bacteria-based cancer therapy because of its immense potential to fill in the gaps left by previous failures, such as the complexity of interactions between tumor bacteria.

Genetically engineered microorganisms and advanced genetic engineering technology can help overcome the above challenges and could be very useful in cancer therapy. Bacterial combinations may be utilized in the future to improve the effectiveness of radiation, chemotherapy, and immunotherapy.

References

Adnan, M., Khan, S., Al-Shammari, E., Patel, M., Saeed, M., & Hadi, S. (2017). In pursuit of cancer metastasis therapy by bacteria and its biofilms: History or future. *Medical Hypotheses, 100*, 78–81.

Alexandroff, A. B., Nicholson, S., Patel, P. M., & Jackson, A. M. (2010). Recent advances in bacillus Calmette–Guerin immunotherapy in bladder cancer. *Immunotherapy, 2*(4), 551–560.

Al-Hilu, S. A., & Al-Shujairi, W. H. (2020). Dual role of bacteria in carcinoma: Stimulation and inhibition. *International Journal of Microbiology, 2020*, 4639761.

Alkabban, F. M., & Ferguson, T. (2023). *Breast cancer. [Updated 2022 Sep 26]. Stat Pearls [Internet]*. StatPearls Publishing.

Al-Madboly, L. A., El-Deeb, N. M., Kabbash, A., Nael, M. A., Kenawy, A. M., & Ragab, A. E. (2020). Purification, characterization, identification, and anticancer activity of a circular bacteriocin from *Enterococcus thailandicus*. *Frontiers in Bioengineering and Biotechnology, 8*, 450.

Alpuim Costa, D., Nobre, J. G., Batista, M. V., Ribeiro, C., Marhold, M., Brito, M., & Cortes, J. (2021). Human microbiota and breast cancer—Is there any relevant link?—A literature review and new horizons toward personalised medicine. *Frontiers in Microbiology, 12*, 584332.

Alshamsan, A., Khan, S., Imran, A., Aljuffali, I. A., & Alsaleh, K. (2017). Prediction of *Chlamydia pneumoniae* protein localization in host mitochondria and cytoplasm and possible involvements in lung cancer etiology: A computational approach. *Saudi Pharm J, 25*(8), 1151–1157.

Armstrong, N., Ryder, S., Forbes, C., Ross, J., & Quek, R. G. (2019). A systematic review of the international prevalence of BRCA mutation in breast cancer. *Clinical Epidemiology, 11*, 543–561.

Ashu, E. E., Xu, J., & Yuan, Z. C. (2019). Bacteria in cancer therapeutics: A framework for effective therapeutic bacterial screening and identification. *Journal of Cancer, 10*(8), 1781.

Azcárate-Peril, M. A., Sikes, M., & Bruno-Bárcena, J. M. (2011). The intestinal microbiota, gastrointestinal environment and colorectal cancer: A putative role for probiotics in prevention of colorectal cancer? *American Journal of Physiology-Gastrointestinal and Liver Physiology, 301*(3), G401–G424.

Baik, S. C., Youn, H. S., Chung, M. H., Lee, W. K., Cho, M. J., Ko, G. H., & Rhee, K. H. (1996). Increased oxidative DNA damage in Helicobacter pylori-infected human gastric mucosa. *Cancer Research, 56*(6), 1279–1282.

Baindara, P., Korpole, S., & Grover, V. (2018). Bacteriocins: Perspective for the development of novel anticancer drugs. *Applied Microbiology and Biotechnology, 102*, 10393–10408.

Balic, J. J., Saad, M. I., Dawson, R., West, A. J., McLeod, L., West, A. C., & Jenkins, B. J. (2020). Constitutive STAT3 serine phosphorylation promotes *Helicobacter*-mediated gastric disease. *The American Journal of Pathology, 190*(6), 1256–1270.

Bassis, C. M., Erb-Downward, J. R., Dickson, R. P., Freeman, C. M., Schmidt, T. M., Young, V. B., & Huffnagle, G. B. (2015). Analysis of the upper respiratory tract microbiotas as the source of the lung and gastric microbiotas in healthy individuals. *MBio, 6*(2), e00037–e00015.

Basu, P., Mehta, A., Jain, M., Gupta, S., Nagarkar, R. V., John, S., & Petit, R. (2018). A randomized phase 2 study of ADXS11-001 listeria monocytogenes–Listeriolysin O immunotherapy with or without cisplatin in treatment of advanced cervical cancer. *International Journal of Gynecologic Cancer, 28*(4), 764.

Bernardo, G., Le Noci, V., Di Modica, M., Montanari, E., Triulzi, T., Pupa, S. M., & Sfondrini, L. (2023). The emerging role of the microbiota in breast cancer progression. *Cells, 12*(15), 1945.

Blanarova, C., Galovicova, A., & Petrasova, D. (2009). Use of probiotics for prevention of radiation-induced diarrhea. *Bratislavske Lekarske Listy, 110*(2), 98–104.

Bleyer, A., & Welch, H. G. (2012). Effect of three decades of screening mammography on breast-cancer incidence. *New England Journal of Medicine, 367*(21), 1998–2005.

Bonnet, M., Buc, E., Sauvanet, P., Darcha, C., Dubois, D., Pereira, B., & Darfeuille-Michaud, A. (2014). Colonization of the human gut by E. coli and colorectal cancer risk. *Clinical Cancer Research, 20*(4), 859–867.

Bowen, J. M., Stringer, A. M., Gibson, R. J., Yeoh, A. S., Hannam, S., & Keefe, D. M. (2007). VSL# 3 probiotic treatment reduces chemotherapy-induced diarrhoea and weight loss. *Cancer Biology & Therapy, 6*(9), 1445–1450.

Bronte-Tinkew, D. M., Terebiznik, M., Franco, A., Ang, M., Ahn, D., Mimuro, H., & Jones, N. L. (2009). *Helicobacter pylori* cytotoxin-associated gene A activates the signal transducer and activator of transcription 3 pathway in vitro and in vivo. *Cancer Research, 69*(2), 632–639.

Budisan, L., Zanoaga, O., Braicu, C., Pirlog, R., Covaliu, B., Esanu, V., & Berindan-Neagoe, I. (2021). Links between infections, lung cancer, and the immune system. *International Journal of Molecular Sciences, 22*(17), 9394.

Bulut, Y., Shimada, K., Wong, M. H., Chen, S., Gray, P., Alsabeh, R., & Arditi, M. (2009). Chlamydial heat shock protein 60 induces acute pulmonary inflammation in mice via the toll-like receptor 4-and MyD88-dependent pathway. *Infection and Immunity, 77*(7), 2683–2690.

Calmels, S., Ohshima, H., & Bartsch, H. (1988). Nitrosamine formation by denitrifying and non-denitrifying bacteria: Implication of nitrite reductase and nitrate reductase in nitrosation catalysis. *Microbiology, 134*(1), 221–226.

Candela, M., Turroni, S., Biagi, E., Carbonero, F., Rampelli, S., Fiorentini, C., & Brigidi, P. (2014). Inflammation and colorectal cancer, when microbiota-host mutualism breaks. *World Journal of Gastroenterology: WJG, 20*(4), 908.

Cano-Garrido, O., Seras-Franzoso, J., & Garcia-Fruitos, E. (2015). Lactic acid bacteria: Reviewing the potential of a promising delivery live vector for biomedical purposes. *Microbial Cell Factories, 14*, 1–12.

Cao, Y., Wu, K., Mehta, R., Drew, D. A., Song, M., Lochhead, P., & Chan, A. T. (2018). Long-term use of antibiotics and risk of colorectal adenoma. *Gut, 67*(4), 672–678.

Carbonero, F., Benefiel, A. C., & Gaskins, H. R. (2012). Contributions of the microbial hydrogen economy to colonic homeostasis. *Nature Reviews Gastroenterology & Hepatology, 9*(9), 504–518.

Castellarin, M., Warren, R. L., Freeman, J. D., Dreolini, L., Krzywinski, M., Strauss, J., & Holt, R. A. (2012). *Fusobacterium nucleatum* infection is prevalent in human colorectal carcinoma. *Genome Research, 22*(2), 299–306.

Cheng, W. T., Kantilal, H. K., & Davamani, F. (2020). The mechanism of Bacteroides fragilis toxin contributes to colon cancer formation. *The Malaysian Journal of Medical Sciences: MJMS, 27*(4), 9.

Chial, H. (2008). Genetic regulation of cancer. Nature. *Education, 1*(1), 67.

Choi, J. (2021). Successful endoscopic resection of residual colonic mucosa-associated lymphoid tissue lymphoma after polypectomy. *Clinical Endoscopy, 54*(5), 759–762.

Chung, L., Orberg, E. T., Geis, A. L., Chan, J. L., Fu, K., Shields, C. E. D., & Housseau, F. (2018). *Bacteroides fragilis* toxin coordinates a pro-carcinogenic inflammatory cascade via targeting of colonic epithelial cells. *Cell Host & Microbe, 23*(2), 203–214.

Correa, P., & Piazuelo, M. B. (2012). The gastric precancerous cascade. *Journal of Digestive Diseases, 13*(1), 2–9.

Cronin, M., Le Boeuf, F., Murphy, C., Roy, D. G., Falls, T., Bell, J. C., & Tangney, M. (2014). Bacterial-mediated knockdown of tumor resistance to an oncolytic virus enhances therapy. *Molecular Therapy, 22*(6), 1188–1197.

Cuevas-Ramos, G., Petit, C. R., Marcq, I., Boury, M., Oswald, E., & Nougayrède, J. P. (2010). *Escherichia coli* induces DNA damage in vivo and triggers genomic instability in mammalian cells. *Proceedings of the National Academy of Sciences, 107*(25), 11537–11542.

Dang, L. H., Bettegowda, C., Huso, D. L., Kinzler, K. W., & Vogelstein, B. (2001). Combination bacteriolytic therapy for the treatment of experimental tumors. *Proceedings of the National Academy of Sciences, 98*(26), 15155–15160.

Danino, T., Prindle, A., Hasty, J., & Bhatia, S. (2013). Measuring growth and gene expression dynamics of tumor-targeted S. typhimurium bacteria. *JoVE (Journal of Visualized Experiments), 77*, e50540.

Dasgupta, D., Kumar, A., Mukhopadhyay, B., & Sengupta, T. K. (2015). Isolation of phenazine 1, 6-di-carboxylic acid from Pseudomonas aeruginosa strain HRW. 1-S3 and its role in biofilm-mediated crude oil degradation and cytotoxicity against bacterial and cancer cells. *Applied Microbiology and Biotechnology, 99*, 8653–8665.

De Martel, C., Ferlay, J., Franceschi, S., Vignat, J., Bray, F., Forman, D., & Plummer, M. (2012). Global burden of cancers attributable to infections in 2008: A review and synthetic analysis. *The Lancet Oncology, 13*(6), 607–615.

Dejea, C., Wick, E., & Sears, C. L. (2013). Bacterial oncogenesis in the colon. *Future Microbiology, 8*(4), 445–460.

Di Domenico, E. G., Cavallo, I., Pontone, M., Toma, L., & Ensoli, F. (2017). Biofilm producing *Salmonella typhi*: Chronic colonization and development of gallbladder cancer. *International Journal of Molecular Sciences, 18*(9), 1887.

Dickson, R. P., Erb-Downward, J. R., Freeman, C. M., McCloskey, L., Beck, J. M., Huffnagle, G. B., & Curtis, J. L. (2015). Spatial variation in the healthy human lung microbiome and the adapted Island model of lung biogeography. *Annals of the American Thoracic Society, 12*(6), 821–830.

Diller, I. C. (1974). Tumor incidence in icr/albino and c57bl/6jnicr male mice injected with organisms cultured from mouse malignant tissues.

Divyashree, M., Prakash, S. K., Aditya, V., Aljabali, A. A., Alzahrani, K. J., Azevedo, V., & Barh, D. (2022). Bugs as drugs: Neglected but a promising future therapeutic strategy in cancer. *Future Oncology, 18*(13), 1609–1626.

Dong, Q. L., & Xing, X. Y. (2018). Cancer cells arise from bacteria. *Cancer Cell International, 18*, 1–9.

Dunne, A., Marshall, N. A., & Mills, K. H. (2011). TLR based therapeutics. *Current Opinion in Pharmacology, 11*(4), 404–411.

Elise, S. L. (1931). Studies in Hodgkin's disease. *Annals of Surgery, 93*(1), 162.

Erdman, S. E., Rao, V. P., Poutahidis, T., Ihrig, M. M., Ge, Z., Feng, Y., & Fox, J. G. (2003). CD4+ CD25+ regulatory lymphocytes require interleukin 10 to interrupt colon carcinogenesis in mice. *Cancer Research, 63*(18), 6042–6050.

Eyvazi, S., Vostakolaei, M. A., Dilmaghani, A., Borumandi, O., Hejazi, M. S., Kahroba, H., & Tarhriz, V. (2020). The oncogenic roles of bacterial infections in development of cancer. *Microbial Pathogenesis, 141*, 104019.

Fan, J. Y., Huang, Y., Li, Y., Muluh, T. A., Fu, S. Z., & Wu, J. B. (2022). Bacteria in cancer therapy: A new generation of weapons. *Cancer Medicine, 11*(23), 4457–4468.

Felgner, S., Kocijancic, D., Frahm, M., Curtiss, R., III, Erhardt, M., & Weiss, S. (2016). Optimizing *Salmonella enterica* serovar Typhimurium for bacteria-mediated tumor therapy. *Gut Microbes, 7*(2), 171–177.

Fu, G. F., Li, X., Hou, Y. Y., Fan, Y. R., Liu, W. H., & Xu, G. X. (2005). *Bifidobacterium longum* as an oral delivery system of endostatin for gene therapy on solid liver cancer. *Cancer Gene Therapy, 12*(2), 133–140.

Gagnière, J., Raisch, J., Veziant, J., Barnich, N., Bonnet, R., Buc, E., & Bonnet, M. (2016). Gut microbiota imbalance and colorectal cancer. *World Journal of Gastroenterology, 22*(2), 501.

Gardlik, R. O. M. A. N., Behuliak, M., Palffy, R., Celec, P., & Li, C. J. (2011). Gene therapy for cancer: Bacteria-mediated anti-angiogenesis therapy. *Gene Therapy, 18*(5), 425–431.

Garza-Morales, R., Rendon, B. E., Malik, M. T., Garza-Cabrales, J. E., Aucouturier, A., Bermúdez-Humarán, L. G., et al. (2020). Targeting melanoma hypoxia with the food-grade lactic acid bacterium *Lactococcus lactis*. *Cancers (Basel), 12*(2), 438.

Gaydos, C. A. (2000). Growth in vascular cells and cytokine production by *Chlamydia pneumoniae*. *The Journal of Infectious Diseases, 181*(Supplement_3), S473–S478.

Gholizadeh, P., Eslami, H., & Kafil, H. S. (2017). Carcinogenesis mechanisms of *Fusobacterium nucleatum*. *Biomedicine & Pharmacotherapy, 89*, 918–925.

Giralt, J., Regadera, J. P., Verges, R., Romero, J., de la Fuente, I., Biete, A., & Guarner, F. (2008). Effects of probiotic Lactobacillus casei DN-114 001 in prevention of radiation-induced diarrhea: Results from multicenter, randomized, placebo-controlled nutritional trial. *International Journal of Radiation Oncology, Biology, Physics, 71*(4), 1213–1219.

Gong, J., Jaiswal, R., Mathys, J.-M., Combes, V., Grau, G. E. R., & Bebawy, M. (2012). Microparticles and their emerging role in cancer multidrug resistance. *Cancer Treatment Reviews, 38*(3), 226–234.

Goodman, B., & Gardner, H. (2018). The microbiome and cancer. *The Journal of Pathology, 244*(5), 667–676.

Górska, A., Przystupski, D., Niemczura, M. J., & Kulbacka, J. (2019). Probiotic bacteria: A promising tool in cancer prevention and therapy. *Current Microbiology, 76*, 939–949.

Gratia, A. (1925). Sur un remarquable example d'antagonisme entre deux souches de colibacille. *Compt Rend Soc Biol, 93*, 1040–1042.

Groizeleau, J., Rybtke, M., Andersen, J. B., Berthelsen, J., Liu, Y., Yang, L., & Tolker-Nielsen, T. (2016). The anti-cancerous drug doxorubicin decreases the c-di-GMP content in *Pseudomonas aeruginosa* but promotes biofilm formation. *Microbiology, 162*(10), 1797–1807.

Gurunathan, S., Kang, M. H., Qasim, M., & Kim, J. H. (2018). Nanoparticle-mediated combination therapy: Two-in-one approach for cancer. *International Journal of Molecular Sciences, 19*(10), 3264.

He, Q. (2020). Comparison of cancer morbidity and mortality between developed countries, developing countries, and China. In *2020 International conference on public health and data science (ICPHDS)* (pp. 364–370). IEEE.

Hess, S., Peters, J., Bartling, G., Rheinheimer, C., Hegde, P., Magid-Slav, M., & Klos, A. (2003). More than just innate immunity: Comparative analysis of *Chlamydophila pneumoniae* and Chlamydia trachomatis effects on host-cell gene regulation. *Cellular Microbiology, 5*(11), 785–795.

Hicks, R. M., Ismail, M. M., Walters, C. L., Beecham, P. T., Rabie, M. F., & El Alamy, M. A. (1982). Association of bacteriuria and urinary nitrosamine formation with *Schistosoma haematobium* infection in the Qalyub area of Egypt. *Transactions of the Royal Society of Tropical Medicine and Hygiene, 76*(4), 519–527.

Higashi, H., Nakaya, A., Tsutsumi, R., Yokoyama, K., Fujii, Y., Ishikawa, S., & Hatakeyama, M. (2004). *Helicobacter pylori* CagA induces Ras-independent morphogenetic response through SHP-2 recruitment and activation. *Journal of Biological Chemistry, 279*(17), 17205–17216.

Hobbs, G. A., Der, C. J., & Rossman, K. L. (2016). RAS isoforms and mutations in cancer at a glance. *Journal of Cell Science, 129*(7), 1287–1292.

Hu, B., Kou, L., Li, C., Zhu, L. P., Fan, Y. R., Wu, Z. W., & Xu, G. X. (2009). Bifidobacterium longum as a delivery system of TRAIL and endostatin cooperates with chemotherapeutic drugs to inhibit hypoxic tumor growth. *Cancer Gene Therapy, 16*(8), 655–663.

Hu, Y. Y., Zheng, M. H., Zhang, R., Liang, Y. M., & Han, H. (2012). Notch signaling pathway and cancer metastasis. *Notch Signaling in Embryology and Cancer*, 186–198.

Hwang, S., Jo, M., Hong, J. E., Park, C. O., Lee, C. G., Yun, M., & Rhee, K. J. (2019). Zerumbone suppresses enterotoxigenic *Bacteroides fragilis* infection-induced colonic inflammation through inhibition of NF-κB. *International Journal of Molecular Sciences, 20*(18), 4560.

Jackson, C. B., Judd, L. M., Menheniott, T. R., Kronborg, I., Dow, C., Yeomans, N. D., & Giraud, A. S. (2007). Augmented gp130-mediated cytokine signalling accompanies human gastric cancer progression. *The Journal of Pathology, 213*(2), 140–151.

Jackson, J. R., Gilmartin, A., Imburgia, C., Winkler, J. D., Marshall, L. A., & Roshak, A. (2000). An indolocarbazole inhibitor of human checkpoint kinase (Chk1) abrogates cell cycle arrest caused by DNA damage. *Cancer Research, 60*(3), 566–572.

Kanazawa, Y., Ohkawa, K., Ueda, K., Mita, E., Takehara, T., Sasaki, Y., et al. (1996). Hammerhead ribozyme-mediated inhibition of telomerase activity in extracts of human hepatocellular carcinoma cells. *Biochemical and Biophysical Research Communications, 225*(2), 570–576.

Kaur, S., & Kaur, S. (2015). Bacteriocins as potential anticancer agents. *Frontiers in Pharmacology, 6*, 165109.

Kaye, J., Porcelli, S., Jones, B., et al. (1994). Treatment of murine colitis by *Lactococcus lactis* secreting IL-10. *Proceedings of National Academic Science USA, 76*(August), 104.

Kerdkumthong, K., Chanket, W., Runsaeng, P., Nanarong, S., Songsurin, K., Tantimetta, P., & Obchoei, S. (2023). Two recombinant bacteriocins, rhamnosin and lysostaphin, show synergistic anticancer activity against gemcitabine-resistant cholangiocarcinoma cell lines. *Probiotics and Antimicrobial Proteins*, 1–13.

Khan, S., Imran, A., Khan, A. A., Abul Kalam, M., & Alshamsan, A. (2016). Systems biology approaches for the prediction of possible role of *Chlamydia pneumoniae* proteins in the etiology of lung cancer. *PLoS One, 11*(2), e0148530.

Klebanoff, C. A., Acquavella, N., Yu, Z., & Restifo, N. P. (2011). Therapeutic cancer vaccines: Are we there yet? *Immunological Reviews, 239*(1), 27–44.

Knorr, J., Ricci, V., Hatakeyama, M., & Backert, S. (2019). Classification of *Helicobacter pylori* virulence factors: Is CagA a toxin or not? *Trends in Microbiology, 27*(9), 731–738.

Kohoutova, D., Smajs, D., Moravkova, P., Cyrany, J., Moravkova, M., Forstlova, M., & Bures, J. (2014). *Escherichia coli* strains of phylogenetic group B2 and D and bacteriocin production are associated with advanced colorectal neoplasia. *BMC Infectious Diseases, 14*, 1–8.

Kolodziejczyk, A. A., Zheng, D., & Elinav, E. (2019). Diet–microbiota interactions and personalized nutrition. *Nature Reviews Microbiology, 17*(12), 742–753.

Koshiol, J., Wozniak, A., Cook, P., Adaniel, C., Acevedo, J., Azócar, L., & Hildesheim, A. (2016). Salmonella enterica serovar Typhi and gallbladder cancer: A case–control study and meta-analysis. *Cancer Medicine, 5*(11), 3310–3235.

Kroft, S. H., & Oyasu, R. (1994). Urinary bladder cancer: Mechanisms of development and progression. Laboratory investigation. *A Journal of Technical Methods and Pathology, 71*(2), 158–174.

Kucerova, P., & Cervinkova, M. (2016). Spontaneous regression of tumour and the role of microbial infection–possibilities for cancer treatment. *Anti-Cancer Drugs, 27*(4), 269–277.

Kumeria, T., Maher, S., Wang, Y., Kaur, G., Wang, L., Erkelens, M., & Losic, D. (2016). Naturally derived iron oxide nanowires from bacteria for magnetically triggered drug release and cancer hyperthermia in 2D and 3D culture environments: Bacteria biofilm to potent cancer therapeutic. *Biomacromolecules, 17*(8), 2726–2736.

Kwok, T., Zabler, D., Urman, S., Rohde, M., Hartig, R., Wessler, S., & Backert, S. (2007). *Helicobacter* exploits integrin for type IV secretion and kinase activation. *Nature, 449*(7164), 862–866.

Laborda-Illanes, A., Sanchez-Alcoholado, L., Dominguez-Recio, M. E., Jimenez-Rodriguez, B., Lavado, R., Comino-Méndez, I., & Queipo-Ortuño, M. I. (2020). Breast and gut microbiota action mechanisms in breast cancer pathogenesis and treatment. *Cancers, 12*(9), 2465.

Laurila, A. L., Anttila, T., Läärä, E., Bloigu, A., Virtamo, J., Albanes, D., & Saikku, P. (1997). Serological evidence of an association between *Chlamydia pneumoniae* infection and lung cancer. *International Journal of Cancer, 74*(1), 31–34.

Lax, A. J. (2005). Bacterial toxins and cancer—A case to answer? *Nature Reviews Microbiology, 3*(4), 343–349.

Le, D. T., Wang-Gillam, A., Picozzi, V., Greten, T. F., Crocenzi, T., Springett, G., & Jaffee, E. M. (2015). Safety and survival with GVAX pancreas prime and *Listeria* monocytogenes–expressing mesothelin (CRS-207) boost vaccines for metastatic pancreatic cancer. *Journal of Clinical Oncology, 33*(12), 1325.

Leblanc, J. G. J., de Moreno, M. A., Perdigon, G. d. V., Miyoshi, A., Rochat, T., et al. (2008). Anti-inflammatory properties of lactic acid bacteria: Current knowledge, applications and prospects. In *Anti-Infective Agents in Medicinal Chemistry* (pp. 148–154). Bentham Science Publishers.

Lee, C. H. (2012). Engineering bacteria toward tumor targeting for cancer treatment: Current state and perspectives. *Applied Microbiology and Biotechnology, 93*, 517–523.

Leschner, S., Westphal, K., Dietrich, N., Viegas, N., Jablonska, J., Lyszkiewicz, M., & Weiss, S. (2009). Tumor invasion of Salmonella enterica serovar Typhimurium is accompanied by strong hemorrhage promoted by TNF-α. *PLoS One, 4*(8), e6692.

Li, N., Zhou, H., Holden, V. K., Deepak, J., Dhilipkannah, P., Todd, N. W., et al. (2023). *Streptococcus pneumoniae* promotes lung cancer development and progression. *iScience, 26*(2), 105923.

Li, X., Fu, G. F., Fan, Y. R., Liu, W. H., Liu, X. J., Wang, J. J., & Xu, G. X. (2003). *Bifidobacterium adolescentis* as a delivery system of endostatin for cancer gene therapy: Selective inhibitor of angiogenesis and hypoxic tumor growth. *Cancer Gene Therapy, 10*(2), 105–111.

Mármol, I., Sánchez-de-Diego, C., Pradilla Dieste, A., Cerrada, E., & Rodriguez Yoldi, M. J. (2017). Colorectal carcinoma: A general overview and future perspectives in colorectal cancer. *International Journal of Molecular Sciences, 18*(1), 197.

Marsh, R. L., Kaestli, M., Chang, A. B., Binks, M. J., Pope, C. E., Hoffman, L. R., & Smith-Vaughan, H. C. (2016). The microbiota in bronchoalveolar lavage from young children with chronic lung disease includes taxa present in both the oropharynx and nasopharynx. *Microbiome, 4*(1), 1–18.

Matsumoto, Y., Marusawa, H., Kinoshita, K., Endo, Y., Kou, T., Morisawa, T., & Chiba, T. (2007). *Helicobacter pylori* infection triggers aberrant expression of activation-induced cytidine deaminase in gastric epithelium. *Nature Medicine, 13*(4), 470–476.

Meng, Z., Ye, Z., Zhu, J., Li, Y., & Meng, L. (2022). New developments and opportunities of microbiota in treating breast cancers. *Frontiers in Microbiology, 13*, 818793.

Menheniott, T. R., Judd, L. M., & Giraud, A. S. (2015). STAT3: A critical component in the response to *Helicobacter pylori* infection. *Cellular Microbiology, 17*(11), 1570–1582.

Mikó, E., Kovács, T., Sebő, É., Tóth, J., Csonka, T., Ujlaki, G., & Bai, P. (2019). Microbiome—Microbial metabolome—Cancer cell interactions in breast cancer—Familiar, but unexplored. *Cells, 8*(4), 293.

Miyake, K., Yamamoto, S., & Iijima, S. (1996). Blocking adhesion of cancer cells to endothelial cell types by *S. agalactiae* type-specific polysaccharides. *Cytotechnology, 22*, 205–209.

Morales, D. K., Grahl, N., Okegbe, C., Dietrich, L. E., Jacobs, N. J., & Hogan, D. A. (2013). Control of Candida albicans metabolism and biofilm formation by *Pseudomonas aeruginosa* phenazines. *MBio, 4*(1), 10–1128.

Morales-Sánchez, A., & Fuentes-Pananá, E. M. (2014). Human viruses and cancer. *Viruses, 6*(10), 4047–4079.

Morris, A., Beck, J. M., Schloss, P. D., Campbell, T. B., Crothers, K., Curtis, J. L., & Weinstock, G. M. (2013). Comparison of the respiratory microbiome in healthy nonsmokers and smokers. *American Journal of Respiratory and Critical Care Medicine, 187*(10), 1067–1075.

Motta, J. P., Wallace, J. L., Buret, A. G., Deraison, C., & Vergnolle, N. (2021). Gastrointestinal biofilms in health and disease. *Nature Reviews Gastroenterology & Hepatology, 18*(5), 314–334.

Murata-Kamiya, N., Kurashima, Y., Teishikata, Y., Yamahashi, Y., Saito, Y., Higashi, H., & Hatakeyama, M. (2007). *Helicobacter pylori* CagA interacts with E-cadherin and deregulates the β-catenin signal that promotes intestinal transdifferentiation in gastric epithelial cells. *Oncogene, 26*(32), 4617–4626.

Nagy, T. A., Frey, M. R., Yan, F., Israel, D. A., Polk, D. B., & Peek, R. M. (2009). *Helicobacter pylori* regulates cellular migration and apoptosis by activation of phosphatidylinositol 3-kinase signaling. *The Journal of Infectious Diseases, 199*(5), 641–651.

Nakayama, M., Hisatsune, J., Yamasaki, E., Isomoto, H., Kurazono, H., Hatakeyama, M., & Hirayama, T. (2009). *Helicobacter pylori* VacA-induced inhibition of GSK3 through the PI3K/Akt signaling pathway. *Journal of Biological Chemistry, 284*(3), 1612–1619.

Nath, G., Gulati, A. K., & Shukla, V. K. (2010). Role of bacteria in carcinogenesis, with special reference to carcinoma of the gallbladder. *World Journal of Gastroenterology: WJG, 16*(43), 5395.

Nguyen, C., & Nguyen, V. D. (2016). Discovery of azurin-like anticancer bacteriocins from human gut microbiome through homology modeling and molecular docking against the tumor suppressor p53. *BioMed Research International, 2016*, 8490482.

Niethammer, A. G., Xiang, R., Becker, J. C., Wodrich, H., Pertl, U., Karsten, G., & Reisfeld, R. A. (2002). A DNA vaccine against VEGF receptor 2 prevents effective angiogenesis and inhibits tumor growth. *Nature Medicine, 8*(12), 1369–1375.

Nistal, E., Fernández-Fernández, N., Vivas, S., & Olcoz, J. L. (2015). Factors determining colorectal cancer: The role of the intestinal microbiota. *Frontiers in Oncology, 5*, 220.

Ohshima, H., & Bartsch, H. (1994). Chronic infections and inflammatory processes as cancer risk factors: Possible role of nitric oxide in carcinogenesis. *Mutation Research/Fundamental and Molecular Mechanisms of Mutagenesis, 305*(2), 253–264.

Parhi, L., Alon-Maimon, T., Sol, A., Nejman, D., Shhadeh, A., Fainsod-Levi, T., & Bachrach, G. (2020). Breast cancer colonization by *Fusobacterium nucleatum* accelerates tumor growth and metastatic progression. *Nature Communications, 11*(1), 3259.

Park, S. J., Park, S. H., Cho, S., Kim, D. M., Lee, Y., Ko, S. Y., & Park, S. (2013). New paradigm for tumor theranostic methodology using bacteria-based microrobot. *Scientific Reports, 3*(1), 3394.

Park, S. J., Lee, Y. K., Cho, S., Uthaman, S., Park, I. K., Min, J. J., & Park, S. (2015). Effect of chitosan coating on a bacteria-based alginate microrobot. *Biotechnology and Bioengineering, 112*(4), 769–776.

Phan, T. X., Nguyen, V. H., Duong, M. T. Q., Hong, Y., Choy, H. E., & Min, J. J. (2015). Activation of inflammasome by attenuated Salmonella typhimurium in bacteria-mediated cancer therapy. *Microbiology and Immunology, 59*(11), 664–675.

Piao, J. Y., Lee, H. G., Kim, S. J., Kim, D. H., Han, H. J., Ngo, H. K. C., & Surh, Y. J. (2016). *Helicobacter pylori* activates IL-6-STAT3 signaling in human gastric cancer cells: Potential roles for reactive oxygen species. *Helicobacter, 21*(5), 405–416.

Podlesek, Z., & Žgur Bertok, D. (2020). The DNA damage inducible SOS response is a key player in the generation of bacterial persister cells and population wide tolerance. *Frontiers in Microbiology, 11*, 1785.

Prasad, S. K., Bhat, S., Shashank, D., Akshatha, C. R., Rachtanapun, P., Devegowda, D., & Sommano, S. R. (2022). Bacteria-mediated oncogenesis and the underlying molecular intricacies: What we know so far. *Frontiers in Oncology, 12*, 836004.

Ray, A., Moore, T. F., Pandit, R., Burke, A. D., & Borsch, D. M. (2023). An overview of selected bacterial infections in cancer, their virulence factors, and some aspects of infection management. *Biology, 12*(7), 963.

Redecke, V., Dalhoff, K., Bohnet, S., Braun, J., & Maass, M. (1998). Interaction of *Chlamydia pneumoniae* and human alveolar macrophages: Infection and inflammatory response. *American Journal of Respiratory Cell and Molecular Biology, 19*(5), 721–727.

Rhee, K. J., Wu, S., Wu, X., Huso, D. L., Karim, B., Franco, A. A., & Sears, C. L. (2009). Induction of persistent colitis by a human commensal, enterotoxigenic *Bacteroides fragilis*, in wild-type C57BL/6 mice. *Infection and Immunity, 77*(4), 1708–1718.

Rodrigues, G., Silva, G. G. O., Buccini, D. F., Duque, H. M., Dias, S. C., & Franco, O. L. (2019). Bacterial proteinaceous compounds with multiple activities toward cancers and microbial infection. *Frontiers in Microbiology, 10*, 465958.

Roy, U. K. B., Rial, N. S., Kachel, K. L., & Gerner, E. W. (2008). Activated K-RAS increases polyamine uptake in human colon cancer cells through modulation of caveolar endocytosis. *Molecular Carcinogenesis: Published in cooperation with the University of Texas MD Anderson Cancer Center, 47*(7), 538–553.

Saleh, A., & Halawa, A. (2020). P1649 is bk virus carcinogenic?! *Nephrology Dialysis Transplantation, 35*(Supplement_3), gfaa142-P1649.

Scanu, T., Spaapen, R. M., Bakker, J. M., Pratap, C. B., Wu, L. E., Hofland, I., & Neefjes, J. (2015). *Salmonella* manipulation of host signaling pathways provokes cellular transformation associated with gallbladder carcinoma. *Cell Host & Microbe, 17*(6), 763–774.

Schmitz-Winnenthal, F. H., Hohmann, N., Schmidt, T., Podola, L., Friedrich, T., Lubenau, H., & Beckhove, P. (2018). A phase 1 trial extension to assess immunologic efficacy and safety of prime-boost vaccination with VXM01, an oral T cell vaccine against VEGFR2, in patients with advanced pancreatic cancer. *Oncoimmunology, 7*(4), e1303584.

Schoppmann, S. F., Birner, P., Stöckl, J., Kalt, R., Ullrich, R., Caucig, C., & Kerjaschki, D. (2002). Tumor-associated macrophages express lymphatic endothelial growth factors and are related to peritumoral lymphangiogenesis. *The American Journal of Pathology, 161*(3), 947–956.

Schwabe, R. F., & Jobin, C. (2013). The microbiome and cancer. *Nature Reviews Cancer, 13*(11), 800–812.

Schwabe, R. F., & Wang, T. C. (2012). Bacteria deliver a genotoxic hit. *Science, 338*(6103), 52–53.

Sears, C. L. (2009). Enterotoxigenic *Bacteroides fragilis*: A rogue among symbiotes. *Clinical Microbiology Reviews, 22*(2), 349–369.

Sears, C. L., & Garrett, W. S. (2014). Microbes, microbiota, and colon cancer. *Cell Host & Microbe, 15*(3), 317–328.
Sedighi, M., Zahedi Bialvaei, A., Hamblin, M. R., Ohadi, E., Asadi, A., Halajzadeh, M., & Kouhsari, E. (2019). Therapeutic bacteria to combat cancer; current advances, challenges, and opportunities. *Cancer Medicine, 8*(6), 3167–3181.
Segal, L. N., Alekseyenko, A. V., Clemente, J. C., Kulkarni, R., Wu, B., Chen, H., & Weiden, M. D. (2013). Enrichment of lung microbiome with supraglottic taxa is associated with increased pulmonary inflammation. *Microbiome, 1*, 19.
Sender, R., Fuchs, S., & Milo, R. (2016). Revised estimates for the number of human and bacteria cells in the body. *PLoS Biology, 14*(8), e1002533.
Shankar, A., Saini, D., Dubey, A., Roy, S., Bharati, S. J., Singh, N., & Rath, G. K. (2019). Feasibility of lung cancer screening in developing countries: Challenges, opportunities and way forward. *Translational Lung Cancer Research, 8*(Suppl 1), S106.
Shen, M., Cai, L., Jiang, K., et al. (2018). The therapeutic role of inhibition of miR-328 on pulmonary carcinoma induced by chlamydia pneumoniae through targeting histone H2AX. *Cancer Biomarkers: Section A of Disease Markers.* https://doi.org/10.3233/cbm-181999
Sheweita, S. A., & Alsamghan, A. S. (2020). Molecular mechanisms contributing bacterial infections to the incidence of various types of cancer. *Mediators of Inflammation, 2020*, 4070419.
Shi, L., Sheng, J., Chen, G., Zhu, P., Shi, C., Li, B., et al. (2020). Combining IL-2-based immunotherapy with commensal probiotics produces enhanced antitumor immune response and tumor clearance. *Journal for Immunotherapy of. Cancer, 8*(2).
Shi, Q., Haenen, G. R., Maas, L., Arlt, V. M., Spina, D., Vasquez, Y. R., & Godschalk, R. W. L. (2016). Inflammation-associated extracellular β-glucuronidase alters cellular responses to the chemical carcinogen benzo [a] pyrene. *Archives of Toxicology, 90*, 2261–2273.
Singh, R. K., Gutman, M., Radinsky, R., Bucana, C. D., & Fidler, I. J. (1994). Expression of interleukin 8 correlates with the metastatic potential of human melanoma cells in nude mice. *Cancer Research, 54*(12), 3242–3247.
Soler, A. P., Miller, R. D., Laughlin, K. V., Carp, N. Z., Klurfeld, D. M., & Mullin, J. M. (1999). Increased tight junctional permeability is associated with the development of colon cancer. *Carcinogenesis, 20*(8), 1425–1432.
Staedtke, V., Roberts, N. J., Bai, R. Y., & Zhou, S. (2016). Clostridium novyi-NT in cancer therapy. *Genes & Diseases, 3*(2), 144–152.
Stern, C., Kasnitz, N., Kocijancic, D., Trittel, S., Riese, P., Guzman, C. A., & Weiss, S. (2015). Induction of CD 4+ and CD 8+ anti-tumor effector T cell responses by bacteria mediated tumor therapy. *International Journal of Cancer, 137*(8), 2019–2028.
Sung, H., Ferlay, J., Siegel, R. L., Laversanne, M., Soerjomataram, I., Jemal, A., & Bray, F. (2021). Global cancer statistics 2020: GLOBOCAN estimates of incidence and mortality worldwide for 36 cancers in 185 countries. *CA: a Cancer Journal for Clinicians, 71*(3), 209–249.
Suzuki, M., Mimuro, H., Kiga, K., Fukumatsu, M., Ishijima, N., Morikawa, H., & Sasakawa, C. (2009). *Helicobacter pylori* CagA phosphorylation-independent function in epithelial proliferation and inflammation. *Cell Host & Microbe, 5*(1), 23–34.
Tabassam, F. H., Graham, D. Y., & Yamaoka, Y. (2009). Helicobacter pylori activate epidermal growth factor receptor-and phosphatidylinositol 3-OH kinase-dependent Akt and glycogen synthase kinase 3β phosphorylation. *Cellular Microbiology, 11*(1), 70–82.
Thanjavur, N., Sangubotla, R., Lakshmi, B. A., Rayi, R., Mekala, C. D., Reddy, A. S., & Viswanath, B. (2022). Evaluating the antimicrobial and apoptogenic properties of bacteriocin (nisin) produced by *Lactococcus lactis*. *Process Biochemistry, 122*, 76–86.
Torres, W., Lameda, V., Olivar, L. C., Navarro, C., Fuenmayor, J., Pérez, A., & Bermudez, V. (2018). Bacteria in cancer therapy: Beyond immune stimulation. *Journal of Cancer Metastasis and Treatment, 4*(4), 1–25.
Tricker, A. R., Stickler, D. J., Chawla, J. C., & Preussmann, R. (1991). Increased urinary nitrosamine excretion in paraplegic patients. *Carcinogenesis, 12*(5), 943–946.

Tsay, J. C. J., Wu, B. G., Badri, M. H., Clemente, J. C., Shen, N., Meyn, P., & Segal, L. N. (2018). Airway microbiota is associated with upregulation of the PI3K pathway in lung cancer. *American Journal of Respiratory and Critical Care Medicine, 198*(9), 1188–1198.

Tsutsumi, R., Takahashi, A., Azuma, T., Higashi, H., & Hatakeyama, M. (2006). Focal adhesion kinase is a substrate and downstream effector of SHP-2 complexed with *Helicobacter pylori* CagA. *Molecular and Cellular Biology, 26*(1), 261–276.

Umbenhauer, D., Wild, C. P., Montesano, R., Saffhill, R., Boyle, J. M., Huh, N., & Lu, S. H. (1985). 06-Methyldeoxyguanosine in oesophageal dna among individuals at high risk of oesophageal cancer. *International Journal of Cancer, 36*(6), 661–665.

Upadhayay, A., Pal, D., & Kumar, A. (2022). *Salmonella typhi* induced oncogenesis in gallbladder cancer: Co-relation and progression. *Advances in Cancer Biology-Metastasis, 4*, 100032.

Urbaniak, C., Gloor, G. B., Brackstone, M., Scott, L., Tangney, M., & Reid, G. (2016). The microbiota of breast tissue and its association with breast cancer. *Applied and Environmental Microbiology, 82*(16), 5039–5048.

Wang, Y., Hu, P. C., Ma, Y. B., Fan, R., Gao, F. F., Zhang, J. W., & Wei, L. (2016). Sodium butyrate-induced apoptosis and ultrastructural changes in MCF-7 breast cancer cells. *Ultrastructural Pathology, 40*(4), 200–204.

Westphal, K., Leschner, S., Jablonska, J., Loessner, H., & Weiss, S. (2008). Containment of tumor-colonizing bacteria by host neutrophils. *Cancer Research, 68*(8), 2952–2960.

White, R. G. (1976). The adjuvant effect of microbial products on the immune response. *Annual Review of Microbiology, 30*(1), 579–600.

Xu, M., Yamada, M., Li, M., Liu, H., Chen, S. G., & Han, Y. W. (2007a). FadA from *Fusobacterium nucleatum* utilizes both secreted and non-secreted forms for functional oligomerization for attachment and invasion of host cells. *Journal of Biological Chemistry, 282*(34), 25000–25009.

Xu, Y. F., Zhu, L. P., Hu, B., Fu, G. F., Zhang, H. Y., Wang, J. J., & Xu, G. X. (2007b). A new expression plasmid in Bifidobacterium longum as a delivery system of endostatin for cancer gene therapy. *Cancer Gene Therapy, 14*(2), 151–157.

Yamaoka, Y. (2010). Mechanisms of disease: *Helicobacter pylori* virulence factors. *Nature Reviews Gastroenterology & Hepatology, 7*(11), 629–641.

Zappa, C., & Mousa, S. A. (2016). Non-small cell lung cancer: Current treatment and future advances. *Translational Lung Cancer Research, 5*(3), 288.

Zhao, Y., Zhang, J., Cheng, A. S., Yu, J., To, K. F., & Kang, W. (2020). Gastric cancer: Genome damaged by bugs. *Oncogene, 39*(17), 3427–3442.

Zhao, M., Chen, X., Yang, Z., Yang, X., & Peng, Q. (2022). Bacteria and tumor: Understanding the roles of bacteria in tumor genesis and immunology. *Microbiological Research, 261*, 127082.

Zhu, Q., Gao, R., Wu, W., & Qin, H. (2013). The role of gut microbiota in the pathogenesis of colorectal cancer. *Tumor Biology, 34*, 1285–1300.

Zhu, H., Shen, Z., Luo, H., Zhang, W., & Zhu, X. (2016). Chlamydia trachomatis infection-associated risk of cervical cancer: A meta-analysis. *Medicine, 95*(13), e3077.

Viral Infections and Different Cancers

4

Sidra Abbas and Shaghufta Perveen

Abstract

Nearly 20% of worldwide cancers, encompassing millions of individuals, stem from viral infections. The human T-cell leukemia virus type 1 (HTLV-1), the human herpes virus 8 (HHV-8), the Epstein-Barr virus (EBV) which is associated with nasopharyngeal carcinoma and Burkitt's lymphoma, the hepatitis C virus (HCV), the hepatitis B virus (HBV), and the human papillomavirus (HPV), which is a major cause of cervical cancer, are just a few of the seven viruses that have been conclusively concomitant with oncogenesis. These viral foes, despite their limited genetic repertoire, exploit sophisticated mechanisms to manipulate crucial cellular pathways. They hijack processes like cell growth (e.g., through E7 oncoprotein in HPV), differentiation (e.g., by disrupting pRB tumor suppressor), and evasion of the host immune system. Sentinel systems monitor these cellular activities and make sure that abnormal cells are eliminated from the pool of proliferating cells. These regulations are governed via complex, redundant regulatory networks. Viruses that want to be packaged within infectious structures have to target cellular regulatory nodes with little redundancy and deactivate the surveillance mechanisms that normally detect and remove these abnormal cells since they have to fit into relatively small genomes. These regulations governed via highly intricate regulatory networks, with built-in redundancy. To fit into their packaging, viruses must target crucial cellular control points with minimal redundancy, effectively disabling the body's ability to detect and eliminate these rogue cells. Important proteins within these similar regulatory networks are frequently mutated in malignancies and illnesses that are not virally related. Oncogenic viruses have been instrumental in unraveling cellular control networks. These insights include identifying genes that promote cancer and

S. Abbas (✉) · S. Perveen
Fatima Jinnah Women University, Rawalpindi, Pakistan

N. Masood, A. Yasmin (eds.), *The Microbiome and Cancer*,
https://doi.org/10.1007/978-981-97-6948-3_4

those that prevent it, understanding pathways crucial for maintaining healthy DNA, and elucidating the processes by which the immune system identifies and eliminates abnormal cells. Viruses employ diverse strategies throughout the multistep process of carcinogenesis, from initiating mutations through integration and DNA damage to hijacking cellular processes and evading immune defenses. This chapter delves into these mechanisms explored through cell culture, animal models, and molecular analyses, providing insights into the complex journey from viral infection to cancer development.

Keywords

Human T-cell leukemia virus type 1 (HTLV-1) · Hepatitis C virus (HCV) · Human papillomavirus (HPV) · Hepatitis B virus (HBV) · Epstein-Barr virus (EBV) · Kaposi's sarcoma-associated herpesvirus (KSHV)/Human herpes virus 8 (HHV-8)

4.1 Introduction

Cancer casts a long shadow, claiming an estimated 6.7 million lives and leaving 10.8 million newly diagnosed each year. This devastating impact is felt by individuals, families, and healthcare systems worldwide (Parkin et al., 2005). Unraveling the mysteries of cancer cells and how they arise is paramount in the fight against this disease. It holds the key to developing effective treatments and preventive measures. Carcinogenesis is generally understood to be a multifaceted, intricate process. The past three decades have witnessed a surprising link between viruses and neoplasms. Up to one-fifth of all cancers may be triggered by viral infections (Parkin, 2006; zur Hausen, 2006). Parkin (2006) stated that cancer-causing viruses can intervene at various points in the path toward cancer. The influence of a particular virus on a specific cancer type shows significant variation, ranging from a minor contribution (15%) to being the sole cause (100%). Research on oncogenic viruses has yielded unexpected dividends. Not only has it helped us in understanding the specific origins of certain cancers, but it has also revealed critical cellular pathways that frequently disrupted during cancer development in general. This unexpected benefit provides valuable insights applicable to a broader fight against the disease.

In 1911, Peyton Rous's groundbreaking experiments demonstrated that filterable substances {later named Rous sarcoma virus (RSV)} from chicken tumors could induce tumors in healthy birds. This discovery paved the way for the idea that viruses can contribute to cancer development (zur Hausen, 2002). This retrovirus's discovery revolutionized the science of tumor virology by indicating that certain types of cancer may have an infectious cause and ultimately paving the way for the identification of oncogenes (Weiss & Vogt, 2011). The 1930s witnessed a breakthrough in cancer research with the identification of two tumor viruses in mammals, hinting at the possibility of viruses causing human cancers (Javier & Butel, 2008). One such virus, the mouse mammary tumor virus (MMTV), also known as the "milk factor," is passed from mothers to their progeny through milk and can cause cancer in mice (Bittner, 1936). Following the identification of the first mammalian tumor viruses in the 1930s, the 1950s saw a surge in discoveries. Kappapapilloma virus 2 was found in cottontail rabbits with skin cancer (Shope, 1932), while the following decade brought the identification of mouse polyomavirus (Stewart et al., 1958) and mouse leukemia virus (Gross, 1951). These findings further solidified the link between viruses and cancer development in mammals.

The 1960s and 1970s marked the birth of human tumor virology with the identification of the human cancer-causing viruses (Pagano, 1999; zur Hausen, 2006). This groundbreaking era began with the discovery of Epstein-Barr virus (EBV), also known as human herpes virus 4 (HHV-$). Using electron microscopy, researchers found EBV present in cells grown from Burkitt's lymphoma (Epstein et al., 1964). Following the identification of EBV in the 1960s, further research revealed key details about the virus and its connection to cancer. Additionally, studies showed that by the age of 20, over 90% of people would have been infected with EBV (Henle et al., 1969). The link between EBV and specific cancers grew stronger. Immunofluorescence assays demonstrated a heightened defense responses to EBV antigens in patients with Burkitt's lymphoma and nasopharyngeal carcinoma (Henle et al., 1970). Furthermore, EBV DNA was directly detected in biopsy tissues from these tumors, solidifying the association between the virus and these tumors (Nonoyama et al., 1975). The EBV carries a genetic blueprint encoded in a linear dsDNA molecule ranging from 168,000 to 184,000 base pairs long. This blueprint contains instructions for building the virus, including around 85 genes. Interestingly, the EBV also has unique sections that repeat at both ends and within its structure (Baer et al., 1984; zur Hausen, 2006). One peculiarity of EBV is its behavior during a specific phase called latency. Unlike most viruses that form a circle with their DNA, EBV stays linear in most cells. However, in some plants and during latency in certain human cells, EBV DNA can form a circular structure called an episome (Adams & Lindahl, 1975). This circular form helps the virus maintain a stable presence within the infected cell. Although it is mostly replicated in lymphocytes, epithelial cells are also susceptible to infection by EBV (Lemon et al., 1977; Sixbey et al., 1984). Additionally, EBV comes in two distinct subtypes, EBV-1 and EBV-2, which can be differentiated base on a specific region of their genetic material called the EBNA locus (zur Hausen, 2006).

Hepatitis B was first recognized in the 1940s as an infectious jaundice carried by serum, and decades later in the 1970s, a breakthrough came with the discovery of culprit: the hepatitis B virus (HBV) (Dane et al., 1970). According to Robinson et al. (1974), HPV is a hepadnavirus that causes cirrhosis, hepatocellular carcinoma (HCC), acute and chronic liver contagions, and liver failure. Its small, circular genome is mostly made up of dsDNA. Epidemiological studies of HBV contagion and HCC in the 1970s connected HBV to the onset of disease; in 1981, conclusive epidemiological data in favor of an etiological role were published (Beasley et al., 1981).

HBV can transmit through blood and other body fluids. These include saliva from home contact, heterosexual intercourse, blood contact in hospital locales, close contact, and perinatal transmission, which is a major way in Asia. Prenatal transmission of HBV is important, but not of HCV, and around 50% of HBV infections are asymptomatic (Wright, 1990). Most people with hepatitis B recover after 3–4 months of incubation; however, 1% of individuals develop meteoric hepatitis B, which can cause severe liver failure and require liver transplantation. Persistent infections cause cirrhosis and chronic active hepatitis in about 5% of cases (Zur Hausen, 2006). The risk of chronic infection is particularly high for infants infected at birth, with 80–90% of such cases in Asia progressing to chronic stages. Similarly, children infected before the age of 6 in Africa have 30–50% chance of developing chronic HBV. The global burden of chronic HBV infection is not evenly distributed. Regions like sub-Saharan Africa and East Asia see a much higher prevalence compared to North America. In these areas, childhood acquisition of the virus is more common, resulting in a higher percentage of adult with chronic infection (5–10%) compared to the 0.5% rate observed in the United States.

The discovery of hepatitis C virus (HCV) in the 1970s shed light on another major cause of chronic liver disease transmitted through transfusions (Feinstone et al., 1975). Hepatitis C virus (HCV) is the major culprit behind liver transplants in the Western world. This is because a staggering 85% of adults infected with HCV develop persistent infections. This can lead to serious complications like cirrhosis and even liver cancer, contrasting with HBV. HCV belongs to a group of viruses called flavivirus. Its genetic makeup is distinct from HBV. It has a ssRNA genome, which is linear and uncapped, and the RNA is positive-sense, encoded for a polyprotein processed into ten proteins (Hoofnagle, 2002). Today, the most common ways to contract HCV are through hypodermal injections and unprotected sexual contact. Additionally, research suggests a higher prevalence among prison inmates (Larney et al., 2013). Both HBV and HCV infections can lead to chronic inflammation in the liver, which is believed to be a key factor in how these viruses contribute to cancer development.

It is well known that human papillomaviruses, or HPVs, are the causative agents of cancer in humans. The connection between HPV and cervical carcinoma was first theorized by Zur Hausen in the 1970s (zur Hausen, 1977; zur Hausen, Meinhof, et al., 1974; zur Hausen, Schulte-Holthausen, et al., 1974). In the 1980s, scientists were able to directly detect specific HPV strains (HPV16 and HPV18) in cervical cancer cells. Further studies in the 1990s solidified the link between these "high-risk" HPVs and cervical cancer development (zur Hausen, 1977). Worldwide, HPV infections account for more than half of all malignancies connected to infections in women. In men, however, HPV cancers are less common, accounting for only about 5% of such cancers. HPV spreads through close skin-to-skin contact, including contact with mucous membrane. Tiny breaks in the skin, called microabrasions, can make transmission easier. Sexual interaction is usually the source of genital HPV infections, which can result in a variety of malignancies. The good news is that most HPV infections in young women clear up on their own. However, certain types of HPV, called "high-risk" hPV, can linger and potentially lead to cancer if left untreated. This understanding of HPV has led to the development of vaccines like Cervarix and Gardasil. These vaccines target the two main "high-risk" HPV types (HPV16 and HPV18) responsible for most cervical cancers (Schiffman & Castle,

2003; zur Hausen, 1977). By getting vaccinated, women can significantly reduce their risk of developing cervical cancer later in life.

In the 1970s, doctors in Southwestern Japan noticed a cluster of unusual leukaemia incidence in adults. This type of leukaemia called adult T-cell leukemia (ATL) became a cause for concern. Scientists looking into the cause isolated a new virus from the blood cells of ATL patients (Yoshida et al., 1982). Interestingly, this wasn't the first time this virus had been seen. Another research group, led by Robert Gallo, had already discovered a similar virus, later named human T-cell leukemia virus type 1 (HTLV-1) (Seiki et al., 1983). While HTLV-1 is most commonly found in southern Japan, it also exists in other areas like Australia and Northern Canada. Its estimated that around 15–20 million people worldwide are infected with this virus (Lairmore et al., 2012). The virus can be passed from person to person through intercourse, parenteral routes such as blood transfusions and contaminated needles, and nursing. Only a tiny percentage of infected people acquire ATL, and the disease usually manifests decades after infection. While HTLV-1 is a worldwide virus, the concentration of ATL cases within Japanese families points to a possible genetic origin that may involve different HLA types (Sonoda et al., 2011).

Thanks to advancements in molecular biology, scientists have recently discovered two new viruses that can cause cancer in humans. One of these discoveries happened in 1994. Researchers led by Chang used a special technique to identify a unique genetic sequence in Kaposi's sarcoma, a rare cancer. This leads them to isolate a new virus called Kaposi's sarcoma-associated herpesvirus (KSHV) (Chang et al., 1994), also known as human herpesvirus 8 (HHV-8). KSHV is linked to several cancers, including Kaposi's sarcoma, primary effusion lymphoma, and a rare disease called multicentric Castleman's disease. Interestingly, KSHV can hide inside infected cells for long periods without causing any symptoms. In this "dormant" state, it produces a protein called LANA. However, the virus can also become active again and start replicating, potentially leading to new infections (Corbellino et al., 1996; Soulier et al., 1995).

The latest human oncogenic virus to be identified is the Merkel cell polyomavirus (MCV), which stands out among other human polyomaviruses as the only one with confirmed carcinogenicity. The 1971 discovery of the human polyomaviruses JC virus and BK virus was made (White et al., 2013). At least eight novel human polyomaviruses have been found in recent years; MCV is one of them. These viruses have been found in patients suffering from Merkel cell carcinoma (MCC), an uncommon but deadly cancer that starts in the skin's mechanoreceptor Merkel cells. According to Bichakjian et al. (2007), weak immunity is a predisposing factor for MCC, which suggests that MCC may have an infectious origin similar to Kaposi's sarcoma. Feng et al. (2007) discovered MCV, a novel circular DNA virus that is clonally incorporated into the cell genome in roughly 80% of MCC patients, after searching for viral sequences in MCC. Integration of MCV DNA at a single place happens prior to tumor formation, suggesting its role in neoplasia. MCV integration in MCC has been validated by molecular research, demonstrating its causal role in oncogenesis. The virus cannot be propagated in tissue culture past the stage of primary low-level viral particle production from cells transfected with the viral DNA, which has made MCV replication investigations difficult (Arora et al., 2012).

Scientist have identified seven viruses that can directly contribute to the development of cancer in humans. This chapter dives deep into our current understanding of how these viruses manipulate our cells on a molecular level, ultimately leading to tumor formation.

4.2 Human Immunodeficiency Virus

HIV can spread through breast milk, blood, vaginal secretions, and semen. Though it does not induce cancer development, it weakened the immune system which allows other oncogenic viruses to thrive and potentially trigger tumor progression. People living with HIV are more susceptible to specific cancers including cervical cancer; anus, liver, lung, mouth, and throat cancer; NHL and HL; and Kaposi's sarcoma (Casper et al., 2017).

4.2.1 How Immunodeficiency Causes Cancer?

Immunodeficiency, acquired through HIV infection or inborn immunity errors, increases the risk of malignancy and infection. HIV-positive patients' CD4+ T-cell lymphopenia and some inborn immunity errors can result in belligerent B-cell NHL and oncogenic viruses such as EBV, HPV, MCV, and Kaposi's sarcoma herpes virus (KSHV). Defects in DNA repair impede the growth of T and B cells, resulting in malignancies that are hematologic or nonhematologic. HIV infection and cancer have been linked since 1981, according to reports of cases of KS (Kaposi's sarcoma) and pneumocystis pneumonia (Chamseddine et al., 2018). Then, in 1994, the identification of KSHV confirmed that oncogenic viruses are responsible for many, if not all, HIV-related malignancies (Pai et al., 2021).

HIV infection resulted in immunological abnormalities that impeded the immune system's ability to regulate cancer-causing viruses, such as MCV, KSHV, EBV, and HPV. As a result, cancer developed. Individuals with KSHV, HIV, HPVP, HBV, and HCV who smoke excessively are at an increased risk of developing Kaposi's sarcoma, cervical cancer, and anal cancer (Nael et al., 2016).

When three-drug ART was developed in the middle of the 1990s, CD4+ T-cell levels increased, and AIDS became a chronic illness. Malignancies linked to severe immunodeficiency were less common after starting antiretroviral therapy (ART); nevertheless, malignancies with moderate reductions in CD4+ T-cell counts were more common in longer-living HIV-positive individuals. Among HIV-positive individuals receiving ART, cancer is now a prevalent cause of mortality. NHL linked to immunocompromised and Hodgkin's lymphoma is the most prevalent kind of tumor among HIV-positive individuals in the United States (Brandão et al., 2021; Coghill et al., 2018).

The WHO categorizes HIV-based lymphomas into three main classes:

1. Immunocompetent patients' lymphomas, such as Burkitt's lymphomas (BL), mucosa-associated lymphoid tissue (MALT), Hodgkin's lymphoma (HL), diffuse large B-cell lymphoma (DLBCL), and peripheral T-cell lymphoma (PTCL)
2. HIV-positive patients' lymphomas, such as DLBCL, plasmablastic lymphoma (PBL), and primary effusion lymphoma (PEL), which are all positive for the human HBV
3. Polymorphic posttransplant lymphoproliferative disorder (PTLD), a complication that affects individuals with immune suppression (Epeldegui et al., 2010; Linke-Serinsöz et al., 2017)

NHL is often connected with AIDS-associated malignancy and is a leading cause of cancer-associated death in HIV-positive patients (Gessese et al., 2023). Compared to other virus mediators of oncogenesis, EBV is more commonly associated with AIDS-related NHL cases, being present in 30–50% of cases and almost 100% of PCNS cases (Meister et al., 2018). PCNSL (primary central nervous system lymphoma), DLBCL, and BL are the most prevalent AIDS-related NHLs (90%), while PEL and PBL are rare forms (Linke-Serinsöz et al., 2017).

Various stages of the pathogenic mechanism linked to lymphomagenesis related to HIV are described as follows:

1. The ability to infect cancer cells, cripple defense system, and prolong antigenic stimulation is the main cause of HIV-related lymphomagenesis. Immunodeficiency is caused by the loss of CD4 T cells in the course of HIV contagion. This can be brought on by viral proteins (such as Tat, Env, and negative regulatory factor), bystander apoptosis of nearby healthy cells, cell death from excessive immune cell activation, or direct cell destruction due to infection (Gessese et al., 2023).
2. The envelope glycoprotein plays a key role in harming healthy cells through a process known as bystander apoptosis. The binding of envelope glycoprotein (gp120) of HIV to healthy cells depends on whether the cells have the correct docking points, specifically CD4 and either CXCR4/CCR5 receptors. The type of envelope glycoprotein produced by the virus can change depending on the stage of HIV infection., which also affect the virus tropism. Medications that target the envelope glycoprotein can help block this binding, and the strength of the body's immune response also plays a role in preventing the virus from attaching to healthy cells (Gessese et al., 2023).
3. According to Garg and Joshi (2017), macrophages utilize the TNFR1 and Fas pathways to eliminate bystander T cells, and the production of polyclonal antibodies has been connected to immunological imbalance. HIV particles themselves can interact with B cells, hindering their ability to function properly— additionally, HIV's Nef protein can indirectly contribute to this imbalance. HIV infection creates a state of constant immune system activation. This includes factors like CD40 ligand, which overstimulates B cells, complement proteins attaching to HIV particles. This chronic activation can damage B cells over time (Gessese et al., 2023). Certain molecules released during HIV infec-

tion, like CXCL13, can also play a role in the development of NHL (Garg & Joshi, 2017).

4. Viral particles, including the P17 protein, which stimulates key intracellular signaling pathways and encourages HIV-1 infection and replication, also have a direct impact on development of lymphomas. In AIDS-related lymphomas, cancerous lymphocytes adhere to the lining of blood vessels that are infected with HIV. This attachment might accelerate the spread of these malignant cells into surrounding tissues (Gessese et al., 2023). Different types of NHL along with their pathogenesis have been mentioned in Fig. 4.1.

The fundamental mechanism of HIV-1 infection in epithelial cells has been documented in a number of investigations. Some research indicates HIV-1 might incorporate its genetic material into genes that are involved in cancer development or control the cell cycle and contribute to cancer formation (Maldarelli, 2016). There's limited evidence for HIV-1 directly replicating within epithelial cells. Early studies showed that some human uterine epithelial cells could be infected, but this finding hasn't been widely replicated (Isaguliants et al., 2021). HIV-1 can use surface proteins from other viruses to disguise itself and enter cells that normally wouldn't be susceptible, like CD4-negative epithelial cells. Recent research suggests infected T cells might fuse with placental epithelial cells, potentially transferring HIV-1 and creating a viral reservoir in these cells (Tang et al., 2020). This could lead to the

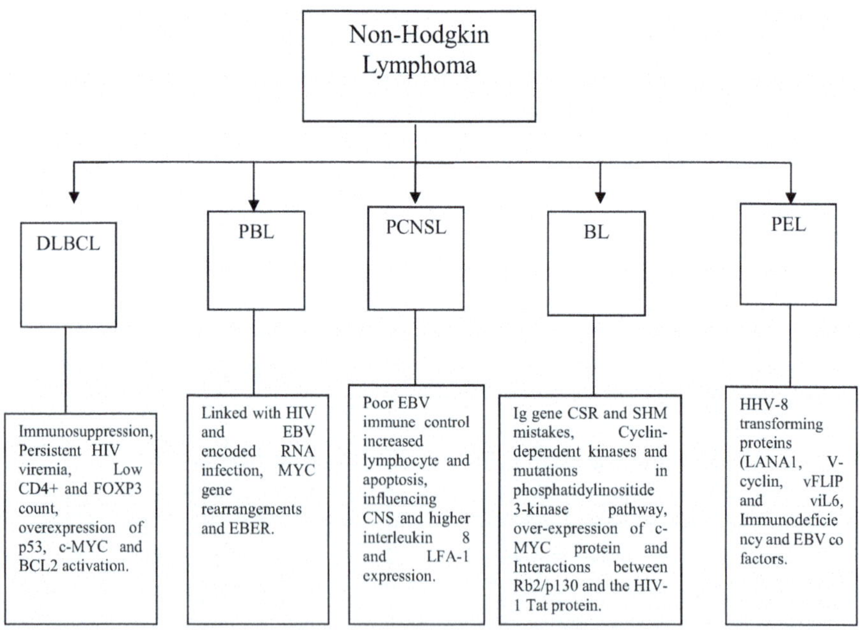

Fig. 4.1 Different types of non-Hodgkin's lymphoma and their pathogenesis. *DLBCL* diffuse large B-cell lymphomas, *PBL* plasmablastic lymphomas, *PCNSL* primary central nervous system lymphomas, *BL* Burkitt's lymphomas, *PEL* primary effusion lymphomas

formation of genetically modified cells susceptible to malignant transformation. In some scenarios, HIV-1 might insert its genetic material into the DNA of epithelial cells. This could disrupt genes involved in normal cell growth, potentially making the cell more susceptible to becoming cancerous. Even without directly infecting epithelial cells, HIV-1 proteins can create an inflammatory environment around them that can damage cells and contribute to tumor development (Anand et al., 2018). HIV-1 might interact with epithelial cells in a way that reduces a protein called E-cadherin. A loss of which is associated with a process called EMT (epithelial-to-mesenchymal transition) that is linked to cancer progression. Interaction of HIV-1 with epithelial cell surface can activate certain signaling pathways within these cells. These pathways, like TGF- β and MAPK, play role in cell growth and division. When these pathways are disrupted, it can lead to uncontrolled cell growth, a hallmark of cancer (Tugizov, 2020).

HIV-1 doesn't just weaken the immune system; it can also team up with other viruses to increase the risk of cancer. This is particularly concerning because some viruses, like EBV, KSHV, HCV, and HPV, are already known to cause cancer. Studies have shown that HIV-1 infection can enhance the cancer-causing effects of these viruses in people infected with both (Isaguliants et al., 2021). In essence, HIV-1 acts like a booster for these other cancer viruses.

1. People living with HIV who are also infected with hepatitis B virus or hepatitis C virus face a higher risk of developing liver cancer. This increased risk is due to complex interactions between the viruses themselves and their effect on liver cells. Research suggests direct communication between HIV and the hepatitis viruses within liver cells. For example, the HBV X protein can interact with the HIV Tat protein, boosting HIV replication. HIV infection can also indirectly promote liver cancer. For instance, HIV can increase the production of proteins from HBV within liver cell, leading to hepatotoxicity. Both HIV and HCV can trigger oxidative stress, which promotes the production of ROS in hepatocytes and liver cells that damage DNA and promote cancer development. HIV proteins further accelerate this process (Anand et al., 2018).
2. The interplay between HIV-1 and HPV may also contribute to the higher incidence of HPV-associated malignancy in PLWH. The decrease of E-cadherin and the increase of vimentin and TGF-b1 expression in PLWH epithelial cells indicate a process called EMT, which is linked to cancer progression. Scientists are still figuring out exactly how HIV and HPV work together, but some examples exist. HIV can increase the production of a protein from HPV type 16 (E6) in cervical cancer cells, potentially boosting cancer development (Isaguliant et al. 2021). Another factor that contributes in virally induced cancer development is the oxidative environment with chronic inflammation. Viral infections induce oxidative stress, leading to DNA damage and alterations that can turn healthy cells cancerous. This damage along with inflammation caused by the immune system's response plays a role in HIV- associated cancers. PLWH are impacted by oxidative stress brought on by HIV-1 contagion, which damages DNA and produces ROS (Barillari et al., 2016).

4.2.2 Oncogenic Effect of HIV-1 Protein

4.2.2.1 Mechanisms of HIV-1 Tat Protein in Tumorigenesis

Tat, a protein, disrupts multiple cellular processes such as cell cycle progression, inhibiting epithelial differentiation, blocking apoptosis, and promoting tumorigenesis and oxidative stress in HeLa cells. It also interacts with Tat-interacting promoter 30(TIP30), enhancing metastasis and EMT in non-small lung cancer. Tat protein further disrupts the cellular processes by modulating VEGF and targeting VEGFRs, promoting angiogenesis, and altering DNA repair which could potentially lead to genomic instability (Dandachi & Morón, 2020; Liu et al., 2018). HIV-1 infected cells lead to cellular oxidative stress via activating spermine oxidase (SMO), mitochondrial dysfunction, and NADPH oxidases. A study found a significant increase in H_2O_2 levels in the endoplasmic reticulum (ER), primarily involved in NOX4 production, which is thought to contribute to HIV-1-associated dementia (Isaguliants et al., 2021).

4.2.2.2 Interactions of HIV-1 Envelope Protein Gp120

Glioma cells interact with the envelope protein Gp120 of HIV-1 resulting in a several effects promoting the tumor growth such as migration, proliferation, and a metabolic shift known as the Warburg effect. The Warburg effect increases the cellular dependence on glycolysis even in oxygen rich environments and enhances the synthesis of lipids and proteins needed for the tumor progression. Furthermore, the Gp120 protein stimulates the EMT and cell migration through pathways involving TGF-B1 and MAPK signaling (Yuan et al., 2019). Beyond the role in tumor growth, Gp120 also contributes to the oxidative stress. Studies have shown that Gp120 increases production of free radicals from macrophages. This protein can trigger oxidative stress in variety of cells, including those of the immune system, blood vessels, brain (astrocytes, neurons), and microglia. It specifically increases the production of reactive oxygen species (ROS) through mechanisms involving Fenton-Weiss-Haber reaction and various enzymes (NOX2, NOX4, and CYP2E1). Gp120 stimulates proline oxidase, another source of ROS in the cancer cells. Interestingly, Gp120 also upregulated the expression of multidrug resistance (MDR) protein 1 and nuclear factor erythroid derived 2-related factor 2 (Nrf2), which activates antioxidant defence enzymes (Isaguliants et al., 2021).

4.2.2.3 Role of Protein Negative Factor (Nef) in HIV-1 Pathogenesis

Nef, a commonly expressed protein in HIV-1, interferes with the signaling in immune cells (CD4+ lymphocytes and macrophages), therefore blocking the apoptotic ability of p53. Nef also works with K1, a cancer protein in Kaposi's sarcoma, to promote the growth of vascular tubes and spread of infected cells. Studies in mice showed that Nef enhances ability of K1 to promote angiogenesis and tumor formation. Nef protein induces phosphorylation and translocation of NADPH oxidase complex p47 (phox) to release superoxide anions from macrophages. Nef indirectly activates NADPH oxidase, increasing ROS production and cell responses to stimuli. Nef further reduces the ability of cells to tolerate hydrogen peroxide, a type of ROS, potentially contributing to blood vessel dysfunction observed in HIV-1 infection (Chelvanambi et al., 2019).

4.2.2.4 Reverse Transcriptase (RT)

A study suggests an interesting role of HIV's RT protein in promoting cancer. The study showed that the constitutive expression of HIV-1 RT in murine mammary carcinoma 4T1 upregulated the genes (Twist and Snail) that promote tumor spread (EMT). This resulted in enhanced tumor growth and spread in mice as compared to other types of reverse transcriptase. In addition, human cells expressing RT show increased reactive oxygen species (ROS) production. Various RT variants, including resistant to drug lysosomal processing variants, cause higher expression of Nqo- and HO-1 proteins (Latanova et al., 2017).

4.2.2.5 Matrix/p17 Protein

HIV's matrix protein, p17, disrupts the immune systems and creates a conductive environment in the lymph nodes that favors the growth and metastasis of lymphoma. This favorable environment is called prolymphangiogenic microenvironment linked with enhanced blood and lymphatic vessel growth, crucial for tumor development. Studies have shown that genetically modified mice with HIV-1 gene exhibited p17 ability to directly induce B-cell lymphomas. HIV-1 proteins, including p17, are oncogenic, promoting cell transformation and aggressive propagation of cancer cells through oxidative stress induction and cellular exit. The p17 protein forms amyloidogenic assemblies toxic to the nerve cells causing neurocognitive disorder. A study conducted on nematode (*Caenorhabditis elegans*) suggested that p17 disrupts feeding behavior and potentially causes neurological disorders. Amyloidogenic proteins, like Aβ, contributes to increased oxidative stress, which may be triggered by p17's amyloidogenic assemblies. This cellular imbalance is linked with tumor development— it is possible that several HIV-1 proteins, including Tat, gp120, Nef, RT, and potentially p17, contribute to the process by inducing oxidative stress (Isaguliants et al., 2021).

4.3 Epstein-Barr Virus (EBV)

Multiple factors such as compromised host immunity, environmental influences, and cellular mutations within the host create an environment where certain types of viruses can induce various forms of cancer. For example, EBV is linked to Burkitt's lymphoma, nasopharyngeal cancer, and some types of Hodgkin's diseases. Epstein-Barr virus (EBV), discovered in 1964, is a well-studied example of cancer- causing virus due to its monoclonal genome. EBV readily infects B cells in humans, alters their genetic makeup, and increases the risk of malignant tumor, especially in individuals with weakened immune system (Lukes & Butler, 1966). Burkitt's lymphoma develops when *myc* gene is overexpressed. This gene controls the activity of various genes crucial for cell cycle advancement, apoptosis (programmed cell death), and transformation of cells. EBV infection can lead to mutations in the *myc* gene causing the B cells to multiply continuously. Different variants of Burkitt's lymphoma may arise depending on the stage of B-cell development (Kennedy et al., 2003). Recent studies suggest that various microorganisms might contribute to the onset of Burkitt's lymphoma. For example, children who have recently been infected with *Plasmodium falciparum*, the malaria- causing parasite, have a higher risk of developing Burkitt's lymphoma.

EBV may directly cause carcinogenesis in people with Hodgkin's disease, especially when myc is persistently expressed (Mundo et al., 2020). The intricate relationship between EBV and Hodgkin's disease is impacted by a number of variables, including age, gender, ethnicity, geography, and histological subtype. EBV infection can also result in various kinds of Burkitt's lymphoma, depending on the stage of B-cell growth during the infection (Shannon-Lowe et al., 2017). According to recent discoveries, Burkitt's lymphoma may have a polymicrobial origin (Bednarska et al., 2024). For instance, children were more likely to acquire Burkitt's lymphoma if they had recently contracted the malaria-causing *Plasmodium falciparum* parasite (Broen et al., 2023). EBV may directly cause carcinogenesis in Hodgkin's disease patients, particularly through constitutive myc expression. There are several variables that contribute to the multifactorial relationship between EBV and Hodgkin's disease, including age, sex, ethnicity, country of residency, and histological subtype. Finally, EBV can result in nasopharyngeal cancer, possibly due to latency of the virus in nasopharyngeal epithelial cells (Flavell & Murray, 2000). Depending on the strain and type of the virus, smoking might result in EBV reactivation and produce genetic alterations related to nasopharyngeal cancer. Thanks to recent developments in viral oncology, pathologists can now see cancer under a microscope (Perri et al., 2023).

4.4 Hepatitis B Virus (HBV)

Globally, an estimated 400 million people are chronically infected with HBV (Ganem & Prince, 2004). Most of these infections don't show any symptoms, and because HBV is not cytopathic, it usually doesn't trigger a strong immune response upon initial infection, probably because entry and replication of HBV doesn't activate cellular gene expression (Wieland & Chisari, 2005). Nevertheless, HBV significantly contributes to the development of HCC— in fact, 15–40% of chronically infected individuals develop chronic active hepatitis (CAH), which then progresses to cirrhosis, failure of liver, or HCC (Feitelson, 1999; Lok, 2004). However, the progress of chronic liver disease can be expedited by exposure to environmental toxins like aflatoxin B, cigarette smoke, and alcohol. CAH damages the liver through cell death (necrosis), inflammation, and fibrosis. This ongoing damage and repair cycle (cirrhosis) is thought to eventually lead to liver cancer (HCC) because the liver cells rapidly regenerate after each injury. This rapid regeneration can lead to mutations and the selection of cells with carcinogenic phenotype. Studies have indicated that individuals with cirrhosis are at a higher risk of developing HCC as compared to those without the disease. While both cirrhosis and chronic active hepatitis (CAH) appear to contribute to liver cancer (HCC), evidence suggests that HBV has a direct oncogenic role in the process. This is exemplified by the correlation between HBV DNA in blood level and increased HCC risk. Consequently, HBV-linked HCC probably arises from a combination of cancer-causing activities of HBV and the synergetic effects of chronic inflammation (Brechot, 2004; Singh & Kumar, 2003; Tanaka et al., 2005).

The DNA of HBV is circular and partially double stranded, comprising four overlapping open reading frames (ORFs) encoding X, polymerase, envelope, and core proteins (Seeger & Mason, 2000). Replication of HBV, unlike that of retroviruses, relies on reverse transcription. HBV can replicate without integrating its genome into the host chromosome— it does facilitate viral persistence (Feitelson & Lee, 2007). The HBV genome is often found integrated into the host chromosome in patients with both chronic active hepatitis (CAH) and hepatocellular carcinoma (HCC), indicating that integration occurs prior to tumor development (Brechot et al., 2000; Minami et al., 2005). The integration of HBV is a dynamic process because integrated viral and surrounding cellular sequences may reorganize because of long-term inflammation and increased hepatocyte proliferation (Brechot et al., 2000). Additionally, the integration process may cause deletions and transpositions of viral sequences between chromosomes (Dandri et al., 2002; Wang et al., 1998). As a result, HBV integration may cause activation of proto-oncogenes (Horikawa & Barrett, 2001) and genomic instability (Hino et al., 2002). Although integration isn't essential for malignancy development, approximately 20% of HBV-linked HCC cases lack proof of integration (Brechot et al., 2000). Analysis of viral DNA sequences in liver cancer (HCC) has revealed additional ways HBV can promote cancer. Researchers have found that most HCC tumor cells contain HBV X protein (HBx) and/or truncated envelope viral protein such as PreS2/S. A novel viral protein called HBSP has been identified made from spliced viral genes in patients with HBV infection. However, the presence of these proteins alone does not definitely establish their role in the development of HCC, necessitating further research to elucidate their possible contribution.

HBx is a regulatory protein consisting of 154 amino acids and is highly conserved among mammalian hepadnaviruses, suggesting its vital role in the life cycle of viruses (Diao et al., 2001). It is expressed consistently throughout cancer development, in cells with integrated HBV genomes as well (Paterlini et al., 1995; Peng et al., 2005; Wollersheim et al., 1988). Studies in cultured cells and transgenic animals indicate HBx's involvement in transformation and its regulatory role in viral replication, essential for infectivity in woodchuck hepatitis virus (WHV). HBx also influences cell proliferation and viability both in vitro and in vivo (Brechot et al., 2001; Klein et al., 1999). Its biological functions include (1) modulating cellular promoters and signaling pathways, impacting cell responses to genotoxic stress; (2) regulating proteasomal function; (3) affecting mitochondrial function; (4) modulating calcium homeostasis; and (5) inducing genomic instability (Forgues et al., 2003; Huh & Siddiqui, 2002; Zhang et al., 2000). Besides HBx, the HBV genome also encodes regulatory proteins called PreS2. These include large surface proteins (LHBs) and truncated middle surface proteins (MHBst), which can activate other viral genes (Hildt et al., 1996; Murakami, 1999). The importance of PreS2 activators, proteins that regulate the virus, is further strengthened by research. Studies have shown that nearly a third of HBV DNA found integrated into liver cancer (HCC) cells associated with HBV infection contain functional MHBst, a specific type of PreS2 activators (Schluter et al., 1994; Zhong et al., 1999). Studies of viral DNA in HCC have identified additional cancercausing mechanisms associated to HBV. These include the expression of HBx protein and small PreS2/S viral envelope proteins, which are found in majority of tumor cells (Soussan et al., 2000). Studies have shown an association between liver fibrosis, occurrence of HBSP antibodies, and viral replication (Soussan et al., 2003).

4.5 Hepatitis C Virus (HCV)

In the 1970s, researchers discovered a new type of hepatitis virus distinct from hepatitis A and hepatitis B. This virus was later named hepatitis C (HVC), which was found to be transmitted through blood transfusions (Feinstone et al., 1975). Later, Choo et al. (1989) isolated cDNA clone from a 104-nucleotide positive-stranded RNA, encoding an antigen linked to non-A and non-B infections from a plasma library of chimpanzee infected with non-A and non-B hepatitis. The continuous ORF encodes for a polyprotein, which splits into ten structural and non-structural proteins by both viral and host cellular proteases (Hoofnagle, 2002). Hepatitis C virus (HCV) is a leading cause of liver disease on a global scale. The virus, which is thought to infect roughly 3% of the global population, is mostly spread by injectable medication usage, blood transfusions, unsafe injection techniques, and other healthcare procedures. About 80% of those infected with HCV have acute hepatitis, which is primarily asymptomatic at first and progresses to chronic hepatitis (Harris, 2005).

Hepatitis C can present as acute or chronic hepatitis. Acute hepatitis C infections are typically asymptomatic, but they can cause cirrhosis and liver damage (Rosen, 2011). There is a high prevalence of infection; 130–200 million cases are reported globally (Gravitz, 2011). Only 10–15% of people experience symptoms of acute hepatitis with jaundice, which can be quite severe; fulminant liver failure is uncommon. In 25–50% of cases with symptomatic infection and 10–15% of cases with asymptomatic infection, spontaneous clearance is seen (Harris, 2005). Unlike HBV, which tends to cause cirrhosis and HCC, 85% of individuals with HCV infection go on to acquire chronic infection. Currently, unsafe sexual contact and intravenous drug use are the main ways that HCV is spread; for example, a high positive rate is seen in prisoners (Larney et al., 2013). Chronic inflammatory responses are the basis for oncogenesis in both HBV and HCV. According to estimates, HCV causes 25% of HCC and 27% of cirrhosis globally.

4.6 Human Papillomavirus (HPV)

Papillomaviruses (PVs) are a large group of small, double-stranded DNA viruses that lack outer envelope. These viruses can infect the skin and other tissues of various animals (de Villiers, 2001). Over 200 human PV types (HPVs) have been isolated and identified causing various skin bumps. These viruses can be categorized as either "mucosal" (infecting moist tissues) or "cutaneous" (infecting the skin). Further, they're classified as low- or high-risk based on their potential to turn cancerous (De Villiers et al., 2004). HPV5 and HPV8, specifically, are considered high-risk because they're linked to a rare genetic dermatological condition called epidermodysplasia verruciformis (EV). This disease can lead to skin cancers, especially areas that are exposed to the sun. Skin cancer in EV patients was the first kind of HPV malignancy linked to HPV infections (Pfister et al., 1981; Ramoz et al., 2002).

Later studies suggest HPV5 and HPV8- infected EV individuals have higher risk of developing nonmelanoma skin cancers (NMSC), especially in individual with compromised immune system (Shamanin et al., 1994; Shamanin et al., 1996). Though common in the population, some of these HPVs, like potentially high-risk HPV5, might even be normal "skin flora" found in hair follicles (Boxman et al., 1999). Studies have indicated that E6 proteins of cutaneous HPVs can target a molecule called Bak, a key player in UV-induced apoptosis, potentially allowing HPV-infected cells to survive DNA damage and contribute to transformation (Akgul et al., 2005, 2007). Studies indicated that transgenic mice expressing early region genes from cutaneous HPVs have shown skin hyperplasia and skin tumors (Dong et al., 2005). However, HPV is often found only in some cancer cells, suggesting it might initiate but not maintain cancer or act indirectly (Akgul et al., 2006).

Mucosal HPVs clearly divide into low-risk (e.g., HPV6/HPV11 causing warts) and high-risk (e.g., HPV16/HPV18 causing precancerous lesions and often cervical cancer). While linked to other cancers, HPV's strongest association is with cervical cancer, with over 99% of cases linked to high-risk types. Strong evidence supports this connection— malignant progression of high-risk HPV-associated lesions is relatively low and influenced by factors like weakened immunity, long-term infection, and additional mutations. Smoking and birth control use are also implicated (Lowy & Howley, 2001). Notably, Fanconi anemia patients, prone to HPV-linked cancers, show higher HPV positivity in oral cancers, suggesting genetic factors might influence progression. Studies on HPV E7's ability to induce instability in these patients offer potential explanations (Kutler et al., 2003). Extensive research has explored the mechanisms through which high-risk HPVs contribute to cervical cancer, revealing numerous interactions between virus and host that contribute to transformation and malignancy (Spardy et al., 2007). The HPV often integrate itself into the host DNA in cervical cancer, which leads to a persistent expression of E6 and E7 viral proteins. This integration disrupts other viral genes, including the transcriptional repressor E2. Integration also leads to increased mRNA stability due to splicing of viral E6/E7 genes with cellular sequences, resulting in dysregulated expression (Jeon et al., 1995). Further, these genes are expressed with cellular sequences, boosting their stability. The combined effect of HPV infection disrupts the normal cell cycle by causing the continuous production of E6 and E7 proteins (Goodwin & DiMaio, 2000; Wells et al., 2000), capable of immortalizing primary human keratinocytes (Munger et al., 1989). When these cells are grown in specific laboratory cultures, they exhibit characteristics similar to precancerous lesion (McCance et al., 1988). Even after immortalization, these cells do not become carcinogenic until long-term passaging in vitro or after the transduction of an additional oncogene (Pei et al., 1993). Prior to this, these cells are not carcinogenic at low stages. Transgenic mice expressing proteins E6 and E7 require low estrogen dose for cervical cancer development, mirroring the progression of HPV-positive cervical lesions, which is a rare process necessitating added host cellular mutations (zur Hausen, 2002).

HPVs exhibited transformative capabilities in various laboratory studies using rodent cell lines, consistent with their clinical categorization as either high-risk HPV or low-risk HPV. E7 is identified as the major driver of transformation, whereas E6 generally doesn't exhibit significant transforming activity (Phelps et al., 1988; Tanaka et al., 1989). Studies suggest that E7 protein from high- risk HPV can work with other factors to promote cancer in kidney cells of baby rats, whereas the E6 protein seems to require ras in kidney cells of baby mice (Storey & Banks, 1993). Moreover, high-risk HPVs through E6 and E7 proteins, prolong the lifespan of primary human genital epithelial cells. These oncoproteins interact with cellular regulatory proteins like p53 and pRB, disrupting their normal functions and posing a risk of cancer progression. Additionally, high-risk HPV E7 acts as mitotic mutator, causing mutation during cell division. This leads to different mitotic irregularities such as multipolar mitosis inducing various mitotic abnormalities, including multipolar mitoses, a signature feature observed in HPV-linked cervical lesions and cancer. This disrupts the normal cell division cycle, contributing in the initiation and progression of cervical cancer (Duensing et al., 2001; Duensing & Munger, 2002).

4.7 Human Herpes Virus 8 (HHV-8)

Herpesvirus-8 (HHV-8) belongs to the family of DNA viruses called *Herpesviridae*, also including Kaposi's sarcoma herpesvirus (KSHV) (Noaimi et al., 2023). HHV-8 is associated with vascular cancer such as Kaposi's sarcoma, as well as B-cell lymphoproliferative disorders like multicentric Castleman disease (MCD) and primary effusion lymphoma (PEL). Notably, risk of these HHV-8-associated cancers is significantly higher in people with HIV.

As the cells are infected with HHV-8, they can take two separate viral processes, the latency replication and lytic replication. During latent infection, the HHV-8 genome remains intact as an episome, with only a limited number of viral genes being expressed. Tumorigenesis, cell proliferation, division, inflammation, and angiogenesis are all aided by the latent-phase proteins (LANA-1; viral (v) cyclin; vFLIP; kaposins A, B, and C; many miRNAs; and maybe ORF K1) in infected spindle cells. But immortalization is not the outcome of latent infection. Though some spindle cells in KS lesions go through lytic replication and release virions, the majority of the cells are latently infected. Continuous lytic replication and cell infection is necessary for HHV-8 to remain persistent within KS tumors. Several aspects of KS pathogenesis, such as inflammation (vGPCR, vIL-6, K15), angiogenesis (vIL-6, vGPCR, K1, vCCL1, vCCL2), cell growth (vIL-6, vGPCR, K1), and inhibition of apoptosis (vCCL1, vCCL2, vBcl2, vIRF1, K1), are mediated by proteins expressed by spindle cells during lytic replication. Comparing plasma from KS patients to controls with asymptomatic infection, HHV-8 DNA is found more often and in larger copy quantities in KS patients' plasma. HHV-8 viremia does not seem to be a result of KS; rather, it seems to be part of the cause. Incident KS is avoided by ganciclovir, which suppresses HHV-8 lytic replication (Gantt & Casper, 2011). HHV-8 performs important role in different cancers.

1. The activated interleukin-6 receptor (IL-6R) signaling pathway, which stimulates angiogenesis and lymphoproliferation, is a characteristic of HHV-8-associated lymphoproliferative disorders (LPDs) (Gonzalez-Farre et al., 2017). Human IL-6 (hIL-6) expression may be upregulated in HHV-8-infected plasmablasts, as evidenced by the production of the HHV-8 gene product LANA. Moreover, HHV-8 enhances an antiapoptotic impact by constitutively activating the transcription factors in the nuclear factor-kappa B (NF-κB) family. It has been demonstrated that NF-κB inhibition causes PEL to undergo apoptosis (Guerrero et al., 2019). When observed instances on the HHV-8-associated LPD spectrum do not meet WHO-defined criteria, diagnostic difficulties may develop (Shivane et al., 2018). Since these disorders are rare, there is potential for more thorough characterisation of these conditions in future classification systems, as the already recognized entities do not fully represent the range of recorded cases. Despite significant progress in understanding HHV-8- associated lymphoproliferative disorders (LPDs), there are some unsolved issues such as why certain individuals develop multiple HHV-8- associated LPDs and what factors contribute to the manifestation of a specific disease with this spectrum. For HIV-positive patients diagnosed with HHV-8-associated LPDs, antiretroviral therapy is an essential part of care. Rituximab therapy should also be taken into consideration for any HHV-8-associated LPD showing CD20 positivity (Guerrero et al., 2019).
2. Breast cancer, a prevalent and often fatal disease with elusive origin, exhibits higher viral infections in breast cancer patients compared to the control cohort. Studies have shown a significant relationship between tumor grade and viral infection frequency, indicating an increased presence of HHV-8 virus genome in breast cancer tissue samples increasing with tumor grade. The virus is associated with different tumors, such as prostate tumors, endothelial cell tumors, and B-cell lymphocytes, demonstrating the capability to immortalize and transform breast cells by inhibiting apoptosis, stimulating cell proliferation and survival, enhancing angiogenesis, and contributing to immune system modulation. Moreover, it can also induce tumorigenesis through production of interleukin homologs (Arabzadeh et al., 2022).

3. According to a study, coinfection of HHV-8 and HPV has been linked to cervical cancer. The frequency of HHV-8 and HPV infection was shown to differ statistically, with HHV-8 being more common in samples that tested positive for HPV than in those that tested negative. HHV-8 infection and HPV synergise to promote cervical carcinogenesis through various mechanisms. The presence of HHV-8 coinfection in SiHa cell lines can elevate levels of interleukin 6 (IL-6), chemokine (C-X-C motif) ligand 1 (CXCL1), chemokine (C–C motif) ligand 5 (CCL5), interleukin 8 (IL-8), macrophage migration inhibitory factor (MIF), and plasminogen activator inhibitor-1 (PAI-) (Chavoshpour-Mamaghani et al., 2021).

4.8 Human T-Cell Lymphotropic Virus 1 (HTLV-1)

Human oncogenic retrovirus HTLV-1 is the cause of HTLV-I-associated myelopathy (HAM/TSP) and adult T-cell leukaemia/lymphoma (ATL). It affects 5–10 million individuals globally and is spread by breast milk, blood, and semen. Viral regulatory proteins Tax and HTLV-1 bZIP factor are important players in the intricate carcinogenic pathway (Marino-Merlo et al., 2023).

The glycoprotein spikes and host-derived lipid envelope that encircle the retrovirus HTLV-I are encoded by its env gene. The HTLV-1 genome shares many characteristics with the HIV genome, including regulatory genes Tax and Rex; structural genes Pol, Env, and Gag; and accessory genes Vif, Vpr, Vpx, Vpu, and Nef (Kalinichenko et al., 2022). The genotype in these supplementary and regulatory areas determines different aspects of pathogenicity. Viral attachment to susceptible cells is mediated by the interaction of certain host cell surface receptors with the surface glycoprotein gp46. The replication of the genome, transcribed genes, translated proteins, assembly, and release of virion are all components of the viral life cycle. A section of the HTLV-I genome codes for distinct gene products, whereas other portions are shared by all replication-competent retroviruses. A key component of viral gene expression and genomic activation that encourages oncogenesis is the viral protein Tax, which is encoded in the pX region (Zhang et al., 2017). Tax promotes or inhibits cellular gene expression through interactions with signaling molecules and cellular transcription factors, which aids in the pathophysiology of HTLV-I-induced illness. Cytokines, receptors, transcription factors, proto-oncogenes, and inhibitors of nucleotide excision repair and apoptosis are among the molecules whose expression is impacted by taxation. The HTLV-I-infected cells' concomitant upregulation of growth-inducing proteins and down regulation of repair and apoptotic factors may promote T-cell transformation and the development of ATL. ATL development is known to be influenced by age, HTLV-1 proviral load (PVL), human leukocyte antigen (HLA) haplotype, and strongyloidiasis; these characteristics may indicate a connection between immunity and carcinogenesis (Bangham, 2023). The seven commonly known oncogenic viruses along with their oncoproteins and binding receptor are presented in Table 4.1, whereas a generalized mechanism for all these viruses is mentioned in Fig. 4.2.

Table 4.1 Human oncoviruses with their binding and oncoproteins

Virus	Family	Genome	Cancer	Viral binding protein	Host binding receptor	Oncoprotein	References
Human immunodeficiency virus (HIV)	*Retroviridae*	ssRNA	Kaposi's sarcoma, non-Hodgkin's lymphoma, and cervical cancer	gp120	CD4, CCR5, or CXCR4	Tat, gp120, Nef, RT, and p17	Isaguliants et al. (2021)
Epstein-Barr virus (EBV)	*Herpesvirus*	dsDNA	Burkitt's lymphoma, Hodgkin's disease, nasopharyngeal carcinoma, stomach and smooth muscle cancer	gp350/220	CD21	EBNA-1, EBNA-2 LMP-1, LMP-2 EBERs, BARTs	Zur Hausen (2006), Perri et al. (2023)
Hepatitis B virus (HBV)	*Hepadnaviridae*	dsDNA	Hepatocellular carcinoma	Glypican 5	NTCP	HBx, HBsAg	Brechot et al. (2001)
Hepatitis C virus (HCV)	*Flaviviridae*	ssRNA +ve	Hepatocellular carcinoma	E1	CD36	Core protein, NS3	McBride (2017)
Human papillomavirus (HPV)	*Papillomaviridae*	dsDNA	Anal cancer, cervical cancer, oropharyngeal cancer, penile cancer, vaginal cancer, and vulvar cancer	L1 and L2	GAGs, HSPG	E6 and E7	McBride (2017)
Human herpes virus 8 (HHV-8)	*Herpeseviridae*	dsDNA	Kaposi's sarcoma and B-cell lymphoproliferative	Envelope Glycoprotein K8	Cell surface heparan sulfate	LANA, vIL-6, vMIP-I, vFLIP, vBCL-2, v cyclin-D, vGPCR, vIRF-1	Gantt and Casper (2011)

(continued)

Table 4.1 (continued)

Virus	Family	Genome	Cancer	Viral binding protein	Host binding receptor	Oncoprotein	References
Human T-cell lymphotropic virus 1 (HTLV-1)	*Retrovirus*	ssRNA+ve	Adult T-cell leukaemia/ lymphoma (ATL)	p8 and p12	Glut-1, NRP-1, BDCA-4, and HSPG	Tax, p12, p30 p8, p13	Marino-Merlo et al. (2023)

GAGs glycosaminoglycans, *HSPG* heparan sulfate proteoglycans, *Glut-1* glucose transporter 1, *NRP-1* neuropilin-1, *NTCP* odium taurocholate cotransporting polypeptide, *CD21* complement receptor 2, *CCR5* chemokine receptor, *CXCR4* vC-X-C motif chemokine receptor 4, *BCDA-4* plasmacytoid dendritic cells, *CD36 and CD4* cluster of differentiation 36 and 4

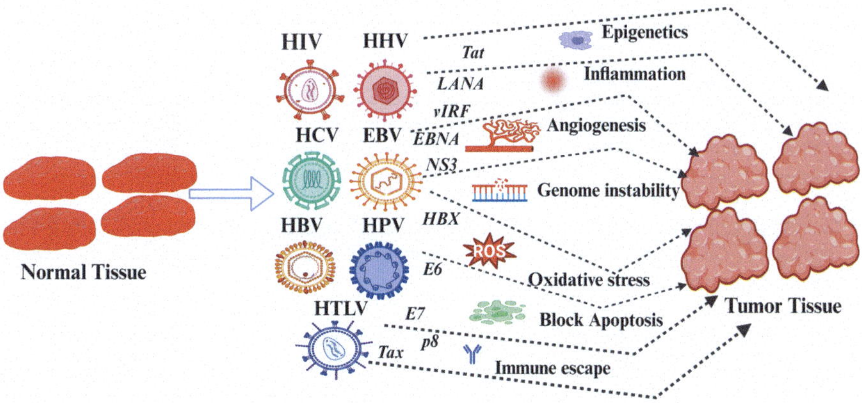

Fig. 4.2 Generalized mechanism of oncogenic viruses (created with BioRender.com)

4.9 Conclusion

Research in the field of viral oncology is continuously advancing. Our increasing understanding of the molecular mechanisms linking oncoviruses to human cancer is driving the development of a broad spectrum of potential treatment. These potential treatments aim to enhance the eradication of infected tumor cells while preserving normal and healthy cells, thus maximizing the therapeutic index. The host signaling pathways that are targeted by the seven known human oncoviruses to cause oncogenesis are relatively similar. Innovative treatments that have proven successful in treating one virus-associated cancer might be extended to treat all virus-related cancers that share the same mechanisms. These innovative approaches hold immense promise eventually leading to the discovery of a cancer cure.

References

Adams, A., & Lindahl, T. (1975). Epstein-Barr virus genomes with properties of circular DNA molecules in carrier cells. *Proceedings of the National Academy of Sciences, 72*(4), 1477–1481.

Akgul, B., Cooke, J. C., & Storey, A. (2006). HPV-associated skin disease. *The Journal of Pathology, 208*, 165–175.

Akgul, B., Garcia-Escudero, R., Ghali, L., Pfister, H. J., Fuchs, P. G., Navsaria, H., & Storey, A. (2005). The E7 protein of cutaneous human papillomavirus type 8 causes invasion of human keratinocytes into the dermis in organotypic cultures of skin. *Cancer Research, 65*, 2216–2223.

Akgul, B., Ghali, L., Davies, D., Pfister, H., Leigh, I. M., & Storey, A. (2007). HPV8 early genes modulate differentiation and cell cycle of primary human adult keratinocytes. *Experimental Dermatology, 16*, 590–599.

Anand, A. R., Rachel, G., & Parthasarathy, D. (2018). HIV proteins and endothelial dysfunction: Implications in cardiovascular disease. *Frontiers in Cardiovascular Medicine, 5*, 1–10. https://doi.org/10.3389/fcvm.2018.00185

Arabzadeh, S. M., Mollaei, H. R., & Malekpour Afshar, R. (2022). Prevalence of human herpes virus-8 infection among women with breast cancer. *Iranian Journal of Virology, 16*(1), 42–47.

Arora, R., Chang, Y., & Moore, P. S. (2012). MCV and Merkel cell carcinoma: A molecular success story. *Current Opinion in Virology, 2*(4), 489–498.

Baer, R., Bankier, A. T., Biggin, M. D., Deininger, P. L., Farrell, P. J., Gibson, T. J., Hatfull, G., Hudson, G. S., Satchwell, S. C., Seguin, C., & Tuffnell, P. S. (1984). DNA sequence and expression of the B95-8 Epstein—Barr virus genome. *Nature, 310*(5974), 207–211.

Bangham, C. R. (2023). HTLV-1 persistence and the oncogenesis of adult T-cell leukemia/lymphoma. *Blood, 141*(19), 2299–2306. https://doi.org/10.1182/blood.2022019332

Barillari, G., Palladino, C., Bacigalupo, I., Leone, P., Falchi, M., & Ensoli, B. (2016). Entrance of the Tat protein of HIV-1 into human uterine cervical carcinoma cells causes upregulation of HPV-E6 expression and a decrease in p53 protein levels. *Oncology Letters, 12*(4), 2389–2394. https://doi.org/10.3892/ol.2016.4921

Beasley, R. P., Hwang, L. Y., Lin, C. C., & Chien, C. S. (1981). Hepatocellular carcinoma and hepatitis B virus. A prospective study of 22,707 men in Taiwan. *Lancet, 2*, 1129–1233.

Bednarska, K., Chowdhury, R., Tobin, J. W., Swain, F., Keane, C., Boyle, S., Khanna, R., & Gandhi, M. K. (2024). Epstein–Barr virus-associated lymphomas decoded. *British Journal of Haematology, 204*(2), 415–433.

Bichakjian, C. K., Lowe, L., Lao, C. D., Sandler, H. M., Bradford, C. R., Johnson, T. M., & Wong, S. L. (2007). Merkel cell carcinoma: Critical review with guidelines for multidisciplinary management. *Cancer, 110*(1), 1–2.

Bittner, J. J. (1936). Some possible effects of nursing on the mammary gland tumor incidence in mice. *Science, 84*(2172), 162–163.

Boxman, I. L., Mulder, L. H., Russell, A., Bouwes Bavinck, J. N., Green, A., & Ter Schegget, J. (1999). Human papillomavirus type 5 is commonly present in immunosuppressed and immunocompetent individuals. *The British Journal of Dermatology, 141*(2), 246–249.

Brandão, M., Bruzzone, M., Franzoi, M. A., De Angelis, C., Eiger, D., Caparica, R., Piccart-Gebhart, M., Buisseret, L., Ceppi, M., Dauby, N., & Carrilho, C. (2021). Impact of HIV infection on baseline characteristics and survival of women with breast cancer. *AIDS, 35*(4), 605–618. https://doi.org/10.1097/QAD.0000000000002810

Brechot, C. (2004). Pathogenesis of hepatitis B virus-related hepatocellular carcinoma: Old and new paradigms. *Gastroenterology, 127*, S56–S61.

Brechot, C., Gozuacik, D., Murakami, Y., & Paterlini-Brechot, P. (2000). Molecular bases for the development of hepatitis B virus (HBV)-related hepatocellular carcinoma (HCC). *Seminars in Cancer Biology, 10*, 211–231.

Brechot, C., Thiers, V., Kremsdorf, D., Nalpas, B., Pol, S., & Paterlini-Brechot, P. (2001). Persistent hepatitis B virus infection in subjects without hepatitis B surface antigen: Clinically significant or purely "occult"? *Hepatology, 34*, 194–203.

Broen, K., Dickens, J., Trangucci, R., Ogwang, M. D., Tenge, C. N., Masalu, N., Reynolds, S. J., Kawira, E., Kerchan, P., Were, P. A., & Kuremu, R. T. (2023). Burkitt lymphoma risk shows geographic and temporal associations with Plasmodium falciparum infections in Uganda, Tanzania, and Kenya. *Proceedings of the National Academy of Sciences, 120*(2), e2211055120.

Casper, C., Crane, H., Menon, M., & Money, D. (2017). HIV/AIDS comorbidities: Impact on cancer, noncommunicable diseases, and reproductive health. In *Major infectious diseases* (3rd ed.). The World Bank.

Chamseddine, A. N., Loulergue, P., & Mir, O. (2018). HIV-associated cancers and related diseases. *The New England Journal of Medicine, 378*(22), 2144–2145. https://doi.org/10.1056/nejmc1804812

Chang, Y., Cesarman, E., Pessin, M. S., Lee, F., Culpepper, J., Knowles, D. M., & Moore, P. S. (1994). Identification of herpesvirus-like DNA sequences in AIDS-associated Kaposi's sarcoma. *Science, 266*(5192), 1865–1869.

Chavoshpour-Mamaghani, S., Shoja, Z., Mollaei-Kandelous, Y., Sharifian, K., & Jalilvand, S. (2021). The prevalence of human herpesvirus 8 in normal, premalignant, and malignant cervical samples of Iranian women. *Virology Journal, 18*(1), 1–7. https://doi.org/10.1186/s12985-021-01614-z

Chelvanambi, S., Gupta, S. K., Chen, X., Ellis, B. W., Maier, B. F., Colbert, T. M., Kuriakose, J., Zorlutuna, P., Jolicoeur, P., Obukhov, A. G., & Clauss, M. (2019). HIV-Nef protein transfer to endothelial cells requires Rac1 activation and leads to endothelial dysfunction implications for statin treatment in HIV patients. *Circulation Research, 125*(9), 805–820. https://doi.org/10.1161/CIRCRESAHA.119.315082

Choo, Q. L., Kuo, G., Weiner, A. J., Overby, L. R., Bradley, D. W., & Houghton, M. (1989). Isolation of a cDNA clone derived from a blood-borne non-A, non-B viral hepatitis genome. *Science, 244*(4902), 359–362.

Coghill, A. E., Engels, E. A., Schymura, M. J., Mahale, P., & Shiels, M. S. (2018). Risk of breast, prostate, and colorectal cancer diagnoses among HIV-infected individuals in the United States. *Journal of the National Cancer Institute, 110*(9), 959–966. https://doi.org/10.1093/jnci/djy010

Corbellino, M., Poirel, L., Aubin, J. T., Paulli, M., Magrini, U., Bestetti, G., Galli, M., & Parravicini, C. (1996). The role of human herpesvirus 8 and Epstein-Barr virus in the pathogenesis of giant lymph node hyperplasia (Castleman's disease). *Clinical Infectious Diseases, 22*(6), 1120–1121.

Dandachi, D., & Morón, F. (2020). Effects of HIV on the tumor microenvironment. In *Tumor microenvironment: State of the science* (pp. 45–54). Springer.

Dandri, M., Burda, M. R., Burkle, A., Zuckerman, D. M., Will, H., Rogler, C. E., Greten, H., & Petersen, J. (2002). Increase in de novo HBV DNA integrations in response to oxidative DNA damage or inhibition of poly(ADP-ribosyl)ation. *Hepatology, 35*, 217–223.

Dane, D. S., Cameron, C. H., & Briggs, M. (1970). Virus-like particles in serum of patients with Australia-antigen-associated hepatitis. *Lancet, 295*(7649), 695–698.

De Villiers, E. M. (2001). Taxonomic classification of papillomaviruses. *Papillomavirus, 12*, 57–63.

De Villiers, E. M., Fauquet, C., Broker, T. R., Bernard, H. U., & zur Hausen, H. (2004). Classification of papillomaviruses. *Virology, 324*, 17–27.

Diao, J., Garces, R., & Richardson, C. D. (2001). X protein of hepatitis B virus modulates cytokine and growth factor related signal transduction pathways during the course of viral infections and hepatocarcinogenesis. *Cytokine & Growth Factor Reviews, 12*, 189–205.

Dong, W., Kloz, U., Accardi, R., Caldeira, S., Tong, W. M., Wang, Z. Q., Jansen, L., Durst, M., Sylla, B. S., Gissmann, L., & Tommasino, M. (2005). Skin hyperproliferation and susceptibility to chemical carcinogenesis in transgenic mice expressing E6 and E7 of human papillomavirus type 38. *Journal of Virology, 79*, 14899–14908.

Duensing, S., Duensing, A., Crum, C. P., & Munger, K. (2001). Human papillomavirus type 16 E7 oncoprotein-induced abnormal centrosome synthesis is an early event in the evolving malignant phenotype. *Cancer Research, 61*, 2356–2360.

Duensing, S., & Munger, K. (2002). The human papillomavirus type 16 E6 and E7 oncoproteins independently induce numerical and structural chromosome instability. *Cancer Research, 62*, 7075–7082.

Epeldegui, M., Vendrame, E., & Martínez-Maza, O. (2010). HIV-associated immune dysfunction and viral infection: Role in the pathogenesis of AIDS-related lymphoma. *Immunologic Research, 48*, 72–83. https://doi.org/10.1007/s12026-010-8168-8

Epstein, M. A., Achong, B. G., & Barr, Y. M. (1964). Virus particles in cultured lymphoblasts from Burkitt's lymphoma. *Lancet, 283*(7335), 702–703.

Feinstone, S. M., Kapikian, A. Z., Purcell, R. H., Alter, H. J., & Holland, P. V. (1975). Transfusion-associated hepatitis not due to viral hepatitis type A or B. *The New England Journal of Medicine, 292*(15), 767–770.

Feitelson, M. A. (1999). Hepatitis B virus in hepatocarcinogenesis. *Journal of Cellular Physiology, 181*, 188–202.

Feitelson, M. A., & Lee, J. (2007). Hepatitis B virus integration, fragile sites, and hepatocarcinogenesis. *Cancer Letters, 252*, 157–170.

Feng, H., Taylor, J. L., Benos, P. V., Newton, R., Waddell, K., Lucas, S. B., Chang, Y., & Moore, P. S. (2007). Human transcriptome subtraction by using short sequence tags to search for tumor viruses in conjunctival carcinoma. *Journal of Virology, 81*(20), 11332–11340.

Flavell, K. J., & Murray, P. G. (2000). Hodgkin's disease and the Epstein-Barr virus. *Molecular Pathology, 53*(5), 262.

Forgues, M., Difilippantonio, M. J., Linke, S. P., Ried, T., Nagashima, K., Feden, J., Valerie, K., Fukasawa, K., & Wang, X. W. (2003). Involvement of Crm1 in hepatitis B virus X protein-induced aberrant centriole replication and abnormal mitotic spindles. *Molecular and Cellular Biology, 23*, 5282–5292.

Ganem, D., & Prince, A. M. (2004). Hepatitis B virus infection–natural history and clinical consequences. *The New England Journal of Medicine, 350*, 1118–1129.

Gantt, S., & Casper, C. (2011). Human herpesvirus 8-associated neoplasms: The roles of viral replication and antiviral treatment. *Current Opinion in Infectious Diseases, 24*(4), 295–301. https://doi.org/10.1097/QCO.0b013e3283486d04

Garg, H., & Joshi, A. (2017). Host and viral factors in HIV-mediated bystander apoptosis. *Viruses, 9*(8), 1–23. https://doi.org/10.3390/v9080237

Gessese, T., Asrie, F., & Mulatie, Z. (2023). Human immunodeficiency virus related non-Hodgkin's lymphoma. *Blood and Lymphatic Cancer Targets and Therapy, 13*, 13–24. https://doi.org/10.2147/BLCTT.S407086

Gonzalez-Farre, B., Martinez, D., Lopez-Guerra, M., Xipell, M., Monclus, E., Rovira, J., Garcia, F., Lopez-Guillermo, A., Colomo, L., Campo, E., & Martinez, A. (2017). HHV8-related lymphoid proliferations: A broad spectrum of lesions from reactive lymphoid hyperplasia to overt lymphoma. *Modern Pathology, 30*(5), 745–760. https://doi.org/10.1038/modpathol.2016.233

Goodwin, E. C., & DiMaio, D. (2000). Repression of human papillomavirus oncogenes in HeLa cervical carcinoma cells causes the orderly reactivation of dormant tumor suppressor pathways. *Proceedings of the National Academy of Sciences of the United States of America, 97*, 12513–12518.

Gravitz, L. (2011). Introduction: A smouldering public-health crisis. *Nature, 474*(7350), S2–S4.

Gross, L. (1951). "Spontaneous" leukemia developing in G3H mice following inoculation, in infancy, with AK-emkemic. *Proceedings of the Society for Experimental Biology and Medicine, 76*(1), 27–32.

Guerrero, C., Jain, T., & Kelemen, K. (2019). HHV-8-associated lymphoproliferative disorders and pathogenesis in an HIV-positive patient. *Case Reports in Hematology, 30*, 745–760. https://doi.org/10.1155/2019/4536157

Harris, H. (2005). A long view of fashions in cancer research. *Bioessays, 27*(8), 833–838.

Henle, G., Henle, W., Clifford, P., Diehl, V., Kafuko, G. W., Kirya, B. G., Klein, G., Morrow, R. H., Munube, G. M., Pike, P., & Tukel, P. M. (1969). Antibodies to Epstein-Barr virus in Burkitt's lymphoma and control groups. *Journal of the National Cancer Institute, 43*(5), 1147–1157.

Henle, W., Henle, G., Zajac, B. A., Pearson, G., Waubke, R., & Scriba, M. (1970). Differential reactivity of human serums with early antigens induced by Epstein-Barr virus. *Science, 169*(3941), 188–190.

Hildt, E., Saher, G., Bruss, V., & Hofschneider, P. H. (1996). The hepatitis B virus large surface protein (LHBs) is a transcriptional activator. *Virology, 225*, 235–239.

Hino, O., Kajino, K., Umeda, T., & Arakawa, Y. (2002). Understanding the hypercarcinogenic state in chronic hepatitis: A clue to the prevention of human hepatocellular carcinoma. *Journal of Gastroenterology, 37*, 883–887.

Hoofnagle, J. H. (2002). Course and outcome of hepatitis C. *Hepatology, 36*(1), 21–29.

Horikawa, I., & Barrett, J. C. (2001). cis-Activation of the human telomerase gene (hTERT) by the hepatitis B virus genome. *Journal of the National Cancer Institute, 93*, 1171–1173.

Huh, K. W., & Siddiqui, A. (2002). Characterization of the mitochondrial association of hepatitis B virus X protein, HBx. *Mitochondrion, 1*, 349–359.

Isaguliants, M., Bayurova, E., Avdoshina, D., Kondrashova, A., Chiodi, F., & Palefsky, J. M. (2021). Oncogenic effects of HIV-1 proteins, mechanisms behind. *Cancer, 13*(2), 1–12. https://doi.org/10.3390/cancers13020305

Javier, R. T., & Butel, J. S. (2008). The history of tumor virology. *Cancer Research, 68*(19), 7693–7706.

Jeon, S., Allen-Hoffmann, B. L., & Lambert, P. F. (1995). Integration of human papillomavirus type 16 into the human genome correlates with a selective growth advantage of cells. *Journal of Virology, 69*, 2989–2997.

Kalinichenko, S., Komkov, D., & Mazurov, D. (2022). HIV-1 and HTLV-1 transmission modes: Mechanisms and importance for virus spread. *Viruses, 14*(1), 1–5. https://doi.org/10.3390/v14010152

Kennedy, G., Komano, J., & Sugden, B. (2003). Epstein-Barr virus provides a survival factor to Burkitt's lymphomas. *Proceedings of the National Academy of Sciences, 100*(24), 14269–14274.

Klein, N. P., Bouchard, M. J., Wang, L. H., Kobarg, C., & Schneider, R. J. (1999). Src kinases involved in hepatitis B virus replication. *The EMBO Journal, 18*, 5019–5027.

Kutler, D. I., Wreesmann, V. B., Goberdhan, A., Ben-Porat, L., Satagopan, J., Ngai, I., Huvos, A. G., Giampietro, P., Levran, O., Pujara, K., Diotti, R., Carlson, D., Huryn, L. A., Auerbach, A. D., & Singh, B. (2003). Human papillomavirus DNA and p53 polymorphisms in squamous cell carcinomas from Fanconi anemia patients. *Journal of the National Cancer Institute, 95*, 1718–1721.

Lairmore, M. D., Haines, R., & Anupam, R. (2012). Mechanisms of human T-lymphotropic virus type 1 transmission and disease. *Current Opinion in Virology, 2*(4), 474–481.

Larney, S., Kopinski, H., Beckwith, C. G., Zaller, N. D., Jarlais, D. D., Hagan, H., Rich, J. D., van den Bergh, B. J., & Degenhardt, L. (2013). Incidence and prevalence of hepatitis C in prisons and other closed settings: Results of a systematic review and meta-analysis. *Hepatology, 58*(4), 1215–1224.

Latanova, A., Petkov, S., Kuzmenko, Y., Kilpeläinen, A., Ivanov, A., Smirnova, O., Krotova, O., Korolev, S., Hinkula, J., Karpov, V., & Isaguliants, M. (2017). Fusion to flaviviral leader peptide targets HIV-1 reverse transcriptase for secretion and reduces its enzymatic activity and ability to induce oxidative stress but has no major effects on its immunogenic performance in DNA-immunized mice. *Journal of Immunology Research, 2017*, 7407136. https://doi.org/10.1155/2017/7407136

Lemon, S. M., Hutt, L. M., Shaw, J. E., Li, J. L., & Pagano, J. S. (1977). Replication of EBV in epithelial cells during infectious mononucleosis. *Nature, 268*(5617), 268–270.

Linke-Serinsöz, E., Fend, F., & Quintanilla-Martinez, L. (2017). Human immunodeficiency virus (HIV) and Epstein-Barr virus (EBV) related lymphomas, pathology view point. *Seminars in Diagnostic Pathology, 34*(4), 352–363.

Liu, Y. P., Chen, C. H., Yen, C. H., Tung, C. W., Chen, C. J., Chen, Y. M., & Huang, M. S. (2018). Human immunodeficiency virus Tat-TIP 30 interaction promotes metastasis by enhancing the nuclear translocation of Snail in lung cancer cell lines. *Cancer Science, 109*(10), 3105–3114. https://doi.org/10.1111/cas.13768

Lok, A. S. (2004). Prevention of hepatitis B virus-related hepatocellular carcinoma. *Gastroenterology, 127*, 303–309.

Lowy, D. R., & Howley, P. M. (2001). *Fields' virology* (pp. 2231–2264). Lippincott Williams and Wilkins.

Lukes, R. J., & Butler, J. J. (1966). The pathology and nomenclature of Hodgkin's disease. *Cancer Research, 26*, 1063–1081.

Maldarelli, F. (2016). The role of HIV integration in viral persistence: No more whistling past the proviral graveyard. *The Journal of Clinical Investigation, 126*(2), 438–447. https://doi.org/10.1172/JCI80564

Marino-Merlo, F., Grelli, S., Mastino, A., Lai, M., Ferrari, P., Nicolini, A., Pistello, M., & Macchi, B. (2023). Human T-cell leukemia virus type 1 oncogenesis between active expression and latency: A possible source for the development of therapeutic targets. *International Journal of Molecular Sciences, 24*(19), 1–15. https://doi.org/10.3390/ijms241914807

McBride, A. A. (2017). Oncogenic human papillomaviruses. *Philosophical Transactions of the Royal Society, B: Biological Sciences, 372*(1732), 1–9. https://doi.org/10.1098/rstb.2016.0273

McCance, D. J., Kopan, R., Fuchs, E., & Laimins, L. A. (1988). Human papillomavirus type 16 alters human epithelial cell differentiation in vitro. *Proceedings of the National Academy of Sciences of the United States of America, 85*, 7169–7173.

Meister, A., Hentrich, M., Wyen, C., & Hübel, K. (2018). Malignant lymphoma in the HIV-positive patient. *European Journal of Haematology, 101*(1), 119–126. https://doi.org/10.1111/ejh.13082

Minami, M., Daimon, Y., Mori, K., Takashima, H., Nakajima, T., Itoh, Y., & Okanoue, T. (2005). Hepatitis B virus-related insertional mutagenesis in chronic hepatitis B patients as an early drastic genetic change leading to hepatocarcinogenesis. *Oncogene, 24*, 4340–4348.

Mundo, L., Del Porro, L., Granai, M., Siciliano, M. C., Mancini, V., Santi, R., Marcar, L., Vrzalikova, K., Vergoni, F., Di Stefano, G., & Schiavoni, G. (2020). Frequent traces of EBV infection in Hodgkin and non-Hodgkin lymphomas classified as EBV-negative by routine methods: Expanding the landscape of EBV-related lymphomas. *Modern Pathology, 33*(12), 2407–2421.

Munger, K., Phelps, W. C., Bubb, V., Howley, P. M., & Schlegel, R. (1989). The E6 and E7 genes of the human papillomavirus type 16 together are necessary and sufficient for transformation of primary human keratinocytes. *Journal of Virology, 63*, 4417–4421.

Murakami, S. (1999). Hepatitis B virus X protein: Structure, function and biology. *Intervirology, 42*, 81–99.

Nael, A., Walavalkar, V., Wu, W., Nael, K., Kim, R., Rezk, S., & Zhao, X. (2016). CD4-positive T-cell primary central nervous system lymphoma in an HIV positive patient. *American Journal of Clinical Pathology, 145*(2), 258–265. https://doi.org/10.1093/ajcp/aqv087

Noaimi, A. A., Hameedi, A. D., Abdullah, M. M., Ibrahim, G. A., & Hassan, H. G. (2023). Association of human herpesvirus type 8 (HHV-8) expression in patients with Kaposi's sarcoma: A clinico-immunohistochemical study. *Journal of Pakistan Association of Dermatologists, 33*(4), 1446–1453.

Nonoyama, M., Kawai, Y., & Pagano, J. S. (1975). Detection of Epstein-Barr virus DNA in human tumors. *Bibliotheca Haematologica, 40*, 577–583.

Pagano, J. S. (1999). Epstein-Barr virus: The first human tumor virus and its role in cancer. *Proceedings of the Association of American Physicians, 111*(6), 573–580.

Pai, S. Y., Lurain, K., & Yarchoan, R. (2021). How immunodeficiency can lead to malignancy. *Hematology, 2021*(1), 287–295. https://doi.org/10.1182/hematology.2021000261

Parkin, D. M. (2006). The global health burden of infection-associated cancers in the year 2002. *International Journal of Cancer, 118*, 3030–3044.

Parkin, D. M., Bray, F., Ferlay, J., & Pisani, P. (2005). Global cancer statistics, 2002. *CA: A Cancer Journal for Clinicians, 55*, 74–108.

Paterlini, P., Poussin, K., Kew, M., Franco, D., & Brechot, C. (1995). Selective accumulation of the X transcript of hepatitis B virus in patients negative for hepatitis B surface antigen with hepatocellular carcinoma. *Hepatology, 21*, 313–321.

Pei, X. F., Meck, J. M., Greenhalgh, D., & Schlegel, R. (1993). Cotransfection of HPV-18 and v-fos DNA induces tumorigenicity of primary human keratinocytes. *Virology, 196*, 855–860.

Peng, Z., Zhang, Y., Gu, W., Wang, Z., Li, D., Zhang, F., Qiu, G., & Xie, K. (2005). Integration of the hepatitis B virus X fragment in hepatocellular carcinoma and its effects on the expression of multiple molecules: A key to the cell cycle and apoptosis. *International Journal of Oncology, 26*, 467–473.

Perri, F., Sabbatino, F., Ottaiano, A., Fusco, R., Caraglia, M., Cascella, M., Longo, F., Rega, R. A., Salzano, G., Pontone, M., & Marciano, M. L. (2023). Impact of Epstein Barr virus infection on treatment opportunities in patients with nasopharyngeal cancer. *Cancer, 15*(5), 1626.

Pfister, H., Nurnberger, F., Gissmann, L., & zur Hausen, H. (1981). Characterization of a human papillomavirus from epidermodysplasia verruciformis lesions of a patient from Upper-volta. *International Journal of Cancer, 27*, 645–650.

Phelps, W. C., Yee, C. L., Munger, K., & Howley, P. M. (1988). The human papillomavirus type 16 E7 gene encodes transactivation and transformation functions similar to those of adenovirus E1A. *Cell, 53*, 539–547.

Ramoz, N., Rueda, L. A., Bouadjar, B., Montoya, L. S., Orth, G., & Favre, M. (2002). Mutations in two adjacent novel genes are associated with epidermodysplasia verruciformis. *Nature Genetics, 32*, 579–581.

Robinson, W. S., Clayton, D. A., & Greenman, R. L. (1974). DNA of a human hepatitis B virus candidate. *Journal of Virology, 14*(2), 384–391.

Rosen, H. R. (2011). Chronic hepatitis C infection. *The New England Journal of Medicine, 364*(25), 2429–2438.

Schiffman, M., & Castle, P. E. (2003). Human papillomavirus: Epidemiology and public health. *Archives of Pathology & Laboratory Medicine, 127*(8), 930–934.

Schluter, V., Meyer, M., Hofschneider, P. H., Koshy, R., & Caselmann, W. H. (1994). Integrated hepatitis B virus X and truncated preS/S sequences derived from human hepatomas encode functionally active transactivators. *Oncogene, 9*, 3335–3344.

Seeger, C., & Mason, W. S. (2000). Hepatitis B virus biology. *Microbiology and Molecular Biology Reviews, 64*, 51–68.

Seiki, M., Hattori, S., Hirayama, Y., & Yoshida, M. (1983). Human adult T-cell leukemia virus: Complete nucleotide sequence of the provirus genome integrated in leukemia cell DNA. *Proceedings of the National Academy of Sciences, 80*(12), 3618–3622.

Shamanin, V., Glover, M., Rausch, C., Proby, C., Leigh, I. M., zur Hausen, H., & de Villiers, E. M. (1994). Specific types of human papillomavirus found in benign proliferations and carcinomas of the skin in immunosuppressed patients. *Cancer Research, 54*, 4610–4613.

Shamanin, V., zur Hausen, H., Lavergne, D., Proby, C. M., Leigh, I. M., Neumann, C., Hamm, H., Goos, M., Haustein, U. F., Jung, E. G., Plewig, G., Wolff, H., & de Villiers, E. M. (1996). Human papillomavirus infections in nonmelanoma skin cancers from renal transplant recipients and nonimmunosuppressed patients. *Journal of the National Cancer Institute, 88*, 802–811.

Shannon-Lowe, C., Rickinson, A. B., & Bell, A. I. (2017). Epstein–Barr virus-associated lymphomas. *Philosophical Transactions of the Royal Society, B: Biological Sciences, 372*(1732), 20160271.

Shivane, A., Pearce, A., Khatib, N., & Smith, M. E. (2018). EBV+ HHV-8+ multicentric Castleman disease with Plasmablastic aggregates in an HIV+ man: An evolving Clinicopathologic entity. *International Journal of Surgical Pathology, 26*(4), 338–341. https://doi.org/10.1177/1066896917745584

Shope, R. E. (1932). A filtrable virus causing a tumor-like condition in rabbits and its relationship to virus myxomatosum. *The Journal of Experimental Medicine, 56*, 803–822.

Singh, M., & Kumar, V. (2003). Transgenic mouse models of hepatitis B virus-associated hepatocellular carcinoma. *Reviews in Medical Virology, 13*, 243–253.

Sixbey, J. W., Nedrud, J. G., Raab-Traub, N., Hanes, R. A., & Pagano, J. S. (1984). Epstein–Barr virus replication in oropharyngeal epithelial cells. *The New England Journal of Medicine, 310*(19), 1225–1230.

Sonoda, S., Li, H. C., & Tajima, K. (2011). Ethnoepidemiology of HTLV-1 related diseases: Ethnic determinants of HTLV-1 susceptibility and its worldwide dispersal. *Cancer Science, 102*(2), 295–301.

Soulier, J., Grollet, L., Oksenhendler, E., Cacoub, P., Cazals-Hatem, D., Babinet, P., d'Agay, M. F., Clauvel, J. P., Raphael, M., Degos, L., & Sigaux, F. (1995). Kaposi's sarcoma-associated herpesvirus-like DNA sequences in multicentric Castleman's disease. *Blood, 86*, 1276–1280.

Soussan, P., Garreau, F., Zylberberg, H., Ferray, C., Brechot, C., & Kremsdorf, D. (2000). In vivo expression of a new hepatitis B virus protein encoded by a spliced RNA. *The Journal of Clinical Investigation, 105*, 55–60.

Soussan, P., Tuveri, R., Nalpas, B., Garreau, F., Zavala, F., Masson, A., Pol, S., Brechot, C., & Kremsdorf, D. (2003). The expression of hepatitis B spliced protein (HBSP) encoded by a spliced hepatitis B virus RNA is associated with viral replication and liver fibrosis. *Journal of Hepatology, 38*, 343–348.

Spardy, N., Duensing, A., Charles, D., Haines, N., Nakahara, T., Lambert, P. F., & Duensing, S. (2007). The human papillomavirus type 16 E7 oncoprotein activates the Fanconi Anemia (FA) pathway and causes accelerated chromosomal instability in FA cells. *Journal of Virology, 81*(23), 13265–13270.

Stewart, S. E., Eddy, B. E., & Borgese, N. (1958). Neoplasms in mice inoculated with a tumor agent carried in tissue culture. *Journal of the National Cancer Institute, 20*(6), 1223–1243.

Storey, A., & Banks, L. (1993). Human papillomavirus type 16 E6 gene cooperates with EJ-ras to immortalize primary mouse cells. *Oncogene, 8*, 919–924.

Tanaka, A., Noda, T., Yajima, H., Hatanaka, M., & Ito, Y. (1989). Identification of a transforming gene of human papillomavirus type 16. *Journal of Virology, 63*, 1465–1469.

Tanaka, H., Iwasaki, Y., Nouso, K., Kobayashi, Y., Nakamura, S., Matsumoto, E., Toshikuni, N., Kaneyoshi, T., Ohsawa, T., Takaguchi, K., Fujio, K., Senoh, T., Ohnishi, T., Sakaguchi, K., & Shiratori, Y. (2005). Possible contribution of prior hepatitis B virus infection to the development of hepatocellular carcinoma. *Journal of Gastroenterology and Hepatology, 20*, 850–856.

Tang, Y., Woodward, B. O., Pastor, L., George, A. M., Petrechko, O., Nouvet, F. J., Haas, D. W., Jiang, G., & Hildreth, J. E. (2020). Endogenous retroviral envelope syncytin induces HIV-1 spreading and establishes HIV reservoirs in placenta. *Cell Reports, 30*(13), 4528–4539. https://doi.org/10.1016/j.celrep.2020.03.016

Tugizov, S. M. (2020). Human immunodeficiency virus interaction with oral and genital mucosal epithelia may lead to epithelial–mesenchymal transition and sequestration of virions in the endosomal compartments. *Oral Diseases, 26*, 40–46. https://doi.org/10.1111/odi.13387

Wang, H. P., Zhang, L., Dandri, M., & Rogler, C. E. (1998). Antisense downregulation of N-myc1 in woodchuck hepatoma cells reverses the malignant phenotype. *Journal of Virology, 72*, 2192–2198.

Weiss, R. A., & Vogt, P. K. (2011). 100 years of Rous sarcoma virus. *The Journal of Experimental Medicine, 208*(12), 2351–2355.

Wells, S. I., Francis, D. A., Karpova, A. Y., Dowhanick, J. J., Benson, J. D., & Howley, P. M. (2000). Papillomavirus E2 induces senescence in HPV-positive cells via pRB- and p21(CIP)-dependent pathways. *The EMBO Journal, 19*, 5762–5771.

White, M. K., Gordon, J., & Khalili, K. (2013). The rapidly expanding family of human polyomaviruses: Recent developments in understanding their life cycle and role in human pathology. *PLoS Pathogens, 9*(3), e1003206.

Wieland, S. F., & Chisari, F. V. (2005). Stealth and cunning: Hepatitis B and hepatitis C viruses. *Journal of Virology, 79*, 9369–9380.

Wollersheim, M., Debelka, U., & Hofschneider, P. H. (1988). A transactivating function encoded in the hepatitis B virus X gene is conserved in the integrated state. *Oncogene, 3*, 545–552.

Wright, R. (1990). Viral hepatitis comparative epidemiology. *British Medical Bulletin, 46*, 548–558.

Yoshida, M., Miyoshi, I., & Hinuma, Y. (1982). Isolation and characterization of retrovirus from cell lines of human adult T-cell leukemia and its implication in the disease. *Proceedings of the National Academy of Sciences of the United States of America, 79*, 2031–2035.

Yuan, Z., Petree, J. R., Lee, F. E., Fan, X., Salaita, K., Guidot, D. M., & Sadikot, R. T. (2019). Macrophages exposed to HIV viral protein disrupt lung epithelial cell integrity and mitochondrial bioenergetics via exosomal microRNA shuttling. *Cell Death & Disease, 10*(8), 1–12. https://doi.org/10.1038/s41419-019-1803-y

Zhang, L. L., Wei, J. Y., Wang, L., Huang, S. L., & Chen, J. L. (2017). Human T-cell lymphotropic virus type 1 and its oncogenesis. *Acta Pharmacologica Sinica, 38*(8), 1093–1103. https://doi.org/10.1038/aps.2017.17

Zhang, Z., Torii, N., Furusaka, A., Malayaman, N., Hu, Z., & Liang, T. J. (2000). Structural and functional characterization of interaction between hepatitis B virus X protein and the proteasome complex. *The Journal of Biological Chemistry, 275*, 15157–15165.

Zhong, S., Chan, J. Y., Yeo, W., Tam, J. S., & Johnson, P. J. (1999). Hepatitis B envelope protein mutants in human hepatocellular carcinoma tissues. *Journal of Viral Hepatitis, 6*, 195–202.

zur Hausen, H. (1977). Human papillomaviruses and their possible role in squamous cell carcinomas. *Current Topics in Microbiology and Immunology, 78*, 1–30.

zur Hausen, H. (2002). Papillomaviruses and cancer: From basic studies to clinical application. *Nature Reviews. Cancer, 2*, 342–350.

zur Hausen, H. (2006). *Infections causing human cancer*. Wiley.

zur Hausen, H., Meinhof, W., Scheiber, W., & Bornkamm, G. W. (1974). Attempts to detect virus-secific DNA in human tumors. I. Nucleic acid hybridizations with complementary RNA of human wart virus. *International Journal of Cancer, 13*, 650–656.

zur Hausen, H., Schulte-Holthausen, H., Wolf, H., Dorries, K., & Egger, H. (1974). Attempts to detect virus-specific DNA in human tumors. II. Nucleic acid hybridizations with complementary RNA of human herpes group viruses. *International Journal of Cancer, 13*, 657–664.

Role of HCV in Hepatocellular Carcinoma 5

Saba Zafar, Farhat Jabeen, and Aiman Sajjad

Abstract

This chapter delves into the intricate relationship between hepatitis C virus (HCV) and the development of hepatocellular carcinoma (HCC). It provides a comprehensive overview of the epidemiological significance of HCV as a major risk factor for HCC worldwide. The molecular mechanisms underlying HCV-induced hepatocarcinogenesis are explored, emphasizing the interplay between viral proteins, host factors, and the cellular microenvironment. Additionally, the impact of HCV infection on liver cirrhosis progression and its subsequent contribution to HCC pathogenesis are discussed. Furthermore, diagnostic and therapeutic strategies targeting HCV-related HCC are reviewed, highlighting the importance of early detection and personalized treatment approaches in improving patient outcomes. Overall, this chapter aims to enhance our understanding of the multifaceted role of HCV in the development and progression of HCC, thereby paving the way for the development of more effective preventive and therapeutic interventions.

Keywords

Hepatitis C virus (HCV) · Hepatocellular carcinoma (HCC) · Hepatocarcinogenesis · Liver cirrhosis

S. Zafar · F. Jabeen (✉) · A. Sajjad
Department of Zoology, Government College University, Faisalabad, Pakistan
e-mail: farhatjabeen@gcuf.edu.pk

N. Masood, A. Yasmin (eds.), *The Microbiome and Cancer*,
https://doi.org/10.1007/978-981-97-6948-3_5

5.1 Introduction

Liver cancer ranks among the most commonly identified cancers and is the primary cause of fatalities related to cancer on a global scale (Bray et al., 2018; Villanueva, 2019). Intrahepatic cholangiocarcinoma (ICC) and hepatocellular carcinoma (HCC) are the main histological subtypes of many malignant diseases that are included in the category of liver cancer (Craig et al., 2020; Massarweh & El-Serag, 2017). According to histology, 10–20% of liver cancer cases are classified as ICC, and 70–85% of cases show signs of HCC (Sia et al., 2017). The morbidity and mortality rates are dramatically caused by liver cancer. It is the seventh most prevalent cancer globally and the third leading cause of cancer-related mortality overall (Goto et al., 2020). Understanding the molecular mechanisms and epidemiological risk factors that drive liver carcinogenesis has advanced significantly in the previous decade due to major discoveries in genomics, molecular and cellular biology, and population-based research (Yang & Roberts, 2010). Males had a higher risk of developing hepatocellular carcinoma than females (HCC), with a ratio of 2.4 to 1. Where the prevalence was higher, this difference was particularly evident in Eastern and Southern Asia, Middle and Western Africa, Malaysia, and Micronesia/Polynesia (Ferlay et al., 2010).

Between 2000 and 2016, the total number of deaths in the USA from liver cancer increased by 43%. Liver tumors have an 18% 5-year survival rate, which makes it the second most deadly malignancy after pancreatic cancer, despite this alarming trend. However, the outlook for patients in Asian nations such as China, whose 5-year survival rates have reportedly dropped to as low as 12%, is even more worrisome (Zheng et al., 2018). Due to an increase in incidence over the past few years, hepatocellular carcinoma (HCC) is one of the most frequently diagnosed new malignancies and the most common cause of cancer-related fatalities in the USA (Siegel et al., 2018). The most recent data collected from population-based cancer registries indicate a continued rise in HCC cases, highlighting the urgent need for improved prevention, early detection, and treatment strategies. According to monitoring, epidemiology, and final outcomes, the rate of hepatocellular carcinoma (HCC) has increased about four times in the past 40 years. From 1.6 per 100,000 in 1975–1977 to 4.8 per 100,000 in 2005–2007, the rates increased. Furthermore, there are worries that the incidence might keep on increasing in the future (Petrick et al., 2016). Furthermore, the diagnosis for those with liver cancer patient is dismal in the long run, which confirms the disease's standing as a leading cause of mortality in cancer patients (Siegel et al., 2018).

The onset of hepatocellular carcinoma is a multiphase, complex disorder that is closely related to food, lifestyle, and environmental factors. It usually appears in more than 90% of individuals with cirrhosis resulting from various factors. While any kind of liver disease raises the possibility of evolving hepatocellular carcinoma, liver failure carried on by persistent diseases like hemochromatosis or hepatitis C or B germs carries the highest risk (Anzola, 2004).

Hepatitis C virus (HCV) and hepatitis B virus (HBV) chronic diseases are considered to be responsible for over 80% of liver cancer cases (de Martel et al., 2015). Hepatocellular carcinoma (HCC) is the most common cancer type which has been linked to several risk factors, such as consuming aflatoxin, drinking water contaminated with algae that produce hepatotoxins, chewing betel nut, having diabetes, non-alcoholic fatty liver disease, consuming alcohol, and smoking cigarettes (Wild & Montesano, 2009). Aflatoxin exposure, which primarily increases the carcinogenicity of the infection with the hepatitis B virus (HBV), has dropped as a result of improved grain storage and changes in diets in a number of developing nations. However, there has been a significant rise in obesity and diabetes. These disorders are on the rise in both industrialized and developing nations, and they were mostly related to the development of hepatocellular carcinoma (HCC) in individuals infected with the hepatitis C virus (HCV) (de Martel et al., 2015).

Typically, HBV is acquired during childhood or at birth. On the other hand, HCV infection can occur to individuals at any age through contaminated blood or needles, and because exposure to these risks increases with age, the prevalence of HCV infection rises continuously. Medical treatments that are unique to each nation have a significant effect on the transmission of HCV, especially with regard to the security of injections and blood transfusions. This can lead to high-prevalence countries being adjacent to low-prevalence countries, especially given the importance of intravenous drug use as a mode of transmission. When exposure occurred at any age, HCV has a significant chance of developing into a chronic infection, unlike HBV infection, where chronic carriage is rare if exposure comes after puberty (Gower et al., 2014; Nelson et al., 2011).

Hepatocellular carcinoma (HCC) development appears to be strongly influenced by the viral pathogenesis resulting from HBV and HCV infections. Implementing hepatitis virus vaccine programs, taking steps to minimize aflatoxin contact, and decreasing the harmful effects of inevitable interactions could all have a significant impact on the disease's incidence in the field of public health. Initiatives targeted on treating folks who are presently infected and stopping viral transmission, consequently, are anticipated to have a major effect on the incidence of hepatocellular carcinoma (HCC) in addition to intrahepatic cholangiocarcinoma (ICC) in the future (Goto et al., 2020).

The most important goal of this chapter is to summarize contemporary research that demonstrates the function of HCV in hepatocellular cancer and explains the counseled mechanisms of action. The objective is to inform researchers and working experts about novel treatments that may be successful to cure HCC. This chapter aims to study the most recent traits in focused remedy and compare how they have an impact on the management of hepatocellular carcinoma (HCC). Moreover, we provide a comprehensive overview of the major areas in HCC research, highlighting emerging medical trials and technological structures. These efforts have the capability to permit good sized traits in the approaching 10 years. A wide variety of digital databases, which include Science Direct, Elsevier, and PubMed, had been used to search the literature. The key phrases used for the exploration of information have been "hepatitis C virus," "hepatitis B virus," "hepatocellular carcinoma," "structure of hepatitis C virus," "HCC variability," "HCV triggered hepatocellular carcinoma," "HCV regulation pathway prompted HCC," "inflammation of the liver fibrosis," "angiogenesis, HCV prognosis," "biomarkers for HCV-related HCC detection," and "medical strategies for HCV-associated HCC."

5.2 Types of Viral Hepatitis

There are at least six distinct hepatotropic viruses that are known to cause viral hepatitis. HCV and HBV are the primary causes of long-term hepatitis among them. About 75% of infections caused by HCV and 5–10% of HBV cases ultimately turn persistent. The specific process behind persistent condition is yet unknown (Waris & Siddiqui, 2003).

5.3 Hepatitis B Virus

The DNA virus that causes prolonged non-inflammatory intestinal disorders is the hepatitis B virus (HBV). Hepatocellular carcinoma (HCC) is the disorder that leads to the development of mutations in liver cells (McGlynn et al., 2015). A double-stranded, circular DNA molecule with eight genotypes identified from A to H characterizes HBV. Genotypes A and D are more common across Europe and the Middle East, whereas genotypes B and C are more frequently identified in Asia (Bruix & Sherman, 2010). Hepatocellular carcinoma (HCC) risk has been associated with genotype C, as instead of A, B, and D. Hepatitis B is transmitted through a IV injections, impure transfusions of blood, and sexual activity. Notably, the most common cause of HBV infection globally is vertical transfer from mother to fetus. It is essential to mention that the worldwide hepatitis B infection rate is roughly 5% (Ott et al., 2012).

Hepatocellular carcinoma (HCC) has no connection with active HBV infection, particularly if HBV DNA levels are more than $>10^5$/mL viral copies. Over an 8–10-year follow-up period, such raised viral copies are linked to a 2.5–3 times greater chance of HCC development (Chen et al., 2006). The development of liver cancer is not exclusively related with the occurrence of the surface antigen of hepatitis B (HBsAg) as a hematological marker. There is still a possibility that HCC will develop in those who test positive for anti-HBc, or hepatitis B core antibody, but are HBsAg-negative (Hosaka et al., 2013). Hepatocellular carcinoma is more likely to occur in HBV carriers due to a variety of cofactors. These includes demographic traits such older age, male sex, African and Asian history, as well as a family history of HCC. Viral factors that affect results include a high rate of replication of the HBV, HBV genetic makeup, duration of infection, and mutual infection with either HIV or the hepatitis C virus (HCV). Contributing factors include alcohol, smoke, obesity, diabetes, and environmental exposures to aflatoxin (El-Serag, 2012).

Antiviral therapy for hepatitis B can significantly reduce the hepatocarcinogenic potential of HBV. HCC rates were successfully decreased over a period of 5 years from 13.7% (in control groups) to 3.7% which can be achieved by controlling the virus; patients with cirrhosis show a significant decrease in rate (Hosaka et al., 2013). The prevalence of HBV-associated HCC has decreased significantly after the introduction of the HBV vaccine. The East Asian premature babies vaccination program is predicted to reduce the incidence of hepatitis B-related HCC by 70–85% (Chiang et al., 2013).

5.4 Hepatitis C Virus

The viral infection hepatitis C virus belongs to the *Hepacivirinae* species and is part of the *Flaviviridae* family of positive strand RNA viruses. The genetic material of HCV is a single-stranded, linear RNA with positive-sense virus that is wrapped that varies in size from 30 to 60 μm. Another major risk factor for HCC is persistent HCV infection, which increases the risk by 10–20 times. Being an RNA virus, HCV cannot incorporate into the host DNA like HBV behaves. Therefore, the chances of HCC is mainly decreased to those people who acquire cirrhosis or sustain permanent damage to the liver with bridging fibrosis (Kanwal et al., 2017). Six distinct HCV genotypes have been identified. While genotype IV is more common in the Middle East, genotypes I, II, and III are more widespread in Western countries and the Far East. At 18%, Egypt has the highest rates of chronic hepatitis C infection; rates in Europe are around 0.5% and 2.5%, the USA is 1.8%, and Canada is 0.8% (Bostan & Mahmood, 2010).

Among the factors that cause HCC in individuals with present HCV infections are male sex, Hispanic background, hepatitis C genotype 3, prolonged infection time, associations with HBV or the HIV virus, resistance to insulin, diabetes, obesity, and both tobacco and alcohol use (El-Serag & Kanwal, 2014). The use of direct-acting antiviral (DAA) drugs has seen a greater proportion of people with HCV infection achieve sustained virologic response (SVR) through effective treatment (Kanwal et al., 2017; Sancho et al., 2022). Obtaining sustained virological response (SVR) with antiviral drugs is the major factor decreasing the incidence of HCC. Direct-acting antiviral medication has been the subject to several large-scale studies. Systematic reviews of these investigations have shown that whereas risk of HCC cannot be eliminated completely, it is reduced by 50–80% in those who are able to achieve sustained virological response (SVR) (Waziry et al., 2017).

5.5 Structure of HCV

HCV is categorized as an encased virus, possessing a positively sensed RNA genome of 9.6 kb. The single polyprotein present in the HCV virus's DNA is ultimately broken into ten mature proteins. The structural proteins are found close to the polyprotein's 5′ end, whereas the nonstructural proteins are found near the polyprotein's 3′ end. The identification and binding of cells by HCV is significantly influenced by the envelope glycoproteins E1 and E2. E1 is specifically essential for the virus to combine with the target cell membrane. The mature protein is created by two cleavages of the core protein. Together with the glycoproteins E1 and E2, this protein forms the nucleocapsid and combines to build the envelope. In addition, p7 is considered to be essential for the development of infectious viruses and could act as a viroporin that influences the permeability of membranes (Levrero, 2006; Penin et al., 2004).

Another serine proteinase that is found in the N region of NS3 emerges as a result of the autocatalytic cleavage of NS2 and NS3. The additional nonstructural proteins

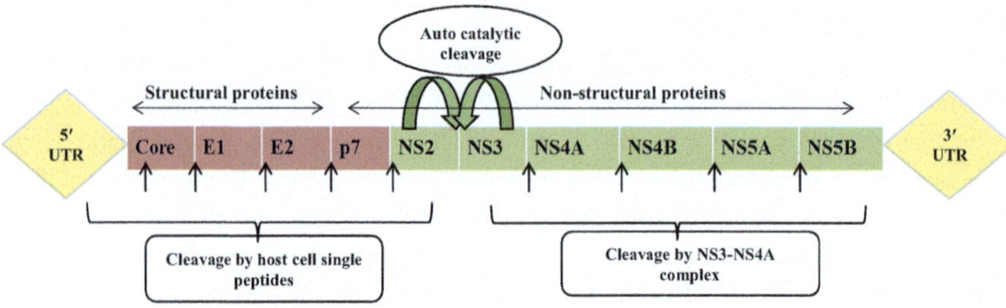

Fig. 5.1 Structure of hepatitis C virus

are then broken down by NS3 and its cofactor NS4A, generating NS4B, NS5A, and NS5B after that (Penin et al., 2004). As an RNA helicase, the C-terminal end of NS3 helps in the unwinding of double-stranded RNA during the replication process (Fig. 5.1). NS4B generates a membranous web that is necessary to the HCV replication complex. With its mixture of structural and nonstructural proteins, this web supports the structure of the replicating HCV complex. It is believed that NS5A involves in the replication complex and links with cellular pathways in its functions. Its specific function is still unknown. However, NS5B functions as the RNA-dependent RNA polymerase during the HCV replication process (Penin et al., 2004; Tran, 2008).

5.6 HCC Variability

Hepatitis B (HBV) and C (HCV) are the two main causes for HCC. Three major pathogenic and clinico-biological characteristics of HCV set it distinct from HBV and are related to hepatocarcinogenesis (Farazi & DePinho, 2006). Ten percent of HBV cases lead to chronic infection—however, 60–80% of HCV cases can do. This is most likely the result of HCV quasi-species immune suppression, which is caused by a high incidence of mistakes in replication. The second j fundamental factor is that HCV is a greater possibility than HBV to cause liver cirrhosis. Between 5 and 10% of HCV-infected individuals experience liver failure after 10 years of infection, which is approximately 10–20 times more common than HBV. Third, since HCV is an RNA virus without a DNA intermediate, it is unable to integrate into host genomes (Castello et al., 2010; Farazi & DePinho, 2006). Since HBV can incorporate into the host's DNA and disrupt cellular communication and growth control, HCC is most frequently observed in cirrhotic hepatitis patients, though it can also occur in HBV individuals without severe liver disorders. It can cause HCC both directly and indirectly. HCV cannot be incorporated into the genome of its host; therefore, HCC is primarily produced via indirect routes: long-term inflammation, cell apoptosis, and proliferation cirrhosis (Lorusso & Rüegg, 2008).

Taking the previous discussion into account, this chapter emphasizes on HCV-related HCC, highlights the significance of pro-tumoral milieu and chronic inflammation in hepatocarcinogenesis, and suggests different ideas for medical treatment.

The virus can be eradicated by the immune system on its own, and immunity generated by cells and antibodies that neutralize specific to the virus have been identified (Castello et al., 2010).

5.7 HCV-Induced Hepatocellular Carcinoma

Because HCV cannot be incorporated into the genome of the host, due to the disruption of epigenetic patterns, it is most likely a major contributing factor in the growth of cancer and the behavior of tumors. The synthesis of viral nucleic acid and viral pathogenesis may lead to oncogenesis through cirrhosis and liver cell regeneration; however, there is no DNA involved in the replication of the HCV genome. On the other hand, certain HCV proteins are associated with the development of liver cancer. For instance, the basic protein and the NS proteins NS4A and NS4B restrict the posttranscriptional synthesis of $p21^{WAF1}$. Translational suppression and possibly enhanced cellular breakdown of proteins are both caused by NS4A and NS4B (Florese et al., 2002). Currently, hepatocarcinogenesis has been associated with several HCV proteins: the core, NS3, and NS5A proteins.

5.8 Core Protein

The structural viral protein known as the hepatitis C virus core protein covers the infectious genomic RNA (Kanda et al., 2008). Various cellular regulatory mechanisms are also regulated by HCV core. In fact, it has been demonstrated that the HCV basic protein, through its physical connection to p53, can cause proliferation and apoptosis by regulating the synthesis of the cyclin-dependent regulator p21WAF1 Q (Kwun & Jang, 2003). One of p53's main targets is the p21WAF1 gene, and the p21WAF1 protein controls the functions of cyclin/cyclin-dependent kinase complexes that contribute in tumor formation and control cell division. Less transcription of the p21WAF1 gene and a shorter half-life of the p21WAF1 protein induce inhibition of p21WAF1 expression (Anzola, 2004). Amino acids 1–191 are the first amino acids which form the core protein. After further processing, these amino acids transform into the adult form (amino acids 1–173). The mature form's subcellular location is regulated partially by the innate form. The mature form, found in the nucleus, reduces p21WAF1 expression, whereas the innate form, located in the cytoplasm, increases p21WAF1 production. Therefore, the core protein in the nucleus reduces the quantity of p21WAF1 through a p53-independent mechanism, while the core protein in the cytoplasm enhances the proportion of p21WAF1 by activating p53. The core protein's subcellular location and control of p21WAF1 expression may determine whether infected cells proliferate or undergo apoptosis. As a result, the innate form controls the location and activities of core protein in addition to being a precursor to the mature form (Anzola, 2004; Yamanaka et al., 2002). Moreover, the HCV core protein has been found to regulate p73, a p53 family member. In addition to its involvement in neurogenesis and the natural immune response, p73 seems to play an important role in the development and preservation of cancer. HCV core protein is nuclear transferred as a result of the p73/core interaction when the tumor-inhibiting proteins p73 or p73 b are present. Moreover, interacting with the HCV core protein inhibits p73 a-induced cell growth arrest in a p53-dependent manner, but not p73 b-induced cell growth arrest. The different p73 activities may be directly affected by HCV core protein, contributing to HCV etiology (Alisi et al., 2003; Tran, 2008).

5.9 NS3 Protein

The protein NS3 of the hepatitis C virus can cause hepatocytes to change early in the hepatocarcinogenic process by acting on host cells through an endogenous mechanism. It has been demonstrated that the NS3 protein and wt-p53 produce a complex. The oligomerization region has been found to be present in a portion of wt-p53 closer to the C-terminus (amino acids 301–360), which is essential for forming a complex with NS3 (Anzola, 2004). In addition, the promoter function of p21WAF1 can be particularly suppressed by NS3 protein in a manner that depends on dosage. When paired to HCV core protein, the is additive impact and not specific to only one cell type (Kwun et al., 2001).

5.10 NS5A Protein

The 56–58 kDa phosphoprotein known as the NS5A protein, encoded by the hepatitis C virus, is synthesized during the breakdown of viral polyprotein. NS5A's role in the transmission of viruses is unknown, despite the fact that it is linked to other virally encoded proteins and is a component of the viral replicase complex, which is located on the cytoplasmic portion of the endoplasmic reticulum. Phosphorylation and possible proteolytic degradation to lower molecular weight molecules that are able to move to the nucleus are examples of posttranslational changes of NS5A. While NS5A was recently shown to interact pleiotropically with cellular proteins, condensed forms of the protein can function as transcriptional regulators in HCV-host contacts (Khabar & Polyak, 2002). Interferon-a (IFN-a) signaling and activation are not directly influenced by NS5A; however, a variety of interactions with aspects of the transcription activating technology, developmental factor protein receptor-bound, p53, p21WAF1, and cyclins, also members of the cellular communicating apparatus, have been reported (Reyes, 2002). NS5A may take part in causing prolonged liver disorders and HCC associated with HCV infection, as demonstrated by the fact that a variety of these interactions suppress the apoptotic cellular response to chronic HCV infection (Tran, 2008).

5.11 DNA Methylation

Cancer development has a close relationship with the outcome of DNA methylation, which comprises both hyper- and hypomethylation. A study demonstrated that the tumor suppressor gene's promoter sequence hypermethylation greatly serves as the carcinogenic activities of HCV-induced HCC (Suhail et al., 2022). HCV viral proteins may induce nuclear signaling in a host's DNA because they are present in the cell's nucleus (Levin et al., 2014). The discovery that the core protein greatly enhances the production of histone deacetylase HDAC1 and DNMT-1 indicates it has the ability to epigenetically restrict the expression of genes that suppress cancer, hence inhibiting the development and spread of tumors (Toh et al., 2019). Through inhibiting SFRP, this protein induces EMT and alters Wnt/β-Catenin signaling (Khalaf et al., 2018). It has been discovered that 11 genes known as tumor suppressors directly contribute to the occurrence of HCC caused by HCV. RASAL1, CSMD1, EGLN3, CDKN2A, RUNX3, SFRP1, ZNF382, RB1, LOX, BCORL1, and P73 are the names of these protein (Zheng et al., 2019).

5.12 Histone Modifications

Histone alterations have been efficiently studied in a variety of cancer models and is essential for the development of HCC. Histone methylation, sumoylation, phosphorylation, acetylation, and ubiquitination are common changes. Elevated JmjC histone demethylase member KDM5B/JARID1B levels in HCV-induced HCC cases regulate downstream E2F1 and E2F2 genes, causing HCC cells to proliferate. The occurrence of HCC is caused by of the regulation of histone acetylation in HCV-infected individuals. HATs and HDACs are the enzymes that control this process (Zhao et al., 2021).

Many cellular pathways are usually involved in the histone acetylation mechanisms linked to HCC. The discovery of virally induced acetylation at histone 3's lysine 27 position (H3K27Ac) in liver tissue infected with HCV suggests that the virus influences the illness through histone deacetylation (Jühling et al., 2021). H3K27Ac functions as an activation mark to differentiate poised or active enhancer elements from inactive ones during embryonic stages. Associated gene transcription is made possible by this differential acetylation (Creyghton et al., 2010). Furthermore, the association between the gene expression predisposed to development of cancer and the genome-wide modifications in H3K27Ac caused by HCV supports the virus's contribution to the formation of cancer through the modification of histone (Hamdane et al., 2019).

Similar findings to H3K27Ac, histone H3 Lysine 9 (H3K9Ac) is also involved in HCV-mediated acetylation in cancer (Perez et al., 2019; Zhou et al., 2015). The reason for the discovery of these epigenetic modifications after the viral therapy has been associated to their determination, which produces a memory or evidence (Hoshida et al., 2014). There is no doubt that in HCV-HCC patients, the presence of this epigenetic signature has been linked with severe tumor behavior. Even after therapy, there is an elevation of cancer risk gene expression, according to experimental studies and findings conducted with HCV patients with hepatitis (Perez et al., 2019).

5.13 Noncoding RNAs in HCV-Induced HCC

Transcriptome analysis at a high rate of production has identified several forms of noncoding RNAs (ncRNAs) in eukaryotes, particularly miRNAs as well as long noncoding RNAs (lncRNAs). A vital role for these noncoding RNAs is played in numerous biological mechanisms, including those related to HCC. It is clear that the 18 genes that have been identified as relevant in HCC have a close relationship with ncRNAs. Both increasing and decreasing the activities of different genes in HCV-caused HCC, these ncRNAs perform a regulatory role. However, more studies continue to figure out the precise mechanisms enabling these approaches of regulation (Unfried & Fortes, 2020). Changes in the expression of miRNA precipitated by HCV infection adjust the development of liver cancer. HCV additionally affects miR-146a-5p, miR-19a, miR-135-5p and miR-122, which are now being studied as prospective signs of liver damage (Plissonnier et al., 2018).

The miRNAs are elaborated in preventing HCV contamination and replication at some point of the method of the virus-host interplay in hepatocellular carcinoma (HCC) improvement. At present, there are over 1,000 acknowledged miRNAs, with miR-122 being 1 of the maximum numerous in the liver. There is a robust correlation among this unique miRNA and hepatocellular carcinoma (HCC) due to HCV (Zhang et al., 2015).

5.14 HCV Regulation on Cellular Pathway Induced HCC

A number of key cellular pathways that are altered in HCV-induced hepatocellular carcinoma (HCC) include TERT, NF-κB pathway, Wnt/β-catenin/c-Myc pathway, β-catenin, Hippo-YAP/TAZ pathway, p53, chromatin remodeling/epigenetic modifications, Rb, hepatocyte differentiation, PI3K-mTOR pathway, cancer stem cells, NRF2-kelch-like ECH-associated protein (KEAP), angiogenesis, and RTKs (Fig. 5.2).

5.15 Telomerase Reverse Transcriptase

Structures called telomeres are composed of DNA and proteins that are situated at the tip of every chromosome. They are protected by shelterin proteins and contain a tandem repeating six-nucleotide pattern (5′-TTAGGG-3). A telomere's length may exceed up to 15,000 base pairs. The cell experiences apoptosis and eventually dies according to program once the telomere length is too shorter. An enzyme called TERT helps to maintain the length of the telomere. The RNA and protein subunits

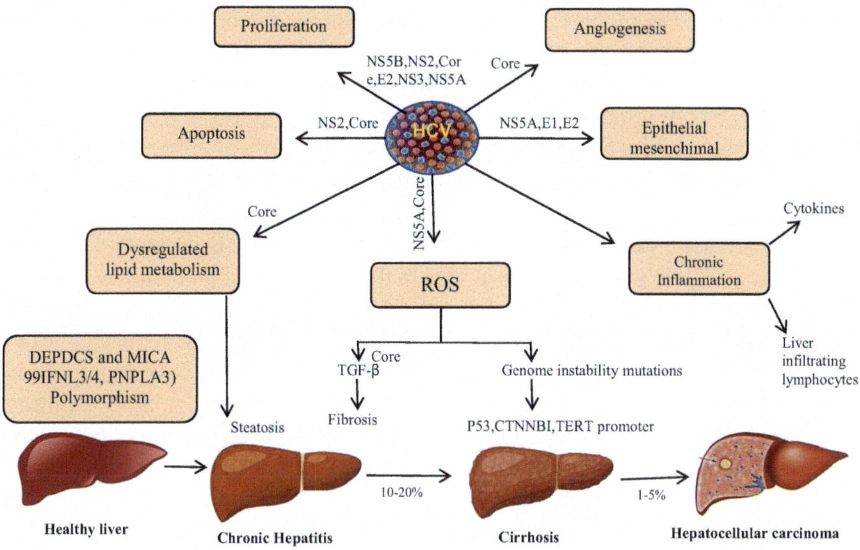

Fig. 5.2 HCV-associated mechanisms of carcinogenesis, from HCV infection to HCC

of telomerase are the enzymes that elongate chromosomes. It does this by marking TTAGGG sequences at the end of chromosomes which already exist. Telomerase and telomeres play a major role in the development of liver cirrhosis and the advancement of liver disease (Nault et al., 2019). Tumor cells have a selected growth benefit due to elevated telomerase levels. Approximately 85% of human cancers, which include those related to HCC, have several activating mutations in the TERT gene promoter regions, according to systematic hepatocellular carcinoma (HCC) gene sequencing (Pezzuto et al., 2017). These mutations are located in the TERT promoter area at nucleotides 146 (G>A) or 124 (often G>A and occasionally G>T), just before the ATG start point. These changes encourage transcription factors to bind to the promoter, which causes TERT to be overproduced and activated. Higher TERT function with prolonged telomeres had been correlated with severe hepatocellular carcinoma, which is characterized by a destructive prognosis (Donaires et al., 2017).

5.16 Wnt/c-Myc/β-Catenin Pathway

The Wnt/β-catenin pathway is altered in 66% of cases of hepatocellular carcinoma (HCC). Of these cases, 51% show that Wnt/β-catenin is activated, which results in changes to gene expression. The Wnt/β-catenin pathway is stimulated to induct when a Wnt ligand binds to frizzled receptors. This binding event triggers the transcription factor β-catenin and stimulates pro-survival genes (Kanda et al., 2019). Wang and colleagues have shown that HCV infection can stimulate the Wnt/β--catenin pathway, with the core protein playing a critical role in this process. To be more precise, the core protein controls transcription factors in the hepatocyte nucleus, which activates the Wnt/β-catenin pathway (Wang et al., 2017).

Furthermore, by blocking antagonists such FZD-related protein 2 and Dickkopf 1 (DDK1), the core protein activates the FZD receptor and low-density lipoprotein receptor-related protein 5/6 (LRP5/6) Liu et al., 2011).

Dickkopf 1 (DDK1) suppression is necessary for the epigenetic silencing of regulators early in an HCV infection. This inhibition facilitates the recruitment of DNA methyltransferase 1 and histone deacetylase. Furthermore, by hypermethylating the E-Cadherin (CDH1) gene activator, it aids in the inhibition of β-catenin accumulation (Mahmoudvand et al., 2019). In contrast, GSK3b regulation is linked to the activity of NS5A, a protein that is part of the HCV complex that replicates. Both direct binding and indirect inhibition via the PI3K-AKT pathway have an impact on its stability. Thus, there is an activation of the Wnt/β-catenin regulatory pathway (Wang et al., 2017). Hepatocellular carcinoma (HCC) has been shown to be caused by the c-Myc oncogene via the Wnt/β-catenin pathways in the HCV transgenic animal approach (Higgs et al., 2013).

5.17 NF-κB Pathway

It is well known that NF-κB contributes to liver tumor growth, fibrosis, swelling, and cancer development, particularly in cases of HCV-induced hepatocellular carcinoma (HCC). According to Shokri et al. (2019), HCV appears to alter the NF-κB pathway in two different ways: it stimulates it and suppresses it. According to earlier studies, people with HCV overexpress NF-κB (Nguyen et al., 2006). As an illustration, core proteins have the ability to activate the TNF receptor (TNFR), which is necessary to start the NF-κB cascade. This stimulation can occur directly or through the replication of cytokines, such as TNF-α, that cause inflammation. Additionally, Th-17 cells, which release cytokines such as CD161, IL-17A, IL-17F, IL-21, and IL-22, can be activated by core proteins. Their connection to TNFR amplifies the activation of NF-κB. Furthermore, there is evidence indicating that the E1 and E2 proteins play a role in controlling the release of TNF-α (Elpek, 2021). On the other hand, recent experimental studies have demonstrated that HCV can suppress the NF-κB pathway by stimulating a number of genes that act as NF-κB inhibitors. Remarkably, TNF-α-induced NF-B activation is inhibited by the NS3 and NS5B proteins (Chen et al., 2015). Given the crucial roles that inflammation and oxidative stress play in liver disorders, it is not surprising that core, E1, E2, NS3, NS4A, NS4B, and NS5A proteins induce oxidative stress to elevate NF-κB. These signals cause the production of reactive oxygen species (ROS) through the activities of the mitochondria (Elpek, 2021).

5.18 Hippo-YAP/TAZ Pathway

There aren't many results currently available regarding the Hippo-YAP-1 pathway in relation to HCV-induced hepatocellular carcinoma (HCC), most of which are from experimental research. More specifically, it has been suggested that the HCV NS5B protein stimulates the epithelial-mesenchymal transition (EMT). This is accomplished by inhibiting the Hippo signaling pathway, which increases Snail activity and makes it easier for PI3K/AKT to initiate (Hu et al., 2017). There is evidence from a number of studies indicating that the HCV E2 protein may resemble CD81, which is the main ligand that attaches to glypican 3. It's interesting to note that this mimicry reduces multiplication by decreasing YAP levels and Hippo stimulation rates (Elpek, 2021; Xue et al., 2018). Since the transcription factor CD81 serves as HCV's entry point, it follows that HCV resistance is conferred on tumor cells by CD81 deletion during the early stages of carcinogenesis. This feature aids in the clonal development of neoplastic cells during the early stages of hepatocellular carcinoma and separates them from non-tumor cells. Therefore, by investigating the prospective effects of these factors on this cancer pathway, it will be feasible to evaluate whether the Hippo pathway could be a target for treatment for hepatocellular carcinoma (Elpek, 2021).

5.19 p53-Rb Pathway

Both of the tumor suppressors, p53 and Rb, are essential for the development of many malignancies. P53 protein levels increase in response to oncogene activation, oxygen deprivation, and DNA alterations. Consequently, p53 regulates the expression of genes that induce cell cycle arrest or apoptosis. The Rb protein prevents cell development by preventing the activation of E2F transcription genes. Stopping the cell cycle's progression into the S-phase and mitosis depends on the critical interaction between the transcription components Rb and E2F (Mitchell et al., 2017). Numerous earlier studies have shown that the minimal synthesis of the NS3, NS5A, and HCV core proteins can alter the p53 suppressor of cancers without using the harmful replication of HCV methods (Mitchell et al., 2017). In Stanley Lemon's laboratory, it was demonstrated that the HCV infection-induced negative regulation of Rb stability of proteins is mediated by the NS5B protein. These results are consistent with information gleaned from human sources, which indicates that mutations in the p53-Rb pathway are involved in more than 70% of cases of hepatocellular carcinoma (HCC). These alterations are connected to issues such aberrant mitosis, cell cycle progression, and genomic instability (Munakata et al., 2005; Munakata et al., 2007).

5.20 mTOR Pathway

Chronic viral infection is linked to the development of hepatocellular carcinoma through activation of the pathway known as mTOR. Immunohistochemistry staining results showed that 33 out of 73 HCC patients, or 45% of the patients, had an elevated level of total S6k. This was linked to the development of the tumor and mTOR stimulation (Villanueva et al., 2008). Stimulation of the mTOR cascade is linked to vascular invasion, tumor differentiation, staging, and phosphoS6 expression in a significant subset of patients with hepatocellular carcinoma (HCC) (Zhou et al., 2010). Numerous stimuli, such as growth hormones, TLR ligands, cytokines, low cellular energy (ATP/AMP ratio), DNA damage, and hypoxia, can activate the mTOR pathway. When mTOR is activated, progenitor cells, also known as cancer stem cells, exhibit a variety of growth advantages, including increased cell division and resistance to programmed cell death brought on by stress signals like malnourishment and low oxygen levels. The NS5A protein can activate mTOR and prevent mortality by interacting with the immunosuppressive FK506-binding protein FKBP38 (Dash et al., 2020; Wouters & Koritzinsky, 2008).

5.21 Inflammation of the Liver Fibrosis

Indeed, swelling and fibrosis of the liver are greatly affected by a number of circumstances. These consist of damage-associated sequences (DAMPs), viral particles, and host-pathogen-associated patterns of molecules (PAMPs). The growth of liver disorders may be facilitated by the combination of several factors. PAMPs and DAMPs are two examples of these genetic variations which lead to macrophage stimulation, which in return induces the beginning of hepatic disease. Based on indicators or aminotransferase stages, observations indicate, in patients with prolonged hepatitis, the reduction in macrophage activity following antiviral therapy. It is noteworthy that individuals with coinfections exhibit a higher prevalence of HCV-induced fibrosis compared to those with mono-infections.

5.22 Angiogenesis

A method by which new blood vessels originate from preexisting ones is connected to the development of liver cancer (HCC) known as angiogenesis. Studies have shown that HCV induces angiogenesis during the advanced phases of liver disease. This participation related to angiogenesis may accelerate the evolution of HCC. In healthy tissues, the angiogenesis process is typically regulated by the balance between proangiogenic (which encourages blood vessel production) and antiangiogenic (which inhibits blood vessel formation) components. Tumor tissues need more oxygen and nutrients, which triggers angiogenesis, the formation of new blood vessels that aid in the growth of the tumors (Li et al., 2010). Hepatocellular carcinoma cells synthesize proangiogenic chemicals because they stimulate endothelial cells via VEGF and FGFs, two fibroblast growth factors. Patients with head and neck cancer had greater expression levels of angiogenic growth factors, including PDGF, VEGF-A, and angiopoietin-2. This upregulation aids in angiogenesis, which encourages the development of blood vessels around tumors (Im et al., 2015). Research indicates that HCV infection controls VEGF and hypoxia-inducible factor-1 alpha (HIF-1α), accelerating the development of hepatocellular carcinoma and promoting hepatic angiogenesis. Many cell survival pathways, including PI3K/Akt, NF-κB, ERK1/2, and STAT3, which support the stability of HIF-1α, can activate VEGF. Together, these pathways support angiogenesis and aid in the development of HCC in the context of HCV infection. Indeed, research has shown that the HCV core protein can actually cause angiogenesis. Vascular endothelial growth factor (VEGF), CD34 expression, and transforming growth factor-beta 2 (TGF-β2) interact in this pathway (Dash et al., 2020).

5.23 HCV Diagnosis

Currently, the HIV infection status is ascertained using a variety of studies, such as biomarkers and molecular, fast, and serological testing. The World Health Organization (WHO) states that a rapid diagnostic test (RDT) or a lab setting can be used to diagnose HCV. A serological assay (antibodies or antibodies/antigen) can be used to identify anti-HCV antibodies. Moreover, the RNA from HCV can be identified using a molecular assay that uses automated real-time PCR and other nucleic acid identification techniques (Chevaliez & Pawlotsky, 2018). The main factors influencing the effectiveness and dependability of fast identification tests are their sensitivity, specificity, and accuracy together with their positive and negative predictive values, which might change based on the disease's prevalence in various populations. Recently, a number of kits for the detection of HCV-RNA have been developed, such as the GeneXpert Omni (Cepheid, Sunnyvale, CA) and small devices like Genedrive® (Genedrive Diagnostics, Manchester, UK). Wearable technology, lab-on-a-chip, NAT-based amplification, biosensors, and smartphone apps are some of the innovative methods for diagnosing HCV (Gupta et al., 2017; Suhail et al., 2022).

5.24 Biomarkers for HCV-Associated HCC Detection

Early identification and treatment of HCC is essential for long-term disease-free life in cirrhosis patients, as it is the primary cause of death in this illness. All scientific societies advise routine ultrasound (US) screening in at-risk patients given the identification of the at-risk group, provided that they would receive treatment if identified as having HCC. This restricts screening to those who do not have serious health conditions and have unchanged function of the liver (Child-Pugh A and B). At the point where identified nodules measure 10 mm, the diagnostic workup ought to continue (Bruix et al., 2014).

Eighty-five percent of cancers that were missed by US had minor HCC nodules (≤2 cm). Furthermore, central obesity impedes ultrasound's capacity to identify tiny tumors. On both CT and MRI images, HCC lesions show increased arterialization and decreased contrast agent presence during the portal phase of screening (washout) (Brahee et al., 2013). When it comes to CT and MRI, the detection rate for tumors greater than 2 cm is 90%; for tumors between 1 and 2 cm, it is 65%; and for tumors smaller than 1 cm, it varies from 10% to 34–71%. It is preferable to use CT and MRI modalities when first US results are unclear (Balogh et al., 2016; Wald et al., 2013).

Alpha-fetoprotein, or AFP, regulates fatty acids in both proliferating mature cells of the liver and embryonic liver cells. There are three different types of AFP: AFP-L1, AFP-L2, and AFP-L3. Chronic hepatitis and liver cirrhosis are connected with elevated AFP-L1 levels, while the growth of hepatocellular carcinoma (HCC) is caused by elevated AFP-L3 levels (Dash et al., 2020). Since 1968, AFP has been utilized as a serum marker to identify HCC. Numerous studies have assessed the sensitivity and specificity of AFP; the ranges for sensitivity are 21–64% and for specificity are 82–93% (Balogh et al., 2016). The lack of clinical usefulness for screening and diagnosis of α-fetoprotein (AFP) and other tumor indicators highlight the need for creative biomarkers to detect early hepatocellular carcinoma (HCC). It is anticipated that the incorporation of these biomarkers will increase the screening efforts' existing efficacy (Bruix et al., 2014).

5.25 Clinical Strategies for HCV-Associated HCC

Hepatocellular carcinoma (HCC) patients with advanced illness usually do not benefit medically from the limited therapies that are available. Small or slowly growing tumors can be effectively treated with surgery or transplantation of the liver, but these procedures are limited in their efficacy because of the scarcity of organ donors and the hepatocellular carcinoma's (HCC) quick and frequent recurrence in the donated liver. Hepatocellular carcinoma (HCC) can be treated with supportive and possibly curative methods, but systemic chemotherapy is not yet an effective treatment for HCC. In addition to the limited benefits of currently available medications, the stage of the disease, the resources available, and the practitioner's expertise all play a role in the decision about HCC therapy (Jeong et al., 2012; Wenger et al., 2011).

5.26 DAA Treatment

In patients with HCV-activated HCC, direct-acting antiviral (DAA) treatment has greatly increased SVR rates. In order to assess the advantages and disadvantages of DAAs in patients with chronic HCV, many studies have been conducted. However, information on DAA therapy's positive outcomes is still limited. Recently, a subtle method known as vibration-controlled transient elastography (VCTE) with a controlled attenuation parameter (CAP) has been used to detect hepatic fibrosis and steatosis 12 weeks after SVR. There was a discernible increase in hepatic steatosis and a drop in fibrosis score after the virus was eradicated by DAAs. Therefore, a long-term follow-up is required in HCV-infected persons to ascertain the degree of hepatic steatosis post-SVR and the probability of increased severity (Dash et al., 2020). A recent investigation was carried out on the efficacy of direct-acting antiviral drugs (DAAs) for patients with HCV infection in the Kingdom of Saudi Arabia (KSA). It was noted that the private sector prescribed DDA therapy, which had an overall 97% cure rate for HCV (Trifan et al., 2022).

5.27 Anticancer Agents for the Treatment of HCV-Induced HCC

Many anticancer drugs, such as sorafenib, have shown promise in helping individuals with HCC receive therapy. Sorafenib is a small molecule multikinase inhibitor having proapoptotic, antiproliferative, and antiangiogenic qualities. Individuals with compensated cirrhosis and advanced HCC may not benefit as much from sorafenib therapy, yet overall survival is increased in these individuals (Yang et al., 2012). Nonetheless, sorafenib's relative efficacy has led to its clinical use as a meaningful medication, either on it alone or when combined with other therapies, in spite of the often observed adverse effects (Huynh et al., 2008; Ibrahim et al., 2012). Furthermore, testing the therapeutic potential of additional small molecules like brivanib and erlotinib, as well as monoclonal antibodies like cetuximab and bevacizumab, in patients diagnosed with hepatocellular carcinoma was inspired by sorafenib's effectiveness as a significant HCC therapy approach (Sanoff et al., 2011). Based on the efficacious clinical development of sorafenib for the treatment of hepatocellular carcinoma, the era of molecularly targeted medicines is currently undergoing active clinical development.

The main signal transduction pathways that control the pathophysiology of HCV-associated HCC are thought to serve as a guide for the creation of clinically applicable treatments or preventative measures for the HCV-associated HCC. As of right now, clinical trials are either developing targeted therapeutics for the pathways outlined in relation to HCV-mediated HCC development, or these medications have previously shown promise in clinical settings. These include therapies that target the pathways for c-MET, Ras/Raf and MEK, Akt/mTOR, and insulin growth factor 1, endothelial growth factor receptor, and vascular endothelial growth factor receptors 1–3. Further research is necessary to determine the clinical significance and therapeutic potential of additional signaling pathways in the treatment of HCV-associated HCC, including TGF-β and Jak-STAT (Selimovic et al., 2012).

5.28 Development of Potential Vaccine for HCC

Despite the fact that DAA-based antiviral medication is delivering a very high percentage of viral cure, the number of new HCV infections is increasing globally. Most new infections go undiagnosed because most HCV infections, acute as well as chronic, are silent. Merely 50% of those infected in the USA know they have HCV. One of the biggest concerns with DAA therapy is the possibility of reinfection, particularly for intravenous drug users (Cox, 2015; Sarrazin et al., 2016). The present DAA-induced viral clearance does not prevent recurrence of liver disease-related consequences or offer protection against reinfection. The largest barrier to developing a highly effective vaccine is the complexity of the virus, which includes genetic diversity, seven genotypes, subtypes, viral variants, and antibody responses. Just two vaccination candidates have been developed thus far utilizing various methods and platforms. They are now in advanced clinical trial stages. The second vaccination employs viral envelope glycoproteins called E1/E2 together with MF59 adjuvant, while the first immunization is based on the use of recombinant replication-defective adenovirus containing entire nonstructural proteins (Suhail et al., 2022). These vaccines produce high rates of polyfunctional responses, eliciting wide responses from CD8 T cells, producing IFN-γ, developing memory cells, and maintaining the HCV-specific T cell response throughout time (Dash et al., 2020).

5.29 Conclusion

Chronic HCV infection can cause hepatocellular carcinoma and cirrhosis. The most recent research papers that have been released, and rapid technical developments—including omic studies—have yielded more comprehensive and in-depth details regarding the involvement of HCV in hepatocellular carcinoma. Viral hepatitis treatment has advanced significantly during the past few decades due to efficient medication. However, HCV infection is the main cause of liver diseases including HCC that have significant death rates. It is anticipated that the use of DAA-based antiviral therapy may lessen the mortality rate due to liver disease and the liver disease development brought on by HCV infection. However, the risk of HCC for individuals with cirrhosis persists even after they get therapy for HCV. It is expected that the number of cirrhosis patients who have been cured of HCV would rise. These individuals are nevertheless susceptible to liver disease progression due to nonviral risk factors such diabetes, obesity, alcoholism, and other autoimmune illnesses. After HCV treatment, cirrhotic individuals' ability to heal over the long term is particularly threatened by fatty liver. This chapter covered the molecular mechanisms behind the hepatic pathological response to viral assaults as well as the immunological pathways associated with the development of hepatocellular carcinoma (HCC) during chronic HCV infection. Immunization, which stops HCV transmission and reinfection, is a vital part of post-cure care and a way to accomplish the worldwide elimination of HCV infection. Therefore, the development of novel targeted therapies based on the inhibition of the signaling pathways, which are directly involved in the regulation of HCV-mediated initiation, progression, and invasion of HCC, may offer a clearer picture of the clinical utility and available treatments for patients with HCV-associated HCC.

References

Alisi, A., Giambartolomei, S., Cupelli, F., Merlo, P., Fontemaggi, G., Spaziani, A., & Balsano, C. (2003). Physical and functional interaction between HCV core protein and the different p73 isoforms. *Oncogene, 22*(17), 2573–2580.

Anzola, M. (2004). Hepatocellular carcinoma: role of hepatitis B and hepatitis C viruses proteins in hepatocarcinogenesis. *Journal of Viral Hepatitis, 11*(5), 383–393.

Balogh, J., Victor, D., III, Asham, E. H., Burroughs, S. G., Boktour, M., Saharia, A., & Monsour, H. P. (2016). Hepatocellular carcinoma: A review. *Journal of Hepatocellular Carcinoma, 3*, 41–53.

Bostan, N., & Mahmood, T. (2010). An overview about hepatitis C: A devastating virus. *Critical Reviews in Microbiology, 36*(2), 91–133.

Brahee, D. D., Ogedegbe, C., Hassler, C., Nyirenda, T., Hazelwood, V., Morchel, H., & Feldman, J. (2013). Body mass index and abdominal ultrasound image quality: A pilot survey of sonographers. *Journal of Diagnostic Medical Sonography, 29*(2), 66–72.

Bray, F., Ferlay, J., Soerjomataram, I., Siegel, R. L., Torre, L. A., & Jemal, A. (2018). Global cancer statistics 2018: GLOBOCAN estimates of incidence and mortality worldwide for 36 cancers in 185 countries. *CA: A Cancer Journal for Clinicians, 68*(6), 394–424.

Bruix, J., Gores, G. J., & Mazzaferro, V. (2014). Hepatocellular carcinoma: Clinical frontiers and perspectives. *Gut, 63*(5), 844–855.

Bruix, J., & Sherman, M. (2010). American Association for the Study of Liver Diseases (AASLD) practical guidelines. Management of hepatocellular carcinoma: An update. *Hepatology, 10*, 1–35.

Castello, G., Scala, S., Palmieri, G., Curley, S. A., & Izzo, F. (2010). HCV-related hepatocellular carcinoma: From chronic inflammation to cancer. *Clinical Immunology, 134*(3), 237–250.

Chen, G., Lin, W., Shen, F., Iloeje, U. H., London, W. T., & Evans, A. A. (2006). Past HBV viral load as predictor of mortality and morbidity from HCC and chronic liver disease in a prospective study. *Official Journal of the American College of Gastroenterology, 101*(8), 1797–1803.

Chen, Y., He, L., Peng, Y., Shi, X., Chen, J., Zhong, J., & Deng, H. (2015). The hepatitis C virus protein NS3 suppresses TNF-α–stimulated activation of NF-κB by targeting LUBAC. *Science Signaling, 8*(403), ra118.

Chevaliez, S., & Pawlotsky, J. M. (2018). New virological tools for screening, diagnosis and monitoring of hepatitis B and C in resource-limited settings. *Journal of Hepatology, 69*(4), 916–926.

Chiang, C. J., Yang, Y. W., You, S. L., Lai, M. S., & Chen, C. J. (2013). Thirty-year outcomes of the national hepatitis B immunization program in Taiwan. *JAMA, 310*(9), 974–976.

Cox, A. L. (2015). Global control of hepatitis C virus. *Science, 349*(6250), 790–791.

Craig, A. J., Von Felden, J., Garcia-Lezana, T., Sarcognato, S., & Villanueva, A. (2020). Tumour evolution in hepatocellular carcinoma. *Nature Reviews Gastroenterology & Hepatology, 17*(3), 139–152.

Creyghton, M. P., Cheng, A. W., Welstead, G. G., Kooistra, T., Carey, B. W., Steine, E. J., & Jaenisch, R. (2010). Histone H3K27ac separates active from poised enhancers and predicts developmental state. *Proceedings of the National Academy of Sciences, 107*(50), 21931–21936.

Dash, S., Aydin, Y., Widmer, K. E., & Nayak, L. (2020). Hepatocellular carcinoma mechanisms associated with chronic HCV infection and the impact of direct-acting antiviral treatment. *Journal of Hepatocellular Carcinoma, 7*, 45–76.

de Martel, C., Maucort-Boulch, D., Plummer, M., & Franceschi, S. (2015). World-wide relative contribution of hepatitis B and C viruses in hepatocellular carcinoma. *Hepatology, 62*(4), 1190–1200.

Donaires, F. S., Scatena, N. F., Alves-Paiva, R. M., Podlevsky, J. D., Logeswaran, D., Santana, B. A., & Martinelli, A. L. (2017). Telomere biology and telomerase mutations in cirrhotic patients with hepatocellular carcinoma. *PLoS One, 12*(8), e0183287.

Elpek, G. O. (2021). Molecular pathways in viral hepatitis-associated liver carcinogenesis: An update. *World Journal of Clinical Cases, 9*(19), 4890.

El-Serag, H. B. (2012). Epidemiology of viral hepatitis and hepatocellular carcinoma. *Gastroenterology, 142*(6), 1264–1273.

El-Serag, H. B., & Kanwal, F. (2014). Epidemiology of hepatocellular carcinoma in the United States: Where are we? Where do we go? *Hepatology, 60*(5), 1767–1775.

Farazi, P. A., & DePinho, R. A. (2006). Hepatocellular carcinoma pathogenesis: From genes to environment. *Nature Reviews Cancer, 6*(9), 674–687.

Ferlay, J., Shin, H. R., Bray, F., Forman, D., Mathers, C., & Parkin, D. M. (2010). Estimates of worldwide burden of cancer in 2008: GLOBOCAN 2008. *International Journal of Cancer, 127*(12), 2893–2917.

Florese, R. H., Nagano-Fujii, M., Iwanaga, Y., Hidajat, R., & Hotta, H. (2002). Inhibition of protein synthesis by the nonstructural proteins NS4A and NS4B of hepatitis C virus. *Virus Research, 90*(1-2), 119–131.

Goto, K., Roca Suarez, A. A., Wrensch, F., Baumert, T. F., & Lupberger, J. (2020). Hepatitis C virus and hepatocellular carcinoma: when the host loses its grip. *International Journal of Molecular Sciences, 21*(9), 3057.

Gower, E., Estes, C., Blach, S., Razavi-Shearer, K., & Razavi, H. (2014). Global epidemiology and genotype distribution of the hepatitis C virus infection. *Journal of Hepatology, 61*(1), S45–S57.

Gupta, E., Agarwala, P., Kumar, G., Maiwall, R., & Sarin, S. K. (2017). Point-of-care testing (POCT) in molecular diagnostics: Performance evaluation of GeneXpert HCV RNA test in diagnosing and monitoring of HCV infection. *Journal of Clinical Virology, 88*, 46–51.

Hamdane, N., Jühling, F., Crouchet, E., El Saghire, H., Thumann, C., Oudot, M. A., & Baumert, T. F. (2019). HCV-induced epigenetic changes associated with liver cancer risk persist after sustained virologic response. *Gastroenterology, 156*(8), 2313–2329.

Higgs, M. R., Lerat, H., & Pawlotsky, J. M. (2013). Hepatitis C virus-induced activation of β-catenin promotes c-Myc expression and a cascade of pro-carcinogenetic events. *Oncogene, 32*(39), 4683–4693.

Hosaka, T., Suzuki, F., Kobayashi, M., Seko, Y., Kawamura, Y., Sezaki, H., & Kumada, H. (2013). Long-term entecavir treatment reduces hepatocellular carcinoma incidence in patients with hepatitis B virus infection. *Hepatology, 58*(1), 98–107.

Hoshida, Y., Fuchs, B. C., Bardeesy, N., Baumert, T. F., & Chung, R. T. (2014). Pathogenesis and prevention of hepatitis C virus-induced hepatocellular carcinoma. *Journal of Hepatology, 61*(1), S79–S90.

Hu, B., Xie, S., Hu, Y., Chen, W., Chen, X., Zheng, Y., & Wu, X. (2017). Hepatitis C virus NS4B protein induces epithelial-mesenchymal transition by upregulation of Snail. *Virology Journal, 14*, 1–9.

Huynh, H., Ngo, V. C., Fargnoli, J., Ayers, M., Soo, K. C., Koong, H. N., & Tran, E. (2008). Brivanib alaninate, a dual inhibitor of vascular endothelial growth factor receptor and fibroblast growth factor receptor tyrosine kinases, induces growth inhibition in mouse models of human hepatocellular carcinoma. *Clinical Cancer Research, 14*(19), 6146–6153.

Ibrahim, N., Yu, Y., Walsh, W. R., & Yang, J. L. (2012). Molecular targeted therapies for cancer: Sorafenib monotherapy and its combination with other therapies. *Oncology Reports, 27*(5), 1303–1311.

Im, Y. K., La Selva, R., Gandin, V., Ha, J. R., Sabourin, V., Sonenberg, N., & Ursini-Siegel, J. (2015). The ShcA adaptor activates AKT signaling to potentiate breast tumor angiogenesis by stimulating VEGF mRNA translation in a 4E-BP-dependent manner. *Oncogene, 34*(13), 1729–1735.

Jeong, S. W., Jang, J. Y., Lee, J. E., Lee, S. H., Kim, S. G., Cha, S. W., & Kim, Y. J. (2012). The efficacy of hepatic arterial infusion chemotherapy as an alternative to sorafenib in advanced hepatocellular carcinoma. *Asia-Pacific Journal of Clinical Oncology, 8*(2), 164–171.

Jühling, F., Hamdane, N., Crouchet, E., Li, S., El Saghire, H., Mukherji, A., & Baumert, T. F. (2021). Targeting clinical epigenetic reprogramming for chemoprevention of metabolic and viral hepatocellular carcinoma. *Gut, 70*(1), 157–169.

Kanda, T., Goto, T., Hirotsu, Y., Moriyama, M., & Omata, M. (2019). Molecular mechanisms driving progression of liver cirrhosis towards hepatocellular carcinoma in chronic hepatitis B and C infections: A review. *International Journal of Molecular Sciences, 20*(6), 1358.

Kanda, T., Steele, R., Ray, R., & Ray, R. B. (2008). Hepatitis C virus core protein augments androgen receptor-mediated signaling. *Journal of Virology, 82*(22), 11066–11072.

Kanwal, F., Kramer, J., Asch, S. M., Chayanupatkul, M., Cao, Y., & El-Serag, H. B. (2017). Risk of hepatocellular cancer in HCV patients treated with direct-acting antiviral agents. *Gastroenterology, 153*(4), 996–1005.

Khabar, K. S., & Polyak, S. J. (2002). Hepatitis C virus-host interactions: The NS5A protein and the interferon/chemokine systems. *Journal of Interferon & Cytokine Research, 22*(10), 1005–1012.

Khalaf, A. M., Fuentes, D., Morshid, A. I., Burke, M. R., Kaseb, A. O., Hassan, M., & Elsayes, K. M. (2018). Role of Wnt/β-catenin signaling in hepatocellular carcinoma, pathogenesis, and clinical significance. *Journal of Hepatocellular Carcinoma, 5*, 61–73.

Kwun, H. J., & Jang, K. L. (2003). Dual effects of hepatitis C virus Core protein on the transcription of cyclin-dependent kinase inhibitor p21 gene. *Journal of Viral Hepatitis, 10*(4), 249–255.

Kwun, H. J., Jung, E. Y., Ahn, J. Y., Lee, M. N., & Jang, K. L. (2001). p53-dependent transcriptional repression of p21waf1 by hepatitis C virus NS3. *Journal of General Virology, 82*(9), 2235–2241.

Levin, A., Neufeldt, C. J., Pang, D., Wilson, K., Loewen-Dobler, D., Joyce, M. A., & Tyrrell, D. L. J. (2014). Functional characterization of nuclear localization and export signals in hepatitis C virus proteins and their role in the membranous web. *PLoS One, 9*(12), e114629.

Levrero, M. (2006). Viral hepatitis and liver cancer: The case of hepatitis C. *Oncogene, 25*(27), 3834–3847.

Li, Y., Chen, J., Wu, C., Wang, L., Lu, M., & Chen, X. (2010). Hepatitis B virus/hepatitis C virus upregulate angiopoietin-2 expression through mitogen-activated protein kinase pathway. *Hepatology Research, 40*(10), 1022–1033.

Liu, J., Ding, X., Tang, J., Cao, Y., Hu, P., Zhou, F., & Tang, N. (2011). Enhancement of canonical Wnt/β-catenin signaling activity by HCV core protein promotes cell growth of hepatocellular carcinoma cells. *PLoS One, 6*(11), e27496.

Lorusso, G., & Rüegg, C. (2008). The tumor microenvironment and its contribution to tumor evolution toward metastasis. *Histochemistry and Cell Biology, 130*, 1091–1103.

Mahmoudvand, S., Shokri, S., Taherkhani, R., & Farshadpour, F. (2019). Hepatitis C virus core protein modulates several signaling pathways involved in hepatocellular carcinoma. *World Journal of Gastroenterology, 25*(1), 42.

Massarweh, N. N., & El-Serag, H. B. (2017). Epidemiology of hepatocellular carcinoma and intrahepatic cholangiocarcinoma. *Cancer Control, 24*(3), 1073274817729245.

McGlynn, K. A., Petrick, J. L., & London, W. T. (2015). Global epidemiology of hepatocellular carcinoma: An emphasis on demographic and regional variability. *Clinics in Liver Disease, 19*(2), 223–238.

Mitchell, J. K., Midkiff, B. R., Israelow, B., Evans, M. J., Lanford, R. E., Walker, C. M., & McGivern, D. R. (2017). Hepatitis C virus indirectly disrupts DNA damage-induced p53 responses by activating protein kinase R. *MBio, 8*(2), 10–1128.

Munakata, T., Liang, Y., Kim, S., McGivern, D. R., Huibregtse, J., Nomoto, A., & Lemon, S. M. (2007). Hepatitis C virus induces E6AP-dependent degradation of the retinoblastoma protein. *PLoS Pathogens, 3*(9), e139.

Munakata, T., Nakamura, M., Liang, Y., Li, K., & Lemon, S. M. (2005). Down-regulation of the retinoblastoma tumor suppressor by the hepatitis C virus NS5B RNA-dependent RNA polymerase. *Proceedings of the National Academy of Sciences, 102*(50), 18159–18164.

Nault, J. C., Ningarhari, M., Rebouissou, S., & Zucman-Rossi, J. (2019). The role of telomeres and telomerase in cirrhosis and liver cancer. *Nature Reviews Gastroenterology & Hepatology, 16*(9), 544–558.

Nelson, P. K., Mathers, B. M., Cowie, B., Hagan, H., Des Jarlais, D., Horyniak, D., & Degenhardt, L. (2011). Global epidemiology of hepatitis B and hepatitis C in people who inject drugs: results of systematic reviews. *The Lancet, 378*(9791), 571–583.

Nguyen, H., Sankaran, S., & Dandekar, S. (2006). Hepatitis C virus core protein induces expression of genes regulating immune evasion and anti-apoptosis in hepatocytes. *Virology, 354*(1), 58–68.

Ott, J. J., Stevens, G. A., Groeger, J., & Wiersma, S. T. (2012). Global epidemiology of hepatitis B virus infection: New estimates of age-specific HBsAg seroprevalence and endemicity. *Vaccine, 30*(12), 2212–2219.

Penin, F., Dubuisson, J., Rey, F. A., Moradpour, D., & Pawlotsky, J. M. (2004). Structural biology of hepatitis C virus. *Hepatology, 39*(1), 5–19.

Perez, S., Kaspi, A., Domovitz, T., Davidovich, A., Lavi-Itzkovitz, A., Meirson, T., & Gal-Tanamy, M. (2019). Hepatitis C virus leaves an epigenetic signature post cure of infection by direct-acting antivirals. *PLoS Genetics, 15*(6), e1008181.

Petrick, J. L., Kelly, S. P., Altekruse, S. F., McGlynn, K. A., & Rosenberg, P. S. (2016). Future of hepatocellular carcinoma incidence in the United States forecast through 2030. *Journal of Clinical Oncology, 34*(15), 1787.

Pezzuto, F., Buonaguro, L., Buonaguro, F. M., & Tornesello, M. L. (2017). Frequency and geographic distribution of TERT promoter mutations in primary hepatocellular carcinoma. *Infectious Agents and Cancer, 12*(1), 1–9.

Plissonnier, M. L., Herzog, K., Levrero, M., & Zeisel, M. B. (2018). Non-coding RNAs and hepatitis C virus-induced hepatocellular carcinoma. *Viruses, 10*(11), 591.

Reyes, G. R. (2002). The nonstructural NS5A protein of hepatitis C virus: an expanding, multifunctional role in enhancing hepatitis C virus pathogenesis. *Journal of Biomedical Science, 9*, 187–197.

Sancho, S. C., Márquez-Coello, M., Illanes-Álvarez, F., Márquez-Ruiz, D., Arizcorreta, A., Galán-Sánchez, F., & Girón-González, J. A. (2022). Hepatitis C: Problems to extinction and residual hepatic and extrahepatic lesions after sustained virological response. *World Journal of Hepatology, 14*(1), 62.

Sanoff, H. K., Bernard, S., Goldberg, R. M., Morse, M. A., Garcia, R., Woods, L., & O'Neil, B. H. (2011). Phase II study of capecitabine, oxaliplatin, and cetuximab for advanced hepatocellular carcinoma. *Gastrointestinal Cancer Research, 4*(3), 78.

Sarrazin, C., Isakov, V., Svarovskaia, E., Hedskog, C., Martin, R., Chodavarapu, K., & Sulkowski, M. S. (2016). Late relapse versus HCV reinfection in patients with sustained virologic response after sofosbuvir-based therapies. *Clinical Infectious Diseases, 64*(1), 44–52.

Selimovic, D., El-Khattouti, A., Ghozlan, H., Haikel, Y., Abdelkader, O., & Hassan, M. (2012). Hepatitis C virus-related hepatocellular carcinoma: An insight into molecular mechanisms and therapeutic strategies. *World Journal of Hepatology, 4*(12), 342.

Shokri, S., Mahmoudvand, S., Taherkhani, R., Farshadpour, F., & Jalalian, F. A. (2019). Complexity on modulation of NF-κB pathways by hepatitis B and C: A double-edged sword in hepatocarcinogenesis. *Journal of Cellular Physiology, 234*(9), 14734–14742.

Sia, D., Villanueva, A., Friedman, S. L., & Llovet, J. M. (2017). Liver cancer cell of origin, molecular class, and effects on patient prognosis. *Gastroenterology, 152*(4), 745–761.

Siegel, R. L., Miller, K. D., & Jemal, A. (2018). Cancer statistics, 2018. *CA: A Cancer Journal for Clinicians, 68*(1), 7–30.

Suhail, M., Sohrab, S. S., Kamal, M. A., & Azhar, E. I. (2022). Role of hepatitis c virus in hepatocellular carcinoma and neurological disorders: An overview. *Frontiers in Oncology, 12*, 913231.

Toh, T. B., Lim, J. J., & Chow, E. (2019). Epigenetics of hepatocellular carcinoma. *Clinical and Translational Medicine, 8*(1), 13.

Tran, G. (2008). The role of hepatitis C virus in the pathogenesis of hepatocellular carcinoma. *Bioscience Horizons, 1*(2), 167–175.

Trifan, A., Stratina, E., Rotaru, A., Stafie, R., Zenovia, S., Nastasa, R., & Stanciu, C. (2022). Changes in liver steatosis using controlled attenuation parameter among patients with chronic hepatitis c infection treated with direct-acting antivirals therapy who achieved sustained virological response. *Diagnostics, 12*(3), 702.

Unfried, J. P., & Fortes, P. (2020). LncRNAs in HCV infection and HCV-related liver disease. *International Journal of Molecular Sciences, 21*(6), 2255.

Villanueva, A. (2019). Hepatocellular carcinoma. *The New England Journal of Medicine, 380*, 1450–1462.

Villanueva, A., Chiang, D. Y., Newell, P., Peix, J., Thung, S., Alsinet, C., & Llovet, J. M. (2008). Pivotal role of mTOR signaling in hepatocellular carcinoma. *Gastroenterology, 135*(6), 1972–1983.

Wald, C., Russo, M. W., Heimbach, J. K., Hussain, H. K., Pomfret, E. A., & Bruix, J. (2013). New OPTN/UNOS policy for liver transplant allocation: standardization of liver imaging, diagnosis, classification, and reporting of hepatocellular carcinoma. *Radiology, 266*(2), 376–382.

Wang, W., Pan, Q., Fuhler, G. M., Smits, R., & Peppelenbosch, M. P. (2017). Action and function of Wnt/β-catenin signaling in the progression from chronic hepatitis C to hepatocellular carcinoma. *Journal of Gastroenterology, 52*, 419–431.

Waris, G., & Siddiqui, A. (2003). Regulatory mechanisms of viral hepatitis B and C. *Journal of Biosciences, 28*, 311–321.

Waziry, R., Hajarizadeh, B., Grebely, J., Amin, J., Law, M., Danta, M., & Dore, G. J. (2017). Hepatocellular carcinoma risk following direct-acting antiviral HCV therapy: A systematic review, meta-analyses, and meta-regression. *Journal of Hepatology, 67*(6), 1204–1212.

Wenger, J. B., Santos, N., Liu, Y., Dallas, J., Subbiah, S., Hochwald, S., & Dang, L. H. (2011). Can we develop effective combination antiangiogenic therapy for patients with hepatocellular carcinoma? *Oncology Reviews, 5*, 177–184.

Wild, C. P., & Montesano, R. (2009). A model of interaction: aflatoxins and hepatitis viruses in liver cancer aetiology and prevention. *Cancer Letters, 286*(1), 22–28.

Wouters, B. G., & Koritzinsky, M. (2008). Hypoxia signalling through mTOR and the unfolded protein response in cancer. *Nature Reviews Cancer, 8*(11), 851–864.

Xue, Y., Mars, W. M., Bowen, W., Singhi, A. D., Stoops, J., & Michalopoulos, G. K. (2018). Hepatitis C virus mimics effects of glypican-3 on CD81 and promotes development of hepatocellular carcinomas via activation of Hippo pathway in hepatocytes. *The American Journal of Pathology, 188*(6), 1469–1477.

Yamanaka, T., Uchida, M., & Doi, T. (2002). Innate form of HCV core protein plays an important role in the localization and the function of HCV core protein. *Biochemical and Biophysical Research Communications, 294*(3), 521–527.

Yang, J. D., & Roberts, L. R. (2010). Hepatocellular carcinoma: A global view. *Nature Reviews Gastroenterology & Hepatology, 7*(8), 448–458.

Yang, Y., Lu, Y., Wang, C., Bai, W., Qu, J., Chen, Y., & Lv, J. (2012). Cryotherapy is associated with improved clinical outcomes of Sorafenib therapy for advanced hepatocellular carcinoma. *Cell Biochemistry and Biophysics, 63*(2), 159–169.

Zhang, H., Zhu, C., Zhao, Y., Li, M., Wu, L., Yang, X., & Zhao, H. (2015). Long non-coding RNA expression profiles of hepatitis C virus-related dysplasia and hepatocellular carcinoma. *Oncotarget, 6*(41), 43770.

Zhao, P., Malik, S., & Xing, S. (2021). Epigenetic mechanisms involved in HCV-induced hepatocellular carcinoma (HCC). *Frontiers in Oncology, 11*, 677926.

Zheng, R., Qu, C., Zhang, S., Zeng, H., Sun, K., Gu, X., & He, J. (2018). Liver cancer incidence and mortality in China: Temporal trends and projections to 2030. *Chinese Journal of Cancer Research, 30*(6), 571.

Zheng, Y., Hlady, R. A., Joyce, B. T., Robertson, K. D., He, C., Nannini, D. R., & Hou, L. (2019). DNA methylation of individual repetitive elements in hepatitis C virus infection-induced hepatocellular carcinoma. *Clinical Epigenetics, 11*(1), 145.

Zhou, L., Huang, Y., Li, J., & Wang, Z. (2010). The mTOR pathway is associated with the poor prognosis of human hepatocellular carcinoma. *Medical Oncology, 27*, 255–261.

Zhou, P., Xia, J., Zhou, Y. J., Wan, J., Li, L., Bao, J., & Bu, H. (2015). Proportions of acetyl-histone-positive hepatocytes indicate the functional status and prognosis of cirrhotic patients. *World Journal of Gastroenterology, 21*(21), 6665.

Cancers Associated with Epstein-Barr Virus

6

Maryam Pervaiz and Parvez Azam Malik

Abstract

Epstein-Barr virus (EBV) is a diverse cancer-causing virus, influencing around 90% of people worldwide. It is also connected to cancers related to both hematopoietic and epithelial cells. The spread of the virus is primarily by mouth secretions; however, spread by organ replacement and blood transfusion has also been stated. EBV develops latent infection in B-cells—there, it shows expression of specific group of proteins known as EBV transcription programs, having six nuclear antigens, three latent membrane proteins, and EBV encoded RNA. These viral proteins have demonstrated efficiency to alter B-cells to lymphoblastic cells. These proteins show varied expression that regulates the incidence of different forms of viral latency in the pathogenesis of a specific type of cancer.

Keywords

Epstein-Barr virus · Epithelial cancers · Burkitt's lymphoma · Hodgkin lymphoma · EBV latency proteins

6.1 Introduction

Epstein-Barr virus or human gammaherpes virus 4 (HHV-4) was the primary herpesvirus shown to cause carcinoma, and the virus is linked with extensive range of human carcinoma related to mesenchymal cells, epithelial cells, and lymphocytes.

M. Pervaiz (✉)
Microbiology and Biotechnology Research Lab, Fatima Jinnah Women University, Rawalpindi, Pakistan

P. A. Malik
Department of Pathology, Rai Medical College, Sargodha, Pakistan

N. Masood, A. Yasmin (eds.), *The Microbiome and Cancer*,
https://doi.org/10.1007/978-981-97-6948-3_6

The growth of EBV-associated neoplasm largely depends on host's genetic susceptibility as well as environmental factors. Viral infection is further connected to genetically susceptible immune deregulation (Kushekhar et al., 2014).

The Epstein-Barr virus (EBV) is a gammaherpesvirus which is reported to develop latent asymptomatic infection in approximately 90% of the adult people worldwide. Most of the EBV infected carriers do not exhibit any medical indication as a result of the virus-related infection— EBV is etiologically connected with up to 2% of all carcinoma, together with multiple types of T, B, and natural killer cell lymphomas, along with cancer of the stomach, nasopharynx, thymus, and parotid (Delecluse et al., 2007).

6.2 Viral Replication

This virus causes the methylation of both the viral and host genome, and these changes facilitate cellular functions that promote and prolong viral persistence and propagation (Tempera & Lieberman, 2014). There are two chief physiological targets that are infected, B lymphocytes and epithelial cells— the virus is transmitted via oral route and exhibits double tropism to start primary infection. This virus also infects other targets like natural killer (NK) cells and T lymphocytes (Saha & Robertson, 2011).

Virus lytic replication happens in epithelial cells; nonetheless, viral cells can create latency after infecting B-cells present in Waldeyer ring's pharyngeal lymphoid tissues. Viral gene either assimilated in genome of the host and then persists as a provirus after entering the B-cells or continues in the nucleus as a nonintegrated circular episome and shows limited group of genes that regulate latency and existence of host cells (Kang & Kieff, 2015). The latency in B-cells progress to B-cell-derived cancers after the conversion of the cells into lymphoblastic lines. The mechanism by which virus can be revived from B-cell latency is still not known. Even though EBV experiences lytic replication in the epithelial cells, wherever lytic genes are elevated, the virus can express latency stage genes and proceed to the alteration of the epithelial cells into permanently multiplying cells, then causing epithelial cell consequential carcinomas (Tsao et al., 2015).

6.3 Genome Structure of Epstein-Barr Virus

The EBV gene is around 172 Kbp having greater than 85 genes that translate to proteins, linear and double stranded DNA. The open reading frames code proteins that regulate DNA replication and gene regulation and maintain gene integrity (Santpere et al., 2014). The open reading frames were regulated according to *Bam* HI restriction fraction map where regarding their size, genes were present in decreasing order. The protein encoding genes have structural and nonstructural role and categorized into lytic and latent genes. The terminal direct repeats upsurge the coding capability of the gene and act as a biomarker to define the ancestral source of Epstein-Barr virus in the viral infected cells (Smatti et al., 2018).

6.4 Genotypes of EBV and Their Worldwide Distribution

EBV is generally categorized into two central genotypes, A and B, or types 1 and 2, grounded on protein sequence deviation in the nuclear proteins, Epstein-Barr nuclear antigen (EBNA)-1, EBNA-2, EBNA-3A, EBNA-3B, and EBNA-3C (Banko et al., 2016). However, EBNA-2 is usually useful for categorization since the protein has the slightest proportion of sequence homology among two genotypes (Münz, 2015). The distribution of two EBV genotypes varied in distribution geographically; EBV type 1 is established worldwide but reported mainly in American, European, Chinese, and Southeast Asian inhabitants, while type 2 is mainly present in Africa (Peh et al., 2002). EBV type 1 is effective in immortalizing B-cells as compared to type 2, and the type 2 possesses greater lytic capability (Klemenc et al., 2006). EBV is further categorized into seven diverse strains— China 1 (C1), China 2 (C2), China 3 (C3), Mediterranean with (Med+) or without (Med-) deletions, Alaskan (AL), and North Carolina— grounded on nucleotide polymorphism, 30 bp deletions, or amino acid alterations in the LMP1 gene in comparison with prototype EBV sequence, B95-8 (Banko et al., 2016). The different strains have diverse tumorigenic activity— if 30bp deletion is reported at the carboxyl terminal, then probably it would have a higher tumorigenic activity and low immunogenicity in comparison with undeleted variant (da Costa et al., 2015).

6.5 EBV-Encoded Latent Gene Proteins

Latent infection by EBV has shown direct relation in development of several malignancies. Expression of limited set of proteins is turned on in establishment of latent infection. Six EBNAs (1, 2, 3A, 3B, 3C, and LP) are turned on in latent host infection (Klein & Ernberg, 2007). Three stages of latency have been detected in virus infected cells. During latency stage 0, EBNA protein is not expressed. During latency stage I and II, EBNA-1 is expressed only; however, in latency stage III, all of the six EBNAs are turned on. These EBNAs take part in transcription regulation and activation of viral genes of LMP-1 and LMP-2 in infected cells (Rowe, 1999).

EBNA-1 is a protein that binds to specific sequences on the genome of virus (Dresang et al., 2009). EBNA-1 guarantees the division of viral genome in duplicating cells and enhances LMP promoter, which eventually leads to cell survival or infected cell immortalization (Klein & Ernberg, 2007).

EBNA-2 attaches to transcription factors and other related proteins to enhance the expression of cellular and viral genes (Kaiser et al., 1999). Genes that are upregulated are involved in cell division, and upregulated viral genes are associated with B-cell immortalization (Klein & Ernberg, 2007).

EBNA-3 consists of a group of three large nuclear phosphoproteins (3A, 3B, 3C). EBNA-3 interacts with different transcription factors and are present in the immortalization of infected cells (Saha & Robertson, 2013).

EBNA-LP is also identified as EBNA-5 which is a phosphorus containing nuclear protein that co-expresses with EBNA-2 when target cell infection takes place (Alfieri et al., 1991).

The co-expression of these proteins (EBNA-2 and EBNA-LP) has been demonstrated to upregulate transcriptional activation of viral and cellular proteins. The EBNA-LP corelates with cellular proteins— some of them are tumor activators and tumor suppressors, cell cycle regulators, and antiapoptotic protein (Kashuba et al., 2005).

Three latent membrane proteins (LMP-1, LMP-2A, LMP-2B) are activated during latency stage II and III in virus infected cells. These proteins show expression as cell surface markers and play an important role in infected cell longevity and transformation into continuously dividing cells (Webster-Cyriaque & Raab-Traub, 1998). LMP-1 gene is a significant oncogene that is constitutively expressed in viral infected tumor cells (Siegler et al., 2003). LMP-2 consists of two variants, LMP-2A and LMP-2B, and is expressed through latency stage II and III in viral infected cells. LMP-2A is necessary for activation, survival, and proliferation of virus infected B-cells which further facilitate long term proliferation of B-cells. LMP-2A has also been involved in activation, growth, and survival of virus infected B-cells at time of initial viral infection— after that, it is essential for B-cells' long term proliferation (Wasil et al., 2013).

The purpose of LMP-2B, however, is not clearly demonstrated, but an earlier work has shown that LMP-2B associates with LMP-2A to regulate the functions of the latter.

6.6 MicroRNAs and EBV

MicroRNAs are noncoding short RNA molecules, generally 18–24 nucleotides long, that transcriptionally downregulate complementary mRNA expression (Dong et al., 2019). Micro RNAs have also been recognized in diverse species of mammals, algae, plants, and worms. EBV was one of the early viruses that were known to possess, express, and regulate these noncoding small RNA molecules (Alptekin et al., 2017).

This virus has shown to possess 44 miRNAs in all types of latency and cancer tissues, modifying viral and cellular genes (Shinozaki-Ushiku, Kunita, & Fukayama, 2015; Shinozaki-Ushiku, Kunita, Isogai, et al., 2015). Micro RNAs facilitate viral escape from immune system by downregulation of immunogenic virus antigens as well as immune proteins of host (Dong et al., 2019). Apart from aiming cellular cytokines, EBV microRNAs also prevent the annihilation of epithelial cell and B-cell-derived tumors through downregulation of tumor suppressor genes; miR-BHRF1-2 has been described to downregulate the expression of PRDM1/Blimp1 that is a tumor suppressor in B-cells, therefore endorsing lymphogenesis (Ma et al., 2016).

Table 6.1 depicts the role of such EBV latency proteins. EBNA-1 and EBNA-2 initiate the expression of LMP-1. EBNA-1 is expressed and present in all EBV-related tumors. EBNA-LP directly activates cyclin D/E and DNA dependent protein kinase to help cell propagation. EBNA-LP also enhances cell longevity by interacting antiapoptotic protein hematopoietic cell-specific protein 1-associated protein X-1. EBNA-2 and EBNA-3 interact with cellular components that eventually leads to unregulated propagation of cells.

6.7 Molecular Diagnosis of Epstein-Barr Virus

Molecular, biochemical, and serological techniques to monitor EBV have gradually become essential in the analysis, evaluation, and monitoring of patients with EBV-related malignancies. Serological techniques that are employed to detect the virus in infected patients include recognition of antibodies against viral specific antigens like viral coat antigen (VCA), EBNA-1, viral nucleic acids, and early antigen (EA) (Chen et al., 2015).

Table 6.1 Biological functions of EBV latency proteins and related carcinomas (Ayee et al., 2020)

EBV latency protein	Category of latency	Biological function and activity	Related carcinomas
EBNA-1	Latency stage I, II, III	Segregation of viral genome in progenies; DNA replication; downregulation of MHC class I; increases p53 breakdown	Burkitt's lymphoma, gastric cancer, breast cancer
EBNA-2	Latency stage III	Upregulation of host and viral proteins; enables immortalization of B-cells	Posttransplant lymphoproliferative disorder
EBNA-3	Latency stage III	Transcription transactivation of host and viral proteins; B-cell immortalization	Posttransplant lymphoproliferative disorder
EBNA-LP	Latency stage III	Transactivation of EBNA-2 to deactivate tumor suppressor genes; vital for B-cell immortalization	Posttransplant lymphoproliferative syndrome
LMP-1/2	Latency stage II/III	Survival of B-cells; activation of antiapoptotic proteins; imitates signaling of CD 40 ligand; continuous cell proliferation; and activation of signaling pathways for cell survival	Hodgkin lymphoma, nasopharyngeal cancer, posttransplant lymphoproliferative disorder, T/NK cell lymphoma, breast cancer
EBV-micro RNAs	Latency stage I, II, III	Host mRNAs involved in apoptosis are targeted; cell proliferation and transformation; overturn antigen presentation and immune cell activation	Gastric cancer, T/NK cell lymphoma, nasopharyngeal cancer

Molecular diagnosis of EBV is attained by various methods as described in Table 6.2—current development has been focusing on the molecular diagnosis and evaluation of EBV.

6.8 Lymphomas Caused by EBV Infection

EBV is related to lymphomas like Hodgkin lymphoma (HL), Burkitt's lymphoma (BL), and posttransplant lymphoproliferative disorder (PTLD). Epithelial carcinomas linked with EBV comprise breast cancer and gastric and nasopharyngeal cancer (Thompson & Kurzrock, 2004). In occasional cases, virus has been exposed to infect natural killer and T cells, triggering nasal-type NK/T-cell lymphoma (Fox, 2011). Different carcinomas caused by EBV are depicted in Fig. 6.1.

Inside host, virus shows tropism for two basic aims, specifically B-cell and epithelial cells along with unusual targets such as natural killer and T cells. EBV in numerous target host cells demonstrates diverse pattern of latent gene regulation, and this governs the nature of cancer that could be established. B-cell-derived lymphomas comprise of Burkitt's lymphoma, Hodgkin lymphoma, and posttransplant lymphoproliferative malignancies that are triggered by the expression of latent I, II, and III, proteins, respectively. Epithelial cell infection by virus may cause cancers of nasopharyngeal, breast, and gastric tissues, instigated by positive expression and regulation of latent I and II proteins.

Table 6.2 Laboratory tests for EBV (Gulley, 2001)

Technique	Diagnostic purpose
In situ hybridization	Identify EBER (Epstein-Barr encoded RNA) transcripts or EBV DNA in infected cells within histologic lesions
EBV clonal assay using Southern blot analysis	Evaluate clonality of tissues with respect to EBV DNA structure; differentiate latent viral infection from replicative infection depending upon the linear vs. episomal structure of the viral genome
EBV DNA amplification	Distinguish viral DNA in patient samples; lacks disease specificity
Viral load detection	Quantify EBV DNA in blood samples or body fluids to evaluate disease progression over time
Immunohistochemistry (LMP-1, EBNA-1, EBNA-2, LMP-2A)	Recognize EBV protein expression in different cell types within histologic samples; differentiate expression profile of latent infection from replicative infection
Culture of EBV or virus infected B lymphocytes	Observe and semiquantitatively calculate infectious virions or latently infected B lymphocytes
Electron microscopy	Detect whole virions depicting replicative EBV infection
Serology	Identify response of antibodies to EBV proteins in serum; differentiation of acute infection from remote infection; evaluate disease progression with time

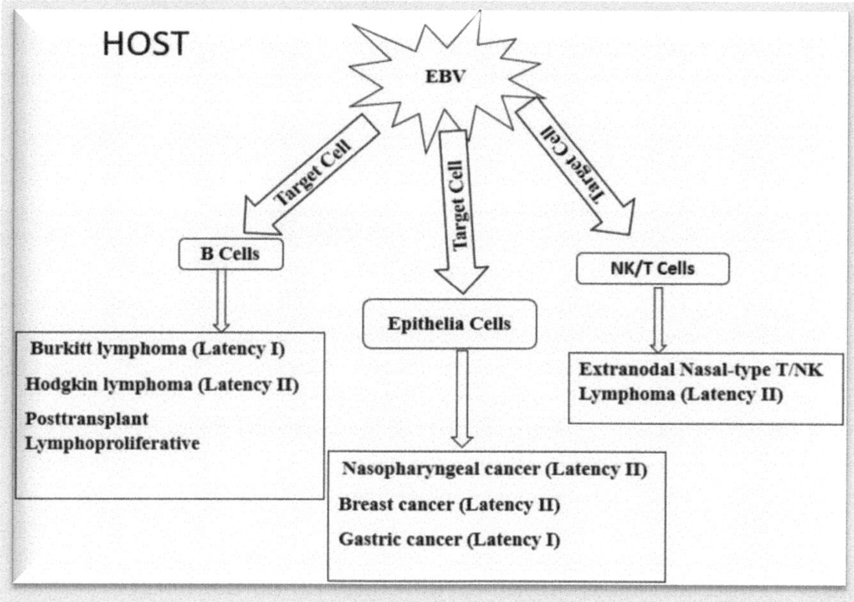

Fig. 6.1 Detailed overview of Epstein-Barr virus-related epithelial and hematopoietic cell-based carcinomas with relevant latency (Ayee et al., 2020)

6.8.1 Burkitt's Lymphoma (BL)

Burkitt's lymphoma (BL) is a pediatric cancer prevalent in sub-Saharan Africa, and its pathogenicity is associated with EBV infection (Rochford & Moormann, 2015). This lymphoma is common in children who live in area where malaria is widespread, and it is speculated that *Plasmodium falciparum* has a part in cancer etiology (Thorley-Lawson et al., 2016). There are three various forms of BL based on clinical interpretations. Endemic BL (eBL) is common in Equatorial Africa and Papua New Guinea— EBV is present in around 95% of all identified patients—common targeted sites are jaws plus the abdomen (Aguilar et al., 2017; Lenze et al., 2011). Sporadic Burkitt's lymphoma (sBL) has an extensive global distribution in less frequency and is usually identified in both children and young adult (Blum et al., 2004). Immunodeficiency-associated Burkitt's lymphoma (iBL) has been established in HIV-effected carriers where lymphoma develops first before progression to AIDS (Gloghini et al., 2013). Compromised immunity that occurs due to HIV infection causes recurrence of virus from latently virus infected B-cells, thus ultimately producing swift development to iBL (Amria et al., 2008).

The viral gene expression and regulation in BL is absolutely latent and limited to type 1 latent default program. Only EBER and EBNA-1 have shown expression in infected cells in BL lymphoma (Tsao et al., 2017). Few reports described the finding of LMP-1 and EBNA-2 proteins in eBL and LMP-1 in several sBL tumors; however, these findings were not reported in all incidences (Kelly et al., 2006). Although the mechanism of EBV activation and regulation in progression of BL is uncertain, nonetheless research has revealed that the upregulation of viral EBNA-1 in BL cell lines encourages cell growth and proliferation via hindering apoptosis (Kirchmaier & Sugden, 1997).

6.8.2 Hodgkin Lymphoma (HL)

Hodgkin lymphoma is a lymphoid neoplasm that initiates from B-cells and categorized by the incidence of some multinucleated huge cells known as Hodgkin/Reed cells that are bounded by a nonneoplastic inflammatory permeate (Mirsadraee et al., 2010). Around 30–40% of Hodgkin lymphoma patients in North America and Europe were described to be positive for EBV, while in Africa, Latin America, and Asia, EBV was present in nearly all HL positive patients (Glaser et al., 1997). Despite the relation of virus to HL infected patients, the part of EBV in the pathogenesis of this cancer remains unidentified (Massini et al., 2009).

6.8.3 Posttransplant Lymphoproliferative Disorder (PTLD)

PTLD is related to life threatening complications related to organ transplant recipients as well as recipients of hematopoietic stem cell allografts (Rasche et al., 2014). This cancer is caused by the increase in B-cell division post transplantation and is mostly caused by EBV infection (Kim et al., 2017).

Occurrence of EBV-related PTLD ranges from 1 to 20%, with prevalence changing because of the factors related to age, type of allograft, and EBV-serostatus of transplant recipient before transplant. PTLD reports are maximum (32%) in small intestine transplant receivers, and 3–12% occurrence is reported in pancreas, liver, lung, and heart transplant recipients. Fewer cases are reported (1–2%) in renal transplant patients (Nijland et al., 2016).

6.8.4 Nasopharyngeal Cancer

Nasopharyngeal cancer (NPC) is a malignancy of epithelial squamous cell that originates from the side wall of the nasopharynx, particularly the fossa of Rosenmüller and superior posterior wall (Carle et al., 2012). Worldwide, this cancer displays extraordinary disparity in racial and terrestrial occurrence, with around 80% of the total of 65,000 new reports that took place globally, present in Southeast Asia and Southern China (Mimi & Yuan, 2002). Classification has been presented for NPC which groups the virus into two histologic variants, called as undifferentiated carcinomas of the nasopharyngeal type (UCNT) and squamous cell carcinomas (SCCs) (Micheau et al., 1978). EBV latent gene protein LMP-1 has been reported in around two-thirds of nasopharyngeal cancer patients, representing the important part this protein holds in tumor progression (Gao et al., 2017).

6.8.5 EBV-Associated Gastric Cancer

Gastric cancer (GC) is the third important reason of cancer-associated deaths globally, with annual frequency of around 950,000 reported cases worldwide. EBV-related gastric cancer (EBVaGC) is a cause for around 10% of all gastric cancer reports (Jácome et al., 2016). EBVaGC is characterized into three histologic types, conventional type adenocarcinoma (CA)-type, carcinoma Crohn's disease-like lymphoid reaction (CLR)-type, and lymphoepithelioma-like carcinoma (LELC)-type (Shinozaki-Ushiku, Kunita, Isogai, et al., 2015). LELC-type is a cancer with dense infiltration of lymphocytes, similar to that of nasopharyngeal cancer.

The genes of EBV (EBER, EBNA-1, BART) are present in the gastric cancer cells and nearby epithelium cells, but they could not be detected in adjoining lymphocytes, stromal cells, intestinal metaplasia, and normal mucosa, consequently suggesting vital part of EBV in gastric cancer pathogenesis (Shinozaki-Ushiku, Kunita, & Fukayama, 2015).

6.8.6 Extranodal Nasal-Type NK/T-Cell Lymphoma

Extranodal nasal-type NK/T-cell lymphomas are rare cancers that have a link with EBV (Neparidze & Lacy, 2014). The cancer is categorized by widespread necrosis and angio-invasion in the upper aerodigestive tract, involving the nasal cavity, paranasal sinus, nasopharynx, and palate (Chan, 2001).

Epstein-Barr virus depicts expression of type II latent stage proteins (EBNA-1, LMP-1, LMP-2) in the extranodal NK/T-cell lymphomas, and EBV occurs as a clonal episome in the virus infected cells (Dojcinov et al., 2018).

Oncogenic LMP-1 protein presence in extranodal NK/T-cell lymphomas has been described to encourage tumor immune escape by the expression of programmed cell death receptor 1 and PD ligand 1 that are both vital checkpoint molecules in the regulation of cancer immunology (Zhang, 2016).

6.8.7 Breast Carcinoma (BC)

Breast carcinoma (BC) is the utmost frequent cancer in females and is the principal cause of deaths in females with global 600,000 mortalities, marking 6.6% of cancer-associated deaths (Shadmani et al., 2017). The progression of cancer depends on genetic and nongenetic factors—genetic factors include breast cancer-associated gene 1 and 2 (*BRCA1/2*) and epidermal growth factor receptor 2 (*HER2*) (Sun et al., 2017). Different viruses like EBV, human papilloma virus, mouse mammary virus, and cytomegalovirus have a part in BC cases—EBV linkage with BC is prevalent in Europe, Asia, and Africa with 30–50% prevalence (Marrão et al., 2014).

EBV gene proteins (LMP-1, LMP-2, EBNA-, and EBER) have also been described to have a role in breast tumorigeneses (Fessahaye et al., 2017), and meta-epidemiological research has depicted that viral infection was greatly linked with the menace of breast cancer progression (Bae & Kim, 2016).

6.9 Conclusion

Epstein-Barr virus is a cancer-causing virus with worldwide occurrence and affects around 90% of the world's human population. This virus has tropism for two basic cell types, specifically B lymphocytes and epithelial cells, though targets like NK cells and T lymphocytes have also been documented. The occurrence of virus in B lymphocytes has been connected to Hodgkin lymphoma, Burkitt's lymphoma, and posttransplant lymphoproliferative disorders. Moreover, epithelial cell infection is associated with pathogenic malignancies like nasopharyngeal, breast, and EBV-associated gastric cancer. The EBV-associated malignancies have pathogenic effect due to the expression of EBV-related latent genes. The proteins are divided into latency I, II, and III stages, and their expression and regulation differ because of the type, differentiation, and stimulation stage of the host cell.

Host and environmental aspects play important roles in the malignancies and tumor progression, thus bringing difficulty in mapping the regulation mechanism of EBV pathogenesis. Recent progress in genome and proteome research and forthcoming work should emphasize on regulation of host genetic components and environmental features that interrelate with viral genes and how they impact the pathogenesis and disease progression of EBV-related disorders.

References

Aguilar, R., Casabonne, D., O'Callaghan-Gordo, C., Vidal, M., Campo, J. J., Mutalima, N., Angov, E., Dutta, S., Gaur, D., Chitnis, C. E., & Chauhan, V. (2017). Assessment of the combined effect of Epstein–Barr Virus and Plasmodium falciparum infections on endemic Burkitt lymphoma using a multiplex serological approach. *Frontiers in Immunology, 8*, 1284.

Alfieri, C., Birkenbach, M., & Kieff, E. (1991). Early events in Epstein-Barr virus infection of human B lymphocytes. *Virology, 181*(2), 595–608.

Alptekin, B., Akpinar, B. A., & Budak, H. (2017). A comprehensive prescription for plant miRNA identification. *Frontiers in Plant Science, 7*, 2058.

Amria, S., Cameron, C., Stuart, R., & Haque, A. (2008). Defects in HLA class II antigen presentation in B-cell lymphomas. *Leukemia & Lymphoma, 49*(2), 353–355.

Ayee, R., Ofori, M. E., Wright, E., & Quaye, O. (2020). Epstein Barr virus associated lymphomas and epithelia cancers in humans. *Journal of Cancer, 11*(7), 1737–1746.

Bae, J. M., & Kim, E. H. (2016). Epstein-Barr virus infection and risk of breast cancer: An adaptive meta-analysis for case-control studies. *Archives of Clinical Infectious Diseases, 11*(3), e34806.

Banko, A. V., Lazarevic, I. B., Folic, M. M., Djukic, V. B., Cirkovic, A. M., Karalic, D. Z., Cupic, M. D., & Jovanovic, T. P. (2016). Characterization of the variability of Epstein-Barr virus genes in nasopharyngeal biopsies: Potential predictors for carcinoma progression. *PLoS One, 11*(4), e0153498.

Blum, K. A., Lozanski, G., & Byrd, J. C. (2004). Adult Burkitt leukemia and lymphoma. *Blood, 104*(10), 3009–3020.

Carle, L. N., Ko, C. C., & Castle, J. T. (2012). Nasopharyngeal carcinoma. *Head and Neck Pathology, 6*, 364–368.

Chan, J. (2001). Extranodal NK/T cell lymphoma, nasal type. In *WHO classification of tumours. Pathology and genetics of tumours of hematopoietic and lymphoid tissues* (pp. 204–208). IARC.

Chen, Y., Zhao, W., Lin, L., Xiao, X., Zhou, X., Ming, H., Huang, T., Liao, J., Li, Y., Zeng, X., & Huang, G. (2015). Nasopharyngeal Epstein-Barr virus load: An efficient supplementary method for population-based nasopharyngeal carcinoma screening. *PLoS One, 10*(7), 121–135.

da Costa, V. G., Marques-Silva, A. C., & Moreli, M. L. (2015). The Epstein-Barr virus latent membrane protein-1 (LMP1) 30-bp deletion and XhoI-polymorphism in nasopharyngeal carcinoma: A meta-analysis of observational studies. *Systematic Reviews, 4*, 1.

Delecluse, H. J., Feederle, R., O'Sullivan, B., & Taniere, P. (2007). Epstein-Barr virus-associated tumours: An update for the attention of the working pathologist. *Journal of Clinical Pathology, 60*(12), 1358–1364.

Dojcinov, S. D., Fend, F., & Quintanilla-Martinez, L. (2018). EBV-positive lymphoproliferations of B-T- and NK-cell derivation in non-immunocompromised hosts. *Pathogens, 7*(1), 28.

Dong, M., Chen, J. N., Huang, J. T., Gong, L. P., & Shao, C. K. (2019). The roles of EBV-encoded microRNAs in EBV-associated tumors. *Critical Reviews in Oncology, 135*, 30–38.

Dresang, L. R., Vereide, D. T., & Sugden, B. (2009). Identifying sites bound by Epstein-Barr virus nuclear antigen 1 (EBNA1) in the human genome: Defining a position-weighted matrix to predict sites bound by EBNA1 in viral genomes. *Journal of Virology, 83*(7), 2930–2940.

Fessahaye, G., Elhassan, A. M., Elamin, E. M., Adam, A. A., Ghebremedhin, A., & Ibrahim, M. E. (2017). Association of Epstein-Barr virus and breast cancer in Eritrea. *Infectious Agents and Cancer, 12*, 1–7.

Fox, C. P. (2011). *Investigating the role of Epstein-Barr virus in the pathogenesis of NK and T cell lymphoproliferations* (Doctoral dissertation, University of Birmingham).

Gao, R., Wang, L., Liu, Q., Zhang, L. F., Ye, Y. F., Xie, S. H., Du, J. L., Chen, S. H., Guo, J., Yang, M. J., & Lin, C. Y. (2017). Evaluation of seven recombinant VCA-IgA ELISA kits for the diagnosis of nasopharyngeal carcinoma in China: A case–control trial. *BMJ Open, 7*(6), e013211.

Glaser, S. L., Lin, R. J., Stewart, S. L., Ambinder, R. F., Jarrett, R. F., Brousset, P., Pallesen, G., Gulley, M. L., Khan, G., O'Grady, J., & Hummel, M. (1997). Epstein-Barr virus-associated Hodgkin's disease: Epidemiologic characteristics in international data. *International Journal of Cancer, 70*(4), 375–382.

Gloghini, A., Dolcetti, R., & Carbone, A. (2013). Lymphomas occurring specifically in HIV-infected patients: From pathogenesis to pathology. *Seminars in Cancer Biology, 23*(6), 457–467.

Gulley, M. L. (2001). Molecular diagnosis of Epstein-Barr virus-related diseases. *The Journal of Molecular Diagnostics, 3*(1), 11–21.

Jácome, A. A., Lima, E. M., Kazzi, A. I., Chaves, G. F., Mendonça, D. C., Maciel, M. M., & Santos, J. S. (2016). Epstein-Barr virus-positive gastric cancer: A distinct molecular subtype of the disease. *Revista da Sociedade Brasileira de Medicina Tropical, 49*, 150–157.

Kaiser, C., Laux, G., Eick, D., Jochner, N., Bornkamm, G. W., & Kempkes, B. (1999). The proto-oncogene c-myc is a direct target gene of Epstein-Barr virus nuclear antigen 2. *Journal of Virology, 73*(5), 4481–4484.

Kang, M. S., & Kieff, E. (2015). Epstein–Barr virus latent genes. *Experimental & Molecular Medicine, 47*(1), 131–143.

Kashuba, E., Yurchenko, M., Szirak, K., et al. (2005). Epstein-Barr virus-encoded EBNA-5 binds to Epstein-Barr virus-induced Fte1/S3a protein. *Experimental Cell Research, 303*, 47–55.

Kelly, G. L., Milner, A. E., Baldwin, G. S., Bell, A. I., & Rickinson, A. B. (2006). Three restricted forms of Epstein–Barr virus latency counteracting apoptosis in c-myc-expressing Burkitt lymphoma cells. *Proceedings of the National Academy of Sciences, 103*(40), 14935–14940.

Kim, H. J., Ko, Y. H., Kim, J. E., Lee, S. S., Lee, H., Park, G., Paik, J. H., Cha, H. J., Choi, Y. D., Han, J. H., & Huh, J. (2017). Epstein-Barr virus–associated lymphoproliferative disorders: Review and update on 2016 WHO classification. *Journal of Pathology and Translational Medicine, 51*(4), 352.

Kirchmaier, A. L., & Sugden, B. (1997). Dominant-negative inhibitors of EBNA-1 of Epstein-Barr virus. *Journal of Virology, 71*(3), 1766–1775.

Klein, G., & Ernberg, I. (2007). Comparative aspects of EBV, HHV-8, and previously known DNA tumor viruses in experimental systems. In *Human herpesviruses* (p. 29). Lyon.

Klemenc, P., Marin, J., Šoba, E., Gale, N., Koren, S., & Strojan, P. (2006). Distribution of Epstein–Barr virus genotypes in throat washings, sera, peripheral blood lymphocytes and in EBV positive tumor biopsies from Slovenian patients with nasopharyngeal carcinoma. *Journal of Medical Virology, 78*(8), 1083–1090.

Kushekhar, K., van den Berg, A., Nolte, I., Hepkema, B., Visser, L., & Diepstra, A. (2014). Genetic associations in classical Hodgkin lymphoma: A systematic review and insights into susceptibility mechanisms. *Cancer Epidemiology, Biomarkers & Prevention, 23*(12), 2737–2747.

Lenze, D., Leoncini, L., Hummel, M., Volinia, S., Liu, C. G., Amato, T., De Falco, G., Githanga, J., Horn, H., Nyagol, J., & Ott, G. (2011). The different epidemiologic subtypes of Burkitt lymphoma share a homogenous micro RNA profile distinct from diffuse large B-cell lymphoma. *Leukemia, 25*(12), 1869–1876.

Ma, J., Nie, K., Redmond, D., Liu, Y., Elemento, O., Knowles, D. M., & Tam, W. (2016). EBV-miR-BHRF1-2 targets PRDM1/Blimp1: Potential role in EBV lymphomagenesis. *Leukemia, 30*(3), 594–604.

Marrão, G., Habib, M., Paiva, A., Bicout, D., Fallecker, C., Franco, S., Fafi-Kremer, S., Simões da Silva, T., Morand, P., Freire de Oliveira, C., & Drouet, E. (2014). Epstein-Barr virus infection and clinical outcome in breast cancer patients correlate with immune cell TNF-α/IFN-γ response. *BMC Cancer, 14*(1), 1.

Massini, G., Siemer, D., & Hohaus, S. (2009). EBV in Hodgkin lymphoma. *Mediterranean Journal of Hematology and Infectious Diseases, 1*(2), e2009013.

Micheau, C., Rilke, F., & Pilotti, S. (1978). Proposal for a new histopathological classification of the carcinomas of the nasopharynx. *Tumori Journal, 64*(5), 513–518.

Mimi, C. Y., & Yuan, J. M. (2002). Epidemiology of nasopharyngeal carcinoma. *Seminars in Cancer Biology, 12*(6), 421–429.

Mirsadraee, S., Farzadnia, M., Heidary, F., Rahimi, M., Jabbari Noghabi, H., & Kakhi, S. (2010). A comparative immunohisto chemical diagnosis of Hodgkin's lymphoma with conventional method. *Medical Journal of Mashhad University of Medical Sciences, 53*(4), 214–221.

Münz, C. (2015). *Epstein Barr virus volume 2: One herpes virus: Many diseases*. Springer.

Neparidze, N., & Lacy, J. (2014). Malignancies associated with Epstein-Barr virus: Pathobiology, clinical features, and evolving treatments. *Clinical Advances in Hematology & Oncology, 12*(6), 358–371.

Nijland, M. L., Kersten, M. J., Pals, S. T., Bemelman, F. J., & Ten Berge, I. J. (2016). Epstein-Barr virus–positive posttransplant lymphoproliferative disease after solid organ transplantation: Pathogenesis, clinical manifestations, diagnosis, and management. *Transplantation direct, 2*(1), e48.

Peh, S. C., Kim, L. H., & Poppema, S. (2002). Frequent presence of subtype A virus in Epstein-Barr virus-associated malignancies. *Pathology, 34*(5), 446–450.

Rasche, L., Kapp, M., Einsele, H., & Mielke, S. (2014). EBV-induced post-transplant lymphoproliferative disorders: A persisting challenge in allogeneic hematopoetic SCT. *Bone Marrow Transplantation, 49*(2), 163–167.

Rochford, R., & Moormann, A. M. (2015). Burkitt's lymphoma. In *Epstein Barr virus volume 1: One herpes virus: Many diseases* (pp. 267–285). Springer.

Rowe, D. T. (1999). Epstein-Barr virus immortalization and latency. *Frontiers in Bioscience, 4*(4), 346–371.

Saha, A., & Robertson, E. S. (2011). Epstein-Barr virus–associated B-cell lymphomas: Pathogenesis and clinical outcomes. *Clinical Cancer Research, 17*(10), 3056–3063.

Saha, A., & Robertson, E. S. (2013). Impact of EBV essential nuclear protein EBNA-3C on B-cell proliferation and apoptosis. *Future Microbiology, 8*(3), 323–352.

Santpere, G., Darre, F., Blanco, S., Alcami, A., Villoslada, P., Mar Albà, M., & Navarro, A. (2014). Genome-wide analysis of wild-type Epstein–Barr virus genomes derived from healthy individuals of the 1000 genomes project. *Genome Biology and Evolution, 6*(4), 846–860.

Shadmani, F. K., Mansori, K., Khazaei, S., Hanis, S. M., Khazaei, S., Sani, M., & Ayubi, E. (2017). Geographic distribution of breast cancer incidence in Iran. *Biomedical Research and Therapy, 4*(5), 1295–1304.

Shinozaki-Ushiku, A., Kunita, A., & Fukayama, M. (2015). Update on Epstein-Barr virus and gastric cancer. *International Journal of Oncology, 46*(4), 1421–1434.

Shinozaki-Ushiku, A., Kunita, A., Isogai, M., Hibiya, T., Ushiku, T., Takada, K., & Fukayama, M. (2015). Profiling of virus-encoded microRNAs in Epstein-Barr virus-associated gastric carcinoma and their roles in gastric carcinogenesis. *Journal of Virology, 89*(10), 5581–5591.

Siegler, G., Kremmer, E., Gonnella, R., et al. (2003). Epstein-Barr virus encoded latent membrane protein 1 (LMP1) and TNF receptor associated factors (TRAF): Colocalisation of LMP1 and TRAF1 in primary EBV infection and in EBV associated Hodgkin lymphoma. *Molecular Pathology, 56*, 156.

Smatti, M. K., Al-Sadeq, D. W., Ali, N. H., Pintus, G., Abou-Saleh, H., & Nasrallah, G. K. (2018). Epstein–Barr virus epidemiology, serology, and genetic variability of LMP-1 oncogene among healthy population: An update. *Frontiers in Oncology, 8*, 211.

Sun, Y. S., Zhao, Z., Yang, Z. N., Xu, F., Lu, H. J., Zhu, Z. Y., Shi, W., Jiang, J., Yao, P. P., & Zhu, H. P. (2017). Risk factors and preventions of breast cancer. *International Journal of Biological Sciences, 13*(11), 1387.

Tempera, I., & Lieberman, P. M. (2014). Epigenetic regulation of EBV persistence and oncogenesis. *Seminars in Cancer Biology, 26*, 22–29.

Thompson, M. P., & Kurzrock, R. (2004). Epstein-Barr virus and cancer. *Clinical Cancer Research, 10*(3), 803–821.

Thorley-Lawson, D., Deitsch, K. W., Duca, K. A., & Torgbor, C. (2016). The link between Plasmodium falciparum malaria and endemic Burkitt's lymphoma—new insight into a 50-year-old enigma. *PLoS Pathogens, 12*(1), e1005331.

Tsao, S. W., Tsang, C. M., & Lo, K. W. (2017). Epstein–Barr virus infection and nasopharyngeal carcinoma. *Philosophical Transactions of the Royal Society, B: Biological Sciences, 372*(1732), 20160270.

Tsao, S. W., Tsang, C. M., To, K. F., & Lo, K. W. (2015). The role of Epstein–Barr virus in epithelial malignancies. *The Journal of Pathology, 235*(2), 323–333.

Wasil, L. R., Tomaszewski, M. J., Hoji, A., & Rowe, D. T. (2013). The effect of Epstein-Barr virus latent membrane protein 2 expression on the kinetics of early B cell infection. *PLoS One, 8*(1), e54010.

Webster-Cyriaque, J., & Raab-Traub, N. (1998). Transcription of Epstein-Barr virus latent cycle genes in oral hairy leukoplakia. *Virology, 248*, 53–65.

Zhang, W. W. (2016). PD-L1 is upregulated by EBV-driven LMP1 through NF-κB pathway and correlates with poor prognosis in natural killer/T-cell lymphoma. *Journal of Hematology & Oncology, 9*(1), 109.

Human Papillomavirus Leading to Urogenital, Head, and Neck Cancer

7

Neelam Batool, Sara Latif, Fatima Ijaz, Fizza Ahmad Tariq, Kiran Saba, Muhammad Suleman Malik, and Mohammad Tahir Waheed

Abstract

Human papillomavirus (HPV) is a group of closely related viruses comprising of more than 200 types and are double-stranded DNA viruses. Several types of HPV cause infections in humans which lead to squamous cell carcinomas and adenocarcinomas. HPV types are categorized into low-risk and high-risk types. Low-risk HPV leads to genital warts and high-risk types causes anogenital, head, and oropharyngeal cancers. High-risk HPV contains HPV types 16 and 18 which are the most pathogenic. HPV has more than 40 genotypes transmitted sexually and are also associated with urinary tract infection that progresses to urogenital cancer. HPV is causatively associated with cervical cancer (100%) and anal cancer (90%) globally. Head and neck cancer, being the sixth most common cancer, is also developed by HPV and it is considered to be involved in 30–60% of oropharyngeal cancer cases as well. The expression of major viral oncoproteins including E6 and E7 dysregulates the cell cycle, degrades the tumor-suppressor P53 and pRb proteins, and promotes cell replication that develops HPV-related malignancies. Downregulating these oncoproteins can be an effective therapy against HPV. Along with radiation therapy and chemotherapy, immunotherapy and vaccines against HPV can provide good measures for controlling cancers, HPV-related cancers.

N. Batool · S. Latif · F. Ijaz · F. A. Tariq · M. S. Malik · M. T. Waheed (✉)
Department of Biochemistry, Quaid-i-Azam University, Islamabad, Pakistan
e-mail: tahirwaheed@qau.edu.pk

K. Saba
Department of Biochemistry, Faculty of Life Sciences, Shaheed Benazir Bhutto Women University, Peshawar, Pakistan

N. Masood, A. Yasmin (eds.), *The Microbiome and Cancer*,
https://doi.org/10.1007/978-981-97-6948-3_7

Keywords

Human papillomavirus · Carcinoma · Urogenital cancer · Head and neck cancer · Cervical cancer · HPV-16 · HPV-18 · E6 · E7

7.1 Introduction

7.1.1 Human Papillomavirus

Human papillomavirus (HPV) is a group of small, non-enveloped, double-stranded (ds) DNA viruses, where the length of the genome is approximately 8 kb (Doorbar et al., 2015). Intensive research on the major types of HPV is primarily driven by HPV-associated pathologies. To date, over 200 types of HPV have been isolated from human infections, fueling the exploration and investigation in this field (Van Doorslaer et al., 2016). Human papillomaviruses are classified into low-risk HPVs (LR-HPVs) that are responsible for anogenital and cutaneous warts, while the other class is high-risk HPVs (HR-HPVs) that cause oropharyngeal (oral, throat, and tonsil) cancers and anogenital cancers such as cervical, anal, vulvar, vaginal, and penile cancers (Kombe Kombe et al., 2021). According to the most recent data published by NIH (National Cancer Institute, USA), there are 14 major types of HR-HPVs including 16, 18, 31, 33, 35, 39, 45, 51, 52, 56, 66, and 68 where types 16 and 18 are most prevalent types in the HPV-related cancers (Waheed et al., 2011). HR-HPV-related cancers are either squamous cell carcinomas or adenocarcinomas depending upon which types of cells are infected by HPV (Rahimi, 2020).

The 8 kb genome of HPV is functionally sectioned into three major parts: "L-region," the late region; "E-region," the early region; and "UR-region," the upstream regulatory region (Graham, 2017). The E and L regions consist of nine open reading frames (ORFs) out of which six are in the E-region (E1, E2, E4, E5, E6, E7) and two are in the L-region (L1 and L2), whereas the URR consists of origin of replication (ORF) as well as all the transcription factor binding sites that are necessary for RNA polymerase-II regulated transcription. Along with these sites, there are two major sets of promoters in the HPV genome, one is P97 (present in HPV-16 and 31) and P105 (present in HPV-18). These promoters lie upstream to the E6-ORF and regulate the expression of early genes. The other set is P670 (present in HPV-16), P811 (present in HPV-18), and P742 (present in HPV-31). This set of promoters lie in E7-ORF and it is vital for the differentiation-dependent late gene expression (Scarth et al., 2021). The major oncoproteins involved in HPV-related pathologies are E5, E6, and E7 although their expression is not sufficient for the progression of infection to cancer, but together they promote the prolonged proliferation and delayed differentiation of keratinocytes in the host cell and provide a suitable environment for viral DNA replication (Moody & Laimins, 2010; Moody, 2022).

7.1.1.1 The Life Cycle of Human Papillomavirus

The life cycle of HPV is intrinsically connected with the programmed differentiation of epithelial cells. For the establishment of infection, virions must have to enter the stratified epithelium through basal lamina, which can happen either through microlesions or by entering the squamocolumnar junction of the cells (Dust et al., 2022; Chen et al., 2023). The binding of L1-capsid protein to heparin sulfate proteoglycans (HSPGs) present on the basement membrane or the surface of basal keratinocytes induces conformational changes in the structure of capsid, exposing the minor capsid protein L2, which is followed by clathrin-independent endocytosis (Morante et al., 2022). The capsid disassembly that occurs in the late endosome is followed by the transfer of the viral genome toward the trans-Golgi complex along with L1 and L2. The L2 protein has an intrinsic property to interact with the components of the retromer complex that is a part of retrograde transport machinery because of the cell-penetrating peptide present at the C-terminal of L2. The viral genome binds to the host mitotic chromosomes with the help of L2 and results in the breakdown of the nuclear envelope during mitosis (McBride, 2021).

Once the viral genome enters the nucleus, it starts amplifying rapidly, forming 50 to 100 copies per cell. This replication happens with the help of interaction between E1 and E2 viral proteins with the host DNA replication process, as HPV lacks any kind of polymerase (Fontan et al., 2020). The copy number is maintained and kept at its minimal level by E2-mediated repression of early (p97 and p105) promoters. This minimal copy number of the virus in the host cell also helps to evade the host's immune response (Bienkowska-Haba et al., 2022). As the cell cycle advances, half of the daughter cells migrate toward the epithelial layers whereas half stay in the basal layer and form an episomal reservoir. The expression of viral oncoproteins (E5, E6, E7) delays the terminal differentiation and halts the cell cycle exit thus providing an extended G2-like environment to provide sustainable conditions for the further viral genome amplification. The virus utilizes the host cellular factor to modulate all these activities as the viral oncoproteins lack any intrinsic enzymatic activity. The life cycle of the virus completes after the required expression of viral oncoproteins L1, L2, and E1^4 occurs, and the terminal differentiation happens in the uppermost layers of the epithelium. The virion gets assembled in the nucleus when the L1 and L2 capsid proteins are imported into the nucleus and the newly formed viral particles are released from the epithelial surface (Wolf, 2023).

7.1.2 HPV and Urogenital Cancer (UGC)

Human papillomavirus is the most common sexually transmitted virus. Therefore, it is the major cause of infection in the anogenital tract. More than 40 genotypes of HPV have been isolated from urogenital tract infections. As discussed earlier, in terms of infection, HPV is divided into low-risk (LR) and high-risk (HR) HPVs, and the mode of infection is highly tissue specific. The infections caused by HPV range from benign (caused by LR-HPVs, i.e., 6, 11, 40, 42, 43, 44, 54, 61, 70, 72, and 81) to invasive tumors (caused by HR-HPVs, i.e., 16, 18, 31, 33, 35, 39, 45, 51, 52, 56,

58, 59, and 66). Along with genital cancers (cervical and penile), HPV also causes urological pathologies such as cancer of the prostate, bladder, testis, and kidney (Rafael et al., 2022).

7.1.2.1 Risk Factors Associated with HPV-Positive UGC

Cancer has an extensive variety of risk factors, but bacterial and viral infections have been identified as significant contributors. Two of the best-known infectious agents associated with cancer worldwide are *Helicobacter pylori*, accounting for 36.3%, and human papillomavirus (HPV), responsible for 31.1% of cases. Additionally, hepatitis B and C infections are also linked to cancer, with hepatitis B accounting for 16.3% and hepatitis C for 7.1% of cases (Plummer et al., 2016).

HPV is a prevalent infection being sexually transmitted affecting both men and women, having a lifetime threat of approximately 80% for sexually active women. This virus is associated with nearly 10% of cancer cases throughout the world, particularly in the subclass of SCC (squamous cell carcinoma) of various body regions, such as the vulva, cervix uteri, anus, penis, and oropharynx. It is hence demonstrated that the virus has an affinity for squamous cells (Kim et al., 2014). In around 90% of cases, the body's effective cell-mediated immune response successfully clears the HPV infection spontaneously. However, the innate immune system has developed defenses against HPV delaying the activation of the adaptive immune system, which is responsible for eliminating the infection. Consequently, about 10% of individuals experience persistent HPV infection, leading to an increased risk of cancer progression, particularly in case of infections with high-risk HPV types (Jørgensen & Jensen, 2020).

HPV infection can affect all sexually active individuals, potentially involving both low- and high-risk strains, usually occurring immediately following the onset of sexual activity. In the majority of cases, the immune system successfully removes the virus. However, in a few individuals, the infection persists, and this can lead to the development of precancerous alterations (Szymonowicz & Chen, 2020). Specifically, cells invaded with HR-HPV strains may experience an increase in their proliferation rate. If evaded by the immune system, these cells can ultimately give rise to precancerous changes or can even lead to tumor formation. Smoking, repeated pregnancies, a weakened immune system, extended use of oral contraceptives, and infection with other sexually transmitted diseases (STDs) can all increase the chance of precancerous cervical cells that could develop into cancer, particularly when associated with high-risk HPV types. Most occurrences of cervical cancer are caused by HPV infection. However, it can also be associated with other varieties of cancer, such as anogenital cancer (including vaginal, vulvar, anal, and penile cancer), head and neck cancers (HNCs), and genital warts, from which both women and men are affected.

Transmission of anogenital warts is frequently via sexual contact. The age at which HPV infections appear is comparable to that of other STDs. Moreover, people infected with the HPV often exhibit an increased incidence of other STDs, and the risk of developing genital warts (condylomata acuminate) are associated with having multiple sexual partners (Schim van der Loeff et al., 2014). Aside from sexual transmission, slight trauma on the infection region may also be involved in HPV transmission, as evidenced by an increased incidence of HPV among individuals involved in activities such as meat handling, like butchers. Interestingly, young children can contract the virus by contracting non-genital lesions on their hands. According to research, the disease may also spread during vaginal delivery and maybe in pregnancy by ascending infection from the mother's genital system. These additional routes of transmission contribute to the spread and persistence of HPV infections in different populations (Deniz et al., 2022).

Among males, the most significant prevalence of HPV infection was seen in HIV-infected men having sexual activities with other men. Among heterosexual men, African males had the highest risk of infection and had three female sexual partners at least (Vardas et al., 2011). Furthermore, the overall prevalence of HPV infection, regardless of the HPV type, was notably higher in men who were HIV positive as compared to men who were HIV negative, with rates of 78.2% and 49.4%, respectively (Olesen et al., 2014).

HPV-associated warts are also detected in individuals who had never been engaged in sexual intercourse, indicating the existence of various routes of horizontal HPV transmission (Sabeena et al., 2017). These routes include autoinoculation, hetero-inoculation, or transmission through inanimate objects known as fomites. HPV infection can spread without involving sexual acts at the genital level. Instead, transmission may occur through contact between infected fingers and the genital area. In case of kids, this occurrence is linked to a parent or custodian who is having hand warts, and unintentionally, the infection is transmitted to the child while washing the anogenital area or during changing diapers. These nonsexual ways of spreading the infection highlight the importance of understanding and preventing HPV infections beyond sexual contexts (Milano et al., 2023).

In addition to other modes of transmission, there exists a recognized hazard of HR-HPV contamination via endo-vaginal ultrasound probes, not being adequately sterilized. This inadequate disinfection renders the probes ineffective in preventing HR-HPV transmission, potentially leading to nosocomial infections during ultrasound procedures (Palma et al., 2021). Another context where HPV transmission has been observed is during laser ablation procedures where the surgical smoke generated is the culprit. This risk of infection has been reported in animal models, where bovine papillomavirus gathered from CO_2 laser smoke during wart treatment in cattle-induced cutaneous fibro papillomas upon reinoculation into the skin of calves. Nevertheless, the risk of infection through surgical smoke in humans remains a subject of controversy (Kalof & Cooper, 2007). As such, further research is needed to ascertain the potential risks associated with surgical smoke and its impact on HPV transmission in human settings (Adams et al., 2014).

7.1.2.2 Progression of HPV-Positive UGC

Persistent cervical HPV infections can lead to the development of premalignant glandular or squamous intraepithelial lesions, known as cervical intraepithelial neoplasia (CIN), and, in some cases, progress to cancer. Histopathologically, CIN is categorized into three grades: CIN 1, indicating mild dysplasia; CIN 2, indicating moderate to marked dysplasia; and CIN 3, indicating severe dysplasia to carcinoma in situ. While many CIN lesions relapse suddenly, it is important to note that over the course of several years, lesions on the cervix may gradually transform into cancerous conditions if left unmonitored or untreated (Li et al., 2019).

CIN diagnosis relies mainly on the identification of nuclear atypia along with the absence of regular squamous maturation (polarity). The exact classification of CIN lesions is crucial as the procedures for treatment and clinical follow-up are significantly dissimilar for the low-grade (CIN 1) and high-grade lesions (CIN 2 and 3) (Barr et al., 2017). Even if the likelihood of progression is increased undoubtedly as the grade of cervical intraepithelial neoplasia rises, a percentage of high-grade lesions could still relapse. Cervical dysplasia that is persistent and/or progressing may be at risk for infection with high-risk HPV subtypes. Additionally, it has been suggested that the disruption of the E1/E2 open reading frames of the HPV genome and consequent loss of the E2-controlled regulation of E6 and E7 are key factors in the cervical carcinogenesis process. As a result, p53 expression is markedly reduced in HPV-associated cancer cells (Abbas & Dutta, 2009). E6 and E7 inactivate the host p53 and retinoblastoma proteins (pRb), respectively, which suppress the cell cycle checkpoints and promote uncontrolled cell division. E6 interacts with p53 leading to its ubiquitination-mediated degradation via E6-associated protein (E6AP) (Huh et al., 2007).

An interesting observation is that for E6 to bind to p53, it needs to first interact with the protein E3-ligase, as none of E6 or E6AP can interact with p53 alone (Hamada et al., 2006). Moreover, only HR-HPV E6 and E7 are capable of binding to pRb39 or p53, indicating that HPV-associated cancers arise exclusively from high-risk HPV types. One of p53's targets, p21 (also known as p21WAF1/Cip1), a cyclin-dependent kinase inhibitor that inhibits cells from entering the S phase, is rendered inactive as a result of the breakdown of p53 by E6. Instead, p21 promotes cell cycle arrest in the G1 phase in response to various stimuli (Jung et al., 2015; Zhen et al., 2020). On the other hand, E7 causes ubiquitination of pRb, resulting in the release of E2F transcription factors, which then promote the transcription of cyclin A, E, and p16INK4A. The p16INK4A protein acts as a CDK4/6 inhibitor, driving the cells to enter premature S phase (Cuesta-Mateos et al., 2018). The mechanism through which E7 mediates pRb ubiquitination is not fully understood, but it is suggested that Cullin2 may interact with E7 and is liable for this ubiquitination process (Wilson, 2014). Additionally, E7 itself can be ubiquitinated by the Cullin 1/ Skp2 complex, though pRb interacts with Skp2 and it is not involved in its ubiquitination (Pérez-Benavente, 2020). The intricate interactions and modifications orchestrated by E6 and E7 (HPV oncoproteins) play a vital role in the development and progression of HPV-related cancers. However, more studies are required to fully elucidate the precise mechanisms involved in these processes. Collectively, p53 and pRb degradation facilitated by high-risk HPV leads to cellular immortalization resulting in the formation of higher grades of dysplasia (CIN 2 and 3), having the possibility of progression into invasive carcinoma (McBride, 2021).

Several studies have reported the significant impact of downregulating HPV E6 and E7 in HPV-related cervical cancer cells. One study utilized adenovirus-mediated transfection to downregulate HPV-16 E6 and E7, leading to suppressed cell growth, accompanied by increased expression of p53 and/or pRb proteins (Hasan et al., 2020). Similarly, another study demonstrated enhanced sensitization of cisplatin treatment by using siRNA against HPV-16 and HPV-18 E6 and E7 in cervical cancer cells. This approach further enhanced the therapeutic effect of cisplatin treatment in these cells. Moreover, suppressed growth of tumors was seen in xenograft mice models and cervical cancer cells by downregulating HPV-16 and HPV-18 E6 and E7, along with radiosensitizing effects (Aggarwal et al., 2019). Targeting of HPV-16 E6/E7 via gRNA resulted in apoptosis in SiHa, i.e., HPV-positive cervical cancer cells. Additionally, another treatment with gRNA targeting PD1 resulted in a significant reduction in the growth of SiHa cell-xenografted humanized SCID mice (Organization WH, 2017).

Collectively, the downregulation of HPV E6 and E7 has a substantial effect on suppressing the growth of cells and restoring the expression of p53 and pRb. Furthermore, this downregulation enhances the susceptibility of tumor cells to chemotherapy and radiation treatment, potentially opening new avenues for targeted therapeutic strategies against HPV-related cervical cancer.

7.1.2.3 Ongoing Clinical Trials on Treatment for HPV-Related Cancers

Currently, immunotherapy has emerged as a promising option for the treatment of cancer patients. It involves two main approaches: First is using checkpoint inhibitors or vaccines for stimulating the immune system of the patient and enhancing its capability to identify and eliminate cancer cells. Second, it utilizes monoclonal-engineered T cells or antibodies that mimic the functions of the immune system's compounds. A total of 14 immunomodulators have been approved by FDA, with 7 of them being checkpoint inhibitors that destroy the PD-1/PD-L1 interaction or block CTLA-4. These agents aid in enhancing the defense system's ability to identify cancer cells and trigger their destruction. Additionally, four cytokines, which act as small molecular messengers among cells, have been approved for treating kidney cancer, lymphoma, leukemia, sarcoma, and melanoma. Over the years, several monoclonal antibodies have received FDA approval for treating various solid tumors and hematological cancers (Schim van der Loeff et al., 2014). Notably, in ongoing phase II clinical studies, ipilimumab, an anti-CTLA4 monoclonal antibody, is under investigation for its effectiveness in treating HPV-related cancers like cervical and head and neck cancer, either alone or combined with radiotherapy (McNeil et al., 2016).

CUE-101 is an innovative E7-pHLA-IL-2-Fc fusion protein designed to activate and increase a specific tumor-specific T-cell population. Its Fc domain is related to the target protein, aiming to eliminate HPV-driven malignancies. Currently, a phase I clinical study (NCT03978689) is underway to assess its safety and determine the optimal dosage by recruiting patients (Palmieri et al., 2017). Avelumab is a monoclonal antibody that targets the protein programmed death-ligand 1 (PD-L1). Recently, it has received FDA approval for treating renal cancer, Merkel cell cancer, and bladder carcinoma. More than 200 clinical studies are examining the effects of nivolumab on various cancer types. In one such clinical trial (NCT03260023), avelumab is being tested in combination with a novel vaccine called TG4001, which contains HPV 16-specific antigens. The study is focusing on cancers associated with HPV infection. Durvalumab is another human immunoglobulin G1 kappa (IgG1K) monoclonal antibody that is approved by the FDA and inhibits the interaction of PD-L1 with PD-1 and is used to treat various types of bladder and lung cancers. Durvalumab is being investigated in combination with the vaccine MEDI0457 for HPV-positive oropharyngeal squamous cell carcinoma (OPSCC) in a phase 2 clinical trial (NCT04001413). The purpose of this study is to evaluate the efficacy of this combination treatment (Cervantes & Doan, 2018).

MEDI0457 is a vaccine containing DNA plasmids that encode the E6 as well as E7 genes of HPV 16 and 18 (VGX-3100) and a DNA plasmid that encodes INO-9012, i.e., human interleukin-12 (IL-12) which stimulates the defense system (Vardas et al., 2011). Some recent studies have shown that MEDI0457 was tolerated well in individuals with cervical cancer who received chemoradiation. Hence, it has opened the possibility for a combination of radiotherapy with tumor-specific vaccines (Olesen et al., 2014). Similarly, individuals having HPV-associated head and neck carcinomas also exhibited good tolerance to MEDI0457, along with increased activity of antigen-specific T cells (Sabeena et al., 2017). Despite these promising safety findings, the vaccine's ultimate effectiveness can only be determined through rigorous evaluation in phase II and phase III clinical trials.

An innovative treatment approach including HPV-specific T cells (HPVST cells) is presently being investigated in a clinical trial that includes individuals with recurrent HPV-related cancers (NCT02379520). These HPVST cells are derived from affected individuals who experienced HPV-related cancers previously. The primary objective of the study is to assess whether these cells can exert an impact on tumors and become resistant to transforming growth factor-beta (TGF-β) formed due to HPV-mediated tumors. Once the optimum dosage of HPVST cells is determined, the study plans to combine them with nivolumab, an FDA-approved human IgG4 monoclonal antibody that inhibits PD-1. The aim is to possibly improve the effectiveness of HPVST cells in patients having HPV-associated cancers. Another phase II study involving patients diagnosed with HPV-positive oropharyngeal squamous cell carcinoma (OPSCC) is being conducted. In this study (NCT03107182, NCT03829722), patients will initially get chemotherapy (carboplatin, nivolumab, and nab-paclitaxel) for 6 months, followed by surgery or radiotherapy. The final step of this study involves the application of nivolumab to assess its impact in combination with the previously administered treatment regimen. The ultimate goal is to improve outcomes for patients with HPV-related cancers using this combination therapy (Deniz et al., 2022). In another clinical study of phase I (NCT03634267), researchers are exploring the use of magnetic resonance imaging (MRI)-guided brachytherapy, i.e., internal radiation therapy to determine if this method can enhance the high-risk clinical target volume (HR-CTV) D90 rate compared to the conventional guidance methods, which involve ultrasound and a freehand technique. The study focuses on patients with stage IB2-IV cervical or stage II-IVA vaginal cancer (Deniz et al., 2022). Proton therapy has also achieved significant attention as a possible treatment approach for various cancer patients. A clinical study is underway currently (NCT03621696) to analyze the use of proton therapy and/or traditional radiotherapy in individuals with HPV-related OPSCC. The goal is to achieve a decline in treatment-related toxicity while upholding treatment effectiveness.

CRISPR/Cas9 technology has become extensively utilized for both in vivo and in vitro studies. Currently, it is the subject of a clinical trial in phase I (NCT03747965) investigating its potential applications. In this trial, individuals having various solid tumors, including cholangiocarcinoma, pancreatic cancer, and ovarian cancer are being enrolled to assess the safety of CRISPR-Cas9-mediated PD-1 gene-knocked out chimeric antigen receptor (CAR) T cells. This novel approach aims to improve the defense response against cancer. Furthermore, another study (NCT03057912) is focusing on evaluating the efficacy and safety of CRISPR/Cas9 technology in targeting HPV E6/E7 genes, specifically disrupting the DNA of HPV 16 and HPV 18. The study involves individuals having HPV-associated cervical intraepithelial neoplasia (CIN) to develop a potential therapeutic strategy to combat this condition (Yanofsky et al., 2012).

Numerous ongoing clinical trials are investigating different aspects and methods of treatment for patients with HPV-related cancers. Among these trials, there are promising approaches utilizing zinc finger nucleases (ZFN-603 and ZFN-758), which specifically slice the E7 oncogenes of HPV 16 and 18. These investigations include patients suffering from HPV-related cervical intraepithelial neoplasia (CIN) (NCT02800369). The trials primarily emphasize unique HPV antigens for vaccination and explore their potential when combined with immune checkpoint agents. These approaches are developed for the enhancement of the immune response against HPV-driven malignancies and hold great promise in the advancement of effective treatments for individuals having HPV-related cancers.

7.1.2.4 Prospects for Treatment

Comprehensive cervical cancer control encompasses multiple aspects of care. It begins with primary prevention, which involves vaccination against HPV, the primary cause of cervical cancer. Secondary prevention focuses on screening and treating precancerous lesions to detect and address potential issues early. Tertiary prevention is based on diagnosing and preventing invasive cervical cancer to provide timely and effective care. Lastly, palliative care is provided to improve the quality of life for those facing advanced stages of the disease, offering supportive measures and symptom management.

The World Health Organization (WHO) recommends HPV vaccines that offer protection against HPV 16 and 18. These vaccines have received approval for use in numerous countries. It is advised that both females and males aged 9–26 years receive the vaccine for HPV. The optimal timing for vaccination is between ages 11 and 12. For individuals aged 9–14 years, two vaccine doses are administered, while those aged 15–45 years require three doses to achieve full immunity. The HPV vaccines provides significant benefits, as it effectively prevents HPV-related genital warts and reduces the risk of various cancers, including cervical, vaginal, vulvar, and anal cancers. Additionally, it lowers the likelihood of developing

HPV-associated throat and penile cancer. These HPV vaccines contain recombinant viral proteins with purified viruslike particles for the target HPV genotypes, offering protection against those HPV types. Essentially, the vaccines are devoid of the DNA component of viruses, rendering them inactive and incapable of infecting cells.

Three vaccines, Cervarix 2v, Gardasil 4v, and Gardasil 9v, are now available in many countries worldwide. These vaccines are categorized as bivalent, quadrivalent, and nonavalent vaccines, respectively. Recombinant *Saccharomyces cerevisiae* is used to manufacture both Gardasil vaccines, while Cervarix is produced via recombinant baculovirus-based system (Jung et al., 2015). Gardasil 9v, a nonavalent vaccine, provides extra protection against various HPV types 31, 33, 45, 52, and 58. Extensive clinical studies and post-marketing scrutiny conducted across different regions have demonstrated the safety of all three vaccines. However, unusual cases of a life-threatening allergic reaction known as anaphylaxis (1.7 cases out of one million doses) have been reported in various parts of the globe (Jung et al., 2015; Zhen et al., 2020).

HPV vaccines have been studied with a focus on potential postvaccination syndromes related to the adjuvants that enhance immunity against specific antigens. Reports of the autoimmune/inflammatory syndrome induced by adjuvants (AIAS) indicate a rate of 3.6 cases out of 100,000 vaccine doses. Nevertheless, it is a matter of debate whether rare cases of death following HPV vaccination can be attributed to the adjuvant or the vaccine itself (Kalof & Cooper, 2007; Adams et al., 2014). Despite these rare occurrences of postvaccination syndromes, the HPV vaccines are generally considered nontoxic and highly effective as a preventive measure. Nevertheless, they are not recommended in pregnancy due to insufficient scientific evidence from clinical trials. Nonetheless, clinical trials done in the past that involved pregnant women did not show any increase in spontaneous abortions in comparison to women who were not pregnant.

The availability of HPV vaccine and Pap tests have a significant contribution to the decline of cervical cancer cases over the years, leading to a decrease in mortality rates (Wilson, 2014). According to the American Cancer Society (ACS), the 5-year overall survival for cervical cancer is relatively high (Szymonowicz & Chen, 2020). However, it is important to note that infection of HPV itself cannot be treated but there are available therapies for HPV-related precancerous cellular changes and cancers. In cases of precancerous lesions, the loop electrosurgical excision procedure (LEEP) is commonly employed for treatment (Cuesta-Mateos et al., 2018). Electrical current is utilized for the removal of abnormal cells during LEEP. Additionally, other treatment options for precancerous lesions include laser therapy, cold knife conization, and cryotherapy (freezing the abnormal tissue). For external genital warts produced due to HPV infection, topical medications are used for treatment (Yanofsky et al., 2012). Therapeutic vaccines are also intended to be utilized for treating the ongoing disease or infection. Therapeutic vaccines evoke an immune response by exposing tumor-specific antigen strengthened by an adjuvant. There are different types of therapeutic vaccines against HPV infections like DNA, protein, peptide and synthetic long peptide, and epitope-based vaccines. E6 and E7 genes of HPV are the most suitable candidates either used separately or in fusion for the development of therapeutic vaccines. All of these are in the lengthy procedure of clinical trials and are currently unavailable for use but they can help us in the future to combat the infections caused by HPV (Khallouf et al., 2014). These various treatment options play a critical role in managing HPV-related conditions and preventing the progression of cervical cancer.

7.1.3 HPV in Head and Neck Cancer

Head and neck cancer (HNC) is the sixth most prevalent type of cancer worldwide, responsible for over 660,000 new cases and approximately 325,000 deaths every year (Kombe Kombe et al., 2021; Waheed et al., 2011). The incidence of HNC is steadily escalating and is projected to surge by 30% annually before 2030 (Sung et al., 2021; Johnson et al., 2020).

HNC also known as head and neck squamous cell carcinoma (HNSCC) and is linked with a set of malignant conditions that affect the mucosal layer across diverse anatomical regions in the upper aerodigestive tract including the larynx, paranasal sinuses, nasopharynx, oropharynx, oral cavity, oropharynx, and hypopharynx. The consumption of tobacco and alcohol is the main conventional risk factor. Over the recent few years, HPV emerged as a new risk factor, which particularly causes oropharyngeal squamous cell carcinoma (OPSCC), carving out a new subset of tumors different from non-HPV ones. Geographical distribution and anatomical location both creates variation in the frequency of HNSCC. Moreover, laryngeal and oral cavity cancers constitute the most widespread forms of HNSCC, with Europe exhibiting higher incidence and mortality rates in comparison to the United States (Bray et al., 2018). Men have a significantly higher predisposition to get HNSCC, with an occurrence ratio of 2:1 to 4:1 compared to women (Ferlay et al., 2015). Typically, diagnosis occurs between the ages of 50 and 70 (Hashim et al., 2019). Global statistics indicate a 36.5% escalation in HNSCC incidence during the past decade (Bray et al., 2018; McDermott & Bowles, 2019).

Oropharyngeal carcinoma which is HPV-positive (HNSCC/OPSCC) usually displays high level of susceptibility to anticancer medications and radiation, resulting in improved prognoses. Roughly one-fourth of all new HNSCC cases comprise OPSCC, with HPV acting as a pivotal causative risk factor (Faraji et al., 2017; Hay & Nixon, 2018). Additionally, infection by HPV is linked to some other subsites such as larynx and oral cavity, of HNSCC, although this information is subject to discrepancies primarily due to variations in anatomical tumor localization and diverse HPV detection methods (Götz et al., 2019). Based on a new meta-analysis, the incidence rates of patients with HPV-positive tumors in head and neck regions, cutting out larynx and oral cavity cancer (OPSCCs), are usually lower than those seen for tumors in the oropharynx (Götz et al., 2019).

The impact of HPV on HNSCC is particularly prominent in the crypts of tonsils present in the oropharynx. Similar to junction cells of cervical squamocolumnar, these crypt cells are arranged in an irregular, single-layered epithelium that is more susceptible to malignant transformation than cells in the mouth cavity (Herfs et al.,

2012). Even though oral cavity infections of HPV are more common than tonsillar crypt infections, only 3.9% of instances of oral cavity squamous cell carcinoma (OCSCC) are caused by HPV, in striking contrast to the 47% of OPSCC cases positive with HPV. Moreover, other viruses including polyomaviruses and the Epstein-Barr virus (EBV) also have a harmful influence on HNSCC, occasionally via coinfection with HPV (Leemans et al., 2018).

7.1.3.1 Association of HPV with Head and Neck Squamous Cell Carcinoma

Squamous cell carcinoma, originating from the lining of epithelia in the pharynx, oral cavity, and larynx accounts for around 90% of the cases of HNCs (Sung et al., 2021). A diverse range of cancers affects the region of the head and neck, each distinctively categorized based on their specific anatomical location, utilizing the World Health Organization (WHO) given classification, under International Classification of Diseases (ICD-10) (Johnson et al., 2020).

Predominantly, consumption of tobacco and alcohol are considered the main risk factors for head and neck squamous cell carcinoma (HNSCC), and they show close linkage to a significant portion of these tumors globally (Gillison et al., 2000a; Andrews et al., 2009). Along with these known risk factors, high-risk strains of human papillomaviruses (HPV), particularly HPV-16, are considered as independent contributors to a subset of HNSCC cases, with the strongest association being observed with oropharyngeal squamous cell carcinomas (OPSCC) (Herrero et al., 2003; Hobbs et al., 2006). While HPV's involvement in the pathogenesis of oral cancer is established, its most robust link to HNSCC is found in the context of oropharyngeal cancer (Chaturvedi et al., 2008).

Recently, a noticeable upsurge in squamous cell carcinoma cases within the region of the neck and head has been observed, garnering global attention (Kobayashi et al., 2018). It is now understood that there exist two distinct subtypes of squamous cell carcinoma within the region of the head and neck. Conventionally, the majority of oropharynx cancers exhibit hyper differentiation and occasionally manifest keratinization. This context of SCC can be categorized as either keratinized or nonkeratinized. The former predominantly affects older males, having a strong association with smoking and alcohol consumption, although HPV does not play its part. In contrast, nonkeratinizing squamous cell carcinoma is most prevalent among men aged 40 to 55, with minimal history of alcohol and tobacco use. In these cases, HPV stands out as the significant hallmark (El-Mofty & Lu, 2003; Chernock et al., 2009; Lewis Jr et al., 2012; El-Mofty et al., 2008).

In the United States, with the decline in smoking rates (Centers for Disease Control and Prevention, Chartbook on Long-term Trends in Health, United States, 2016), the occurrence of oropharyngeal cancer linked to tobacco, which is HPV negative, has diminished. However, the occurrence of oropharyngeal cancer positive with HPV is on the rise (Chaturvedi et al., 2008). The evidence gathered from the Surveillance, Epidemiology, and End Results (SEER) program's repository indicates a 50% reduction in the incidence of cancers that are HPV negative from 1988 to 2004, accompanied by a substantial 225% surge in HPV-positive oropharyngeal carcinoma (Pytynia et al., 2014). A study conducted by Junor and colleagues that included patients from the Edinburgh Cancer Centre revealed that between 1999 and 2001, 41% of head and neck cancer cases were positive for HPV, a percentage that increased to 63% between 2003 and 2005 (Junor et al., 2012).

7.1.3.2 Pathological Molecular Mechanism in Carcinogenesis

Head and neck squamous cell carcinoma (HNSCC) which is not related to HPV commonly displays p53 mutations due to cigarette smoking and alcohol consumption (Carlos de Vicente et al., 2004). The initial stages of the oncogenic process often involve the deletion of 9p21–22, leading to the loss of function of the P16 (tumor-suppressor gene) (Reed et al., 1996). The P16 gene generates P16INK4a, which forms a complex with regulators, i.e., cyclin-dependent kinase 4/cyclin-dependent kinase 6 (CDK4/CDK6), hindering the phosphorylation of Rb followed by inhibiting the activity of the E2F-related cell cycle transcription factor (pRB pathway) (Satyanarayana & Rudolph, 2004). In head and neck cancers associated with HPV, mutations are infrequent (10% or less) as wild-type p53 is typically present. Nevertheless, p53 is inactivated by HPVE6, leading to its reduced functionality. Moreover, P16 deletions are not observed in these tumors. Phosphorylation in Rb is disrupted by HPV E7, a key regulator of host cell cycling, leading to the inhibition of E2F control (Perrone et al., 2006) and an increase in P16 expression. P16 (tumor-suppressor gene) encodes an inhibitor of CDK, blocking cyclin D1, and cyclin-dependent kinase (CDK) 4/6 complex formation. This complex positively regulates cell cycle progression by the release of E2F through the phosphorylating of the Rb protein. Interestingly, the complex Rb protein/E2F also represses the P16 transcription. Thus, when HPV E7 interferes with the Rb protein, overexpression of P16 occurs (Fig. 7.1) (Perrone et al., 2006).

Consequently, there is a marked molecular-level distinction between head and neck malignancies that are positive with HPV and those that are HPV negative. Diverse techniques have been used for the detection of HPV in head and neck malignancies, including serum antibody assays, in situ hybridization, and consensus primer/type-specific PCR, real-time PCR. Similar to the screening of cervical cancer, globally approved guidelines suggest using methods like PCR (GP 5/GP 6) and hybrid capture II (QIAGEN). Immunohistochemistry of P16 serves as a useful marker for the detection of HPV infection, particularly in the cancers of head and neck. With 100% (sensitivity) and 79% (specificity), the immunohistochemistry of P16 is the best standard for detecting HPV, based on HPV 16 E6 and E7 mRNA in specimens of head and neck cancer. It also holds value as a surrogate marker for clinical detection of HPV. Moreover, P16 serves as a valuable molecular marker for assessing prognosis and is included in the WHO classification system (Romagosa et al., 2011).

There is consensus regarding the definition of HPV-associated tumors, requiring the expression of the virus's oncogenic proteins, namely, E6 and E7, which play a pivotal role in transforming infected cells. Notably, the integration of HPV-DNA into the host cell genome is deemed a crucial step for the expression of viral oncogenes in oropharyngeal cancer, mirroring the process seen in cervical cancer.

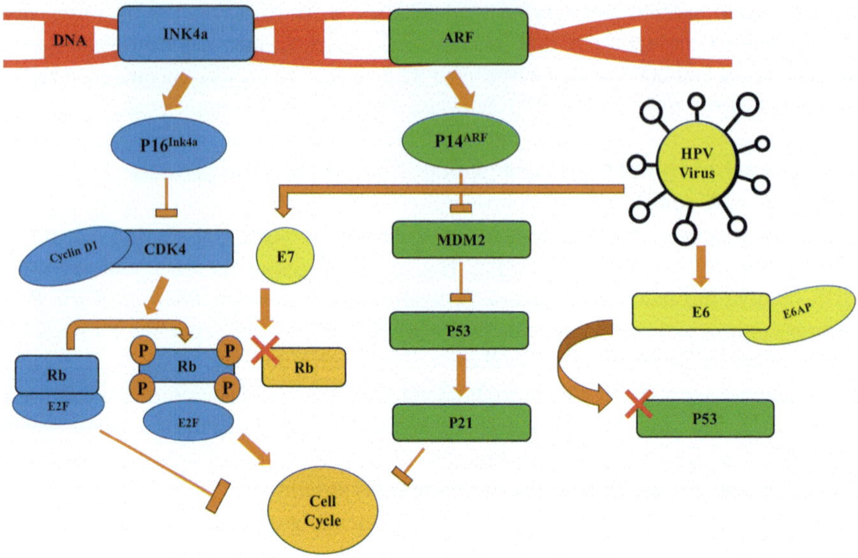

Fig. 7.1 High-risk HPV oncogene signaling pathways. Two known viral oncogenes are encoded by high-risk HPVs. The E6 protein inhibits apoptosis and DNA damage mediated by p53 (tumor suppressor). The E7 protein inhibits the cell cycle regulation mechanism mediated by the tumor-suppressor pRb (modified and redrawn from Kobayashi et al. (Kobayashi et al., 2018) with permission)

Irrespective of the mechanism leading to oncogene expression, the identification of HPV E6/E7 mRNA, whether determined by the expression of protein or DNA, is thought to be a supreme standard for the classification related to HPV, serving patient's stratification and epidemiological purposes (Gao et al., 2014; Snijders et al., 1992). Generally, more accessible methods are favored. In examining pathological specimens, HPV DNA detection through in situ hybridization (ISH) and PCR is often paired with P16 immunohistochemistry. Numerous novel techniques have been explored for the testing of HPV in head and neck malignancies (Morbini & Benazzo, 2016).

When cells infected with HPV develop malignancy, HPV DNA persists in the viral oncoproteins within the nucleus, and particularly E6 is detectable in nearly every case of oropharyngeal cancer driven by HPV (Holzinger et al., 2017; Kuhs et al., 2016). Reports indicate that antibodies of HP-16 E6 and HPV DNA in bodily and oral fluids can serve as tool for detecting head and neck malignancies infected by HPV and predicting the risk of reoccurrence (Mirghani et al., 2018).

In recent years, research using next-generation sequencing (NGS) for HNSCC has emerged. Matthias Lechner conducted a study involving 20 HPV+ and 20 HPV laser-capture microdissected pharyngeal carcinoma cases. These revealed two distinct subgroups for HPV-negative and HPV-positive oropharyngeal cancers. In HPV-negative cases, a TP53 mutation was detected in all instances, and disruption of the G1/S checkpoint due to deficiency of CDKN2A/B and/or amplification of

CCND1 was prominent. A study of 279 HNSCCs in 2015 explored somatic mutations, uncovering deletion of TRAF3, mutations of PIK3CA, and E2F1 amplification in oropharyngeal carcinoma positive with HPV. Subsets of HPV-negative HNSCC recognized concomitant mutations of CASP8 with amplicons on 11q with CCND1, FADD, BIRC2, and YAP1 or with HRAS. Both tumor types exhibited aberrations in NF-κB, cell cycle, cell death, and other cancer pathways (Lechner et al., 2013; Cancer Genome Atlas Network, 2015). Another study involving 92 HNSCC cases using NGS found TP53 to be the most incident mutation in 51% of cases, and it also reports mutation in CDKN2A (25%), CCND1 (24%), and PIK3CA (21%). While changes in TP53, CCND1, and CDKN2A appeared in higher percentages in tumors without HPV, the overall mutation burden was comparable between tumors positive with HPV and those which are without HPV. Tumors caused by HPV showed significant associations with genes related to the immune system. Mutations in NOTCH1, CDKN2A, and TP53 were linked to reduced overall survival. The FAT1 mutation was more prevalent in cisplatin responders, and alterations that are targetable like CDKN2A R58X and PIK3CA E545K have been detected in 15% of patients (Lim et al., 2019).

The evolving liquid biopsy technology involves analyzing a small amount of mRNA and DNA obtained from blood samples. A study by Allen et al. (Allen et al., 2018) introduced a novel in vitro diagnostic approach examining miRNA from cancer cell-exposed sera of HNSCC patients. Of the 377 detected miRNAs, 16 exhibited differential expression when cells are compared which are exposed to HNSCC serum with healthy individuals. The analysis of real-time PCR highlighted the downregulation of five genes linked to carcinogenesis, including SLC2A1 and P53, which are direct targets for the identified miRNAs. This technique holds promise for innovative therapeutic approaches and more precise diagnosis, potentially facilitating the initial detection of primary tumors and personalized treatment (Lindel et al., 2001).

7.1.3.3 Classification and Prognosis of HPV-Positive Head and Neck Cancer

Oropharyngeal carcinoma caused by HPV generally exhibits higher sensitivity to anticancer medications and radiation, leading to a more favorable prognosis in comparison with cancers that have other causes of occurrence than HPV. The development of those oropharyngeal carcinomas that are without the involvement of HPV is often linked to the loss of function due to mutations in the p53 gene (Butz et al., 1996; Bristow et al., 1996; Lindquist et al., 2007). In a subsequent phase II clinical trials by Eastern Cooperative Oncology involving patients with laryngeal and pharyngeal cancer, out of 60 patients, 63% were positive for HPV-16, and all these positive HPV-16 cases were also positive for P16. Notably, the respective response rate for prognosis was 84% in cases with HPV while 57% in cases without HPV. The 2-year survival rates (without progression) were 86% for HPV positive and 53% for cases without HPV, and overall 2-year survival rates were 95% and 62%, all indicating a significantly better prognosis for HPV-positive cancer (Fakhry et al., 2008).

Previous studies have investigated the prognostic implications of HPV in correlation with P16, as well as the impact of tobacco exposure on oropharyngeal cancer outcomes. In a research study of phase III which involved 400 oropharyngeal carcinomas, the 2-year survival rates (without progression) were 72% for HPV positive and 51% for cases without HPV and the overall 2-year survival rates were 88% and 67%. Due to the distinctive prognoses, the new WHO classification categorizes head and neck squamous cell carcinoma. Classification through HPV testing or immunostaining of P16 is suggested. Additionally, some recently suggested classification systems are based on the status of epidermal growth factor receptor (EGFR), dividing cases into two groups: P16-negative squamous cell carcinoma/HPV negative and P16-positive squamous cell carcinoma/HPV positive (Barnes et al., 2005).

A monoclonal antibody, cetuximab that stops EGFR function, has proven effective in cancers, i.e., head and neck (Mirghani et al., 2014; Saltz et al., 2004), colorectal cancer (Cunningham et al., 2004), and head and neck cancer, and it has also demonstrated specificity and cost-effectiveness for the types of cancers stated.

7.1.3.4 Progression of HPV-Positive Head and Neck Squamous Cell Carcinoma

About 15–20% of the patients suffering from HNSCC may comprise a significant number of young men and women who have other potential exposures that cause their disease as they have no documented history of alcohol and tobacco consumption. For instance, Epstein-Barr virus (EBV) and HPV particularly have been linked to the development of head and neck cancer (Ringström et al., 2002; Chen et al., 2005).

Certain head and neck squamous cell carcinomas development are suspected to be caused by mucosotropic high-risk papillomaviruses (HPVs) which are also associated with anogenital and cervical cancers. Although a relationship has been observed between HPV and a group of head and neck cancers since the 1980s, the International Agency for Research on Cancer (IARC) categorized type 16 and 18 HPV in the list of human carcinogens in 1995 (Chen et al., 2005). It has been demonstrated by molecular studies that high-risk HPVs with the expression of E6 and E7 oncoproteins disrupt important processes of the cell including apoptosis and cell cycle by stopping the action of tumor-suppressor proteins pRb and p53 (Azzimonti et al., 2004). It has been shown that these E6 and E7 viral oncoproteins contribute to the genetic development of HNSCC by altering primary human keratinocytes either from epithelia of the upper respiratory tract or from genital epithelia and by interfering with the regulation of the cell cycle (Gillison et al., 2000b). In about 45% of HNSCC cases, there is a mutation in the p53 gene (Sisk et al., 2002), and in HNSCC, pRb regulators including cyclin D and p16 are dysregulated. The infection of HPV can be an alternate molecular mechanism for the tumor development of HNSCC (Gillison et al., 2000b). Infection of HPV-16 may be a contributing factor, particularly for oropharyngeal and tonsillar HNSCC (Niedobitek et al., 1990). According to research, the HPV DNA frequency in HNSCC changes greatly based on the tumor location, i.e., about 15–60%. In 20–90% of cases of HPV-positive HNSCC, HPV-16 has been found, and the types 18, 31, and 33 of HPV are identified in the remaining cases. Moreover, few laryngeal and oral carcinomas have been found to contain HPV types 6 and 11 which are usually considered low-risk types (Chen et al., 2005).

7.1.3.5 Genetic Model for Progression of Head and Neck Cancer

Precursor Lesions in the Mucosal Linings

The most prevalent precursor oral squamous carcinoma lesion is leukoplakia which is a white lesion present in the oral cavity mucosa, and its prevalence ranges from 0.1% to 0.5% (Napier & Speight, 2008; Van der Waal, 2009). The lesion should be treated when feasible and analysis of the biopsy or specimen is done to examine dysplasia by microscopy and a grade is given as severe, mild, or moderate. Even though WHO has established criteria, the classification of dysplasia is challenging due to high assessment variation in intra- and inter-observer. Patients are observed by a wait-and-watch approach following diagnosis and treatment. The proportion of cancer-developing oral leukoplakia is dependent on several factors including research population, the definition of leukoplakia utilized, and the duration of the observation period, but it is acceptable to believe that the transition rate is between 1 and 2 percent every year (Napier & Speight, 2008; Van der Waal, 2009). Potential risk factors for development are gender (female), grade, location, and size of dysplasia. Recent investigations have found that genetic changes are the most accurate indicators of malignant transformation (Zhang et al., 2012). The presence of HPV has been assessed in leukoplakia lesions. However, the results are inconsistent, probably because of sensitive HPV-DNA assays producing false-positive results. The majority of trustworthy studies point to an extremely low incidence of less than 1% (Ha & Califano, 2004).

Field Cancerization

Due the acknowledgeable advancements in molecular science over the past 20 years, field cancerization can be described in molecular terms. Based on the genetic classification of the squamous epithelium morphological variations, the first multistep genetic progression model was proposed in 1996 for HNSCC (Ha & Califano, 2004). In dysplasia, there is loss of heterozygosity at 3p, 9p, and 17p chromosomal positions that usually represent the initial stages of carcinogenesis, whereas additional changes in chromosomes 11q, 4q, and 8 were commonly found in carcinomas and most probably represent the late carcinogenesis. These genetic markers and TP53 mutations were used to demonstrate that 35% of the mucosal epithelium in oropharyngeal and oral carcinomas has these genetic alterations (Tabor et al., 2001). The mucosal epithelium may appear normal when examined macroscopically but shown to be dysplastic after histological examination.

There is some knowledge about what appears to precede the emergence of fields. According to Van Houten et al., small p53-positive localized patches have been seen in mucosal epithelium next to tumors (Tabor et al., 2001). Based on the hypothesized patch-field-tumor-metastasis model for HNSCC progression, these clonal units mutated by p53 formed alongside the genetically specified fields and were thought to represent the first oncogenic alterations in the mucosa (Leemans et al., 2011). This model is supported by recent research. The stem cells along with the patches formed by them have been observed in mouse skin by Axin2 lineage tracing methods (Lim et al., 2013).

7.1.3.6 Clinical Characteristics of HPV-Positive HNSCC Patients

Being the sixth most common cancer, HPV-related head and neck cancer has a global burden of approximately 400,000 per annum. Oral infection caused by HPV plays a vital role in the pathogenesis of head and neck cancers (Cheng et al., 2020). As discussed earlier, HPVs have a significant affinity toward squamous epithelium, and they can induce benign hyperproliferation of the epithelial cells resulting in tumors as well as cancer. The archetypal high-risk types (HPV-16 and HPV-18) confer the ability to transform the epithelia of the genital as well as upper respiratory tract (Gallus et al., 2023). The epithelial transformation of infected cells is mainly the result of overexpression of E6 and E7, the two viral oncoproteins, which systematically inactivate p53 and pRb that are human tumor-suppressor proteins. The expression of E6 and E7 results in cellular proliferation, loss of cell cycle regulation, cellular disintegration, amplified frequency of unprompted and mutagen-induced mutations, and chromosomal variability (Nouel & White, 2022).

HPV not only affects the cells at the genetic level, but it also alters the demographic, histopathological, and clinical presentations of the affected person. HPV mostly affects young, sexually active individuals. Upon comparing the gender, males seem to be at equal risk to females. It is evident that HPV-positive tumors in most cases arise in the lingual and palatine tonsils in the oropharynx, but in a few cases, it also originates from other anatomic sites of the head and neck. Histopathology of HPV-positive tumors, in most studies, shows poorly differentiated and frequently basaloid histology (Aggarwal et al., 2021).

Even though HPV-associated HNSCC has a low recurrence rate, a small population is documented to experience recurrence and death from HNSCC. For these individuals, marker-assisted screening helps in the detection of disease persistence or early disease recurrence. For instance, a cohort study was conducted recently containing 59 affected persons who were screened by histology of their salivary rinsing for the detection of HNSCC recurrence. In this cohort examination, the finding of HPV E6 and E7 copy numbers by RT-PCR of the DNA isolated from salivary rinses showed a specificity of 50% and a sensitivity of 100% for the detection of oropharyngeal cancer (Chuang et al., 2008). However, due to the small size of the cohort, this study was not much appreciated and a further examination with the larger number of patients is needed to validate the outcomes.

Another hypothetically convenient clinical application is in the probing of the site of the primary tumor as a neck mass deprived of a known primary site of origin. Only 1–5% of the HNSCC detection tests comprise the neck mass of an unknown primary site of origin, and in those cases, a comprehensive clinical diagnosis including endoscopy and radiography could not help in the identification of the primary site of origin (Schmalbach & Miller, 2007; Wong, 2008). HPV-associated tumor markers are proven to be the best solution in the localization of the primary site of the tumor as well as the recurrence rate. The fine needle biopsies of HPV-affected neck metastases (Begum et al., 2007; Zhang et al., 2008) and immunohistochemistry for p16 overexpression (Begum et al., 2007) predict the oropharyngeal tumor at its early stage. Furthermore, a recent examination verified that the presence of cystic lymph nodes in in situ hybridization highly correlates with HPV positivity (Goldenberg et al., 2008). These studies validate that the presence of HPV in tumor tissue can help in probing the primary site of tumor origin for patients who present regional metastasis.

Lastly, the introduction of Gardasil® (the vaccine targeting HPV-16, HPV-18, HPV-6, and HPV-11) and the imminent FDA approval of Cervarix® (the vaccine targeting HPV-16 and HPV-18), if properly used, could limit or even converse the tendency of the cumulative prevalence of HNSCC. These vaccines are exceedingly effective at averting cervical intraepithelial neoplasia and cervical cancer (Harper et al., 2004; Villa, 2006), but it is still a matter of interest to see how extensively these will be utilized and whether their use has an impact on the incidence of recurrence of HNSCC.

7.1.3.7 Clinical Significance of HPV-Positive HNSCC

The detection of HPV in tumors isolated from patients with HNSCC provides future inferences for the diagnosis, prognosis, therapeutics, and prevention of HNSCC. Considering the prognosis, patients diagnosed with HPV-positive tumors have a better prognosis as compared to HPV-negative patients (Wichmann et al., 2023). The two distinct clinical cohorts have been conducted to study the epidemiological shift in patients with HPV-related HNSCC in Western countries, and the studies demonstrate that baseline characteristics of the patients vary tremendously. The HPV-positive HNSCC patients were typically young with small primary tumors, along with an advanced degree of lymph nodal disease (Ferreira, 2023). Smoking is also considered a clinical aspect of the patients and it comes out that most of them were current or former smokers (60–70%), whereas nonsmokers also make up a minor percentage of the HPV-positive HNSCC group (Bahig et al., 2022). Gender also contributes to the prognosis of the disease as male-to-female predominance is documented as 3:1, and it is the same as seen for the non-HPV-related HNSCC. However, gender difference remains a poorly understood risk factor in the transmission of HPV. The tumor, node, metastasis (TNM) staging revealed that HPV-positive tumors were locoregionally strikingly advanced as compared to the same staged HPV-negative tumors. As most patients had stage IV disease due to the advanced N-stage, these patients had excellent clinical outcomes (Tawk et al., 2022). Although upshots are enhanced compared with HPV-negative tumors, the inferences for the selection of patients for therapy by their HPV status have not yet been determinedly established (Ghosh et al., 2022).

7.1.3.8 Prospects for Treatment

Given the insights provided, the HPV virus is anticipated to emerge as a therapeutic target in the realm of human cancer treatment (Bonner et al., 2006). It is believed that the prognosis for cervical cancer can experience substantial enhancement through the deployment of early detection methods including HPV screening, cytodiagnosis, and P16 immunostaining (Hellner & Münger, 2011). Moreover, HPV vaccination has the potential to serve as a preventive measure against HPV-related cancers beyond cervical cancer. Notably, a considerable portion of oropharyngeal carcinoma cases, around 90%, are attributed to HPV-16, alongside HPV-18 (Kreimer, 2014). In this context, HPV vaccination holds the promise of exerting a more potent disease-suppressing impact compared to cervical cancer. Rigorous trials (randomized) have furnished robust evidence for the highly effaceable VLP-based vaccines approved by FDA, namely, the HPV16/18 (bivalent) vaccine (Cervarix, from GlaxoSmithKline) and the HPV 6/11/16/18 (quadrivalent) vaccine (Gardasil by Merck), against HPV16/18 infections and related diseases across the cervical, vaginal, and vulvar domains, along with anal HPV16/18 infections in women (Kreimer et al., 2011; Schiller et al., 2012; Muñoz et al., 2010). Along with these two vaccines, a nonavalent vaccine is also available which is created by Merck that targets five additional high-risk HPV types (HPV 31/33/45/52/58) in addition to the quadrivalent vaccine's HPV types 6/11/16/18. This nonavalent HPV vaccine appears to be both safe and efficient at avoiding chronic infection and precancerous lesions caused by HPV (Chatterjee, 2014).

Vaccination is one of the techniques offered for the control of oropharyngeal cancer related to HPV. The Advisory Committee on Immunization Practices (ACIP) of the Centers for Disease Control and Prevention (CDC) has extended the guideline for vaccination of HPV to include females and boys aged 11 to 12, starting at least from 9, with booster doses from age of 26 and 21 years for males and females, respectively. The protective effect of HPV vaccination must last at least two decades for it to be successful in the prevention of cancer in the oropharynx. Current research has shown that there are no declining systemic antibodies after vaccination of 8 years. However, following unsubstantiated reports of harmful events, Japan's Ministry of Health, Labor, and Welfare discontinued proactive HPV vaccine recommendations in June 2013 (Lehtinen et al., 2012).

As previously discussed in Sect. 7.1.2.4, several therapeutic vaccines are in the clinical trials, which could also be considered in relation to head and neck cancer. A peptide-based therapeutic vaccine Vicoryx, targeting E6/E7 fusion against HPV-induced cervical, head, and neck cancers, is under first-phase clinical trials. Vicoryx vaccine containing p16INK4A peptide along with Montanide can elicit an immune response and showed positive results when administered to patients (Reuschenbach et al., 2013). The use of the DNA-based vaccine INO-3112 against head and neck squamous cell carcinoma has proved to be beneficial in phase I and II clinical trials. This vaccine is comprised of three plasmids containing a fusion of E6 and E7 proteins of HPV 16 and 18 along with interleukin II and helps in generating cellular and humoral immunity in patients (Aggarwal et al., 2017). ADXS11–001 is another therapeutic vaccine candidate targeting oropharyngeal cancer caused by different HPV types. In ADXS11–001, a *Listeria monocytogenes* bacterium is used to transfect patients' cancerous cells with HPV antigens to elicit a cytotoxic T-cell response (Rumfield et al., 2020a). PDS0101 vaccine is an example of short peptide-based vaccine therapy; it has epitopes of E6 and E7 proteins of HPV 16 with R-DOTAP as an adjuvant. PDS0101 therapeutic vaccine is administered to the head and neck squamous cell carcinoma patients with positive infection of HPV-16. Most of the phase II clinical trials of PDS0101 use other immune-stimulating agents like pembrolizumab for desired outcomes (Rumfield et al., 2020b). To sum up, many therapeutic vaccines are in clinical trials that can help us in the future to combat the infections, tumors, and metastatic cancers caused by HPV. The availability of these vaccines depends on their safety, efficacy, and the results of detailed clinical trials.

In the cases of oropharyngeal carcinoma related to HPV, light invasive and reduced surgery treatments are being suggested. While it is restricted to T1–T2 stages of oropharynx cancer, transoral robotic surgery which is FDA approved since 2009 has gained prominence (Suzuki & Hosono, 2018). Moreover, numerous postoperative adjuvant therapies and reduction surgeries which are based on pathological staging are under investigation (Weinstein et al., 2012; de Almeida & Genden, 2012). As previously mentioned, EGFR is expressed in HNSCCs. Consequently, there is ongoing exploration into using a combination of cetuximab and radiation as an alternative to standard cisplatin therapy (Dowthwaite et al., 2012; Buglione et al., 2017; Roesler & Schwartsmann, 2013; Keck et al., 2015). Researchers have proposed that some HPV-positive patients could potentially undergo radiation dose reduction safely and feasibly through patient selection involving induction chemotherapy (IC). Three factors support this proposition. Firstly, HNSCCs with HPV are thought to be more responsive to radiation than HPV-negative counterparts (Hu et al., 2015). Secondly, doses similar to the supplementary dosage of radiation might be adequate for treating patients with asymptomatic disease. Lastly, the chemotherapy response can serve as an indicator of the subsequent response toward radiation therapy (Chen et al., 2017).

Chen et al. (Chen et al., 2017) conducted a phase II trial (NCT 01716195) involving 44 patients with stage OPCs that are III/IV P16 positive. These patients received two rounds of IC (paclitaxel and carboplatin) for 21 days, followed by paclitaxel and radiation. Radiation doses were reduced for partial or complete responders (54 Gy, $n = 24$) and for those with half or no response ($n = 20$, 60 Gy instead of the standard 70 Gy). All patients except one completed the IC. After 2 years, survival (progression free) and LGS (local region control) were 92% with 95% CI (77–97) and 95% with 95% CI (80–99). The 2-year incidence of grade 3 or worse mucosal and esophageal adverse events was 85% (95% CI 80–90) in patients treated with 54 Gy. These outcomes were achieved with a reduction of 15–20% in doses compared to the ECOG 2399 trial, which employed the same protocol but with a standard chemoradiotherapy regimen involving a dose of 70 Gy.

Apart from this, chemotherapy removal and other approaches to traditional photon beam treatment are also being explored as novel treatment methods (Mirghani et al., 2015). The differing prognosis and underlying mechanisms of head and neck cancer related to HPV compared to traditional head and neck cancers (mostly without HPV) indicate that the detection of HPV could lead to significant changes in future diagnosis, management, and therapeutic strategies (Junor et al., 2012). HPV's involvement isn't limited to pharyngeal cancer. It is also implicated in 23.5% of the cases of oral cancer and 24% of the cases of laryngeal cancer. This suggests that evidence for HPV vaccination could potentially be expanded to contain oral and laryngeal cancers as well (Fakhry et al., 2008).

References

Abbas, T., & Dutta, A. (2009). p21 in cancer: Intricate networks and multiple activities. *Nature Reviews. Cancer, 9*(6), 400–414.

Adams, A. K., Wise-Draper, T. M., & Wells, S. I. (2014). Human papillomavirus induced transformation in cervical and head and neck cancers. *Cancers (Basel), 6*(3), 1793–1820.

Aggarwal, C., Cohen, R. B., Morrow, M. P., Kraynyak, K., Bauml, J., Weinstein, G. S., … & Bagarazzi, M. L. (2017). Immunogenicity results using human papillomavirus (HPV) specific DNA vaccine, INO-3112 (HPV16/HPV18 plasmids+ IL-12) in HPV+ head and neck squamous cell carcinoma (HNSCCa), 6073–6073.

Aggarwal, C., Cohen, R. B., Morrow, M. P., Kraynyak, K. A., Sylvester, A. J., Knoblock, D. M., et al. (2019). Immunotherapy targeting HPV16/18 generates potent immune responses in HPV-associated head and neck cancer. *Clinical Cancer Research, 25*(1), 110–124.

Aggarwal, N., Yadav, J., Chhakara, S., Janjua, D., Tripathi, T., Chaudhary, A., Chhokar, A., Thakur, K., Singh, T., & Bharti, A. C. (2021). Phytochemicals as potential chemopreventive and chemotherapeutic agents for emerging human papillomavirus–driven head and neck cancer: Current evidence and future prospects. *Frontiers in Pharmacology, 12*, 699044.

Allen, B., Schneider, A., Victoria, B., Nunez Lopez, Y. O., Muller, M., Szewczyk, M., Pazdrowski, J., Majchrzak, E., Barczak, W., Golusinski, W., & Golusinski, P. (2018). Blood serum from head and neck squamous cell carcinoma patients induces altered MicroRNA and target gene expression profile in treated cells. *Frontiers in Oncology, 8*, 217.

Andrews, E., Seaman, W. T., & Webster-Cyriaque, J. (2009). Oropharyngeal carcinoma in non-smokers and non-drinkers: A role for HPV. *Oral Oncology, 45*(6), 486–491.

Azzimonti, B., Pagano, M., Mondini, M., De Andrea, M., Valente, G., Monga, G., et al. (2004). Altered patterns of the interferon-inducible gene IFI16 expression in head and neck squamous cell carcinoma: Immunohistochemical study including correlation with retinoblastoma protein, human papillomavirus infection and proliferation index. *Histopathology, 45*(6), 560–572.

Bahig, H., Huang, S. H., & O'Sullivan, B. (2022). Oligometastatic head and neck cancer: Challenges and perspectives. *Cancers, 14*(16), 3894.

Barnes, L., Eveson, J., Reichart, P., & Sidransky, D. (2005). World Health Organization classification of tumours. Pathology and genetics of head and neck tumours. *International Agency on Research on Cancer, 85*, 75–81.

Barr, A. R., Cooper, S., Heldt, F. S., Butera, F., Stoy, H., Mansfeld, J., et al. (2017). DNA damage during S-phase mediates the proliferation-quiescence decision in the subsequent G1 via p21 expression. *Nature Communications, 8*(1), 14728.

Begum, S., Gillison, M. L., Nicol, T. L., & Westra, W. H. (2007). Detection of human papillomavirus-16 in fine-needle aspirates to determine tumor origin in patients with metastatic squamous cell carcinoma of the head and neck. *Clinical Cancer Research, 13*(4), 1186–1191.

Bienkowska-Haba, M., Zwolinska, K., Keiffer, T., Sapp, M. (2022). Human papillomavirus genome copy number is maintained by S phase amplification, leakage to the cytosol during mitosis and lysosomal degradation in G1 phase; doi:https://doi.org/10.1101/2022.06.24.497574.

Bonner, J. A., Harari, P. M., Giralt, J., Azarnia, N., Shin, D. M., Cohen, R. B., Jones, C. U., Sur, R., Raben, D., Jassem, J., & Ove, R. (2006). Radiotherapy plus cetuximab for squamous-cell carcinoma of the head and neck. *New England Journal of Medicine, 354*(6), 567–578.

Bray, F., Ferlay, J., Soerjomataram, I., Siegel, R. L., Torre, L. A., & Jemal, A. (2018). Global cancer statistics 2018: GLOBOCAN estimates of incidence and mortality worldwide for 36 cancers in 185 countries. *CA: A Cancer Journal For Clinicians, 68*(6), 394–424.

Bristow, R. G., Benchimol, S., & Hill, R. P. (1996). The p53 gene as a modifier of intrinsic radiosensitivity: Implications for radiotherapy. *Radiotherapy and Oncology, 40*(3), 197–223.

Buglione, M., Maddalo, M., Corvò, R., Pirtoli, L., Paiar, F., Lastrucci, L., Stefanacci, M., Belgioia, L., Crociani, M., Vecchio, S., & Bonomo, P. (2017). Subgroup analysis according to human papillomavirus status and tumor site of a randomized phase II trial comparing cetuximab and cisplatin combined with radiation therapy for locally advanced head and neck cancer. *International Journal of Radiation Oncology, Biology, Physics, 97*(3), 462–472.

Butz, K., Geisen, C., Ullmann, A., Spitkovsky, D., & Hoppe-Seyler, F. (1996). Cellular responses of HPV-positive cancer cells to genotoxic anti-cancer agents: Repression of E6/E7-oncogene expression and induction of apoptosis. *International Journal of Cancer, 68*(4), 506–513.

Cancer Genome Atlas Network. (2015). Comprehensive genomic characterization of head and neck squamous cell carcinomas. *Nature, 517*(7536), 576.

Carlos de Vicente, J., Junquera Gutierrez, L. M., Zapatero, A. H., Fresno Forcelledo, M. F., Hernández-Vallejo, G., & Lopez Arranz, J. S. (2004). Prognostic significance of p53 expression in oral squamous cell carcinoma without neck node metastases. Head & Neck: Journal for the sciences and specialties of the. *Head and Neck, 26*(1), 22–30.

Centers for Disease Control and Prevention, Chartbook on Long-term Trends in Health, United States, 2016. Retrieved August 27, 2018, from https://www.cdc.gov/nchs/data/hus/hus16.pdf

Cervantes, J. L., & Doan, A. H. (2018). Discrepancies in the evaluation of the safety of the human papillomavirus vaccine. *Memórias do Instituto Oswaldo Cruz, 113*, e180063.

Chatterjee, A. (2014). The next generation of HPV vaccines: Nonavalent vaccine V503 on the horizon. *Expert Review of Vaccines, 13*(11), 1279–1290.

Chaturvedi, A. K., Engels, E. A., Anderson, W. F., & Gillison, M. L. (2008). Incidence trends for human papillomavirus-related and-unrelated oral squamous cell carcinomas in the United States. *Journal of Clinical Oncology, 26*(4), 612–619.

Chen, R., Aaltonen, L., & Vaheri, A. (2005). Human papillomavirus type 16 in head and neck carcinogenesis. *Reviews in Medical Virology, 15*(6), 351–363.

Chen, A. M., Felix, C., Wang, P. C., Hsu, S., Basehart, V., Garst, J., Beron, P., Wong, D., Rosove, M. H., Rao, S., & Melanson, H. (2017). Reduced-dose radiotherapy for human papillomavirus-associated squamous-cell carcinoma of the oropharynx: A single-arm, phase 2 study. *The Lancet Oncology, 18*(6), 803–811.

Chen, J., Wang, D., Wang, Z., Wu, K., Wei, S., Chi, X., et al. (2023). Critical residues involved in the coassembly of L1 and L2 capsid proteins of human papillomavirus 16. *Journal of Virology, 97*, 3. https://doi.org/10.1128/jvi.01819-22

Cheng, L., Wang, Y., & Du, J. (2020). Human papillomavirus vaccines: An updated review. *Vaccine, 8*(3), 391.

Chernock, R. D., El-Mofty, S. K., Thorstad, W. L., Parvin, C. A., & Lewis, J. S. (2009). HPV-related nonkeratinizing squamous cell carcinoma of the oropharynx: Utility of microscopic features in predicting patient outcome. *Head and Neck Pathology, 3*, 186–194.

Chuang, A. Y., Chuang, T. C., Chang, S., Zhou, S., Begum, S., Westra, W. H., et al. (2008). Presence of HPV DNA in convalescent salivary rinses is an adverse prognostic marker in head and neck squamous cell carcinoma. *Oral Oncology, 44*(10), 915–919.

Cuesta-Mateos, C., Alcaraz-Serna, A., Somovilla-Crespo, B., & Muñoz-Calleja, C. (2018). Monoclonal antibody therapies for hematological malignancies: Not just lineage-specific targets. *Frontiers in Immunology, 8*, 1936.

Cunningham, D., Humblet, Y., Siena, S., Khayat, D., Bleiberg, H., Santoro, A., Bets, D., Mueser, M., Harstrick, A., Verslype, C., & Chau, I. (2004). Cetuximab monotherapy and cetuximab plus irinotecan in irinotecan-refractory metastatic colorectal cancer. *New England Journal of Medicine, 351*(4), 337–345.

de Almeida, J. R., & Genden, E. M. (2012). Robotic surgery for oropharynx cancer: Promise, challenges, and future directions. *Current Oncology Reports, 14*, 148–157.

Deniz, Z., Uraz, S., Holem, R., Ozaras, R., & Tahan, V. (2022). Human papillomavirus infection and oropharyngeal and gastrointestinal cancers: A causal relationship? *Diseases, 10*(4), 94.

Doorbar, J., Egawa, N., Griffin, H., Kranjec, C., & Murakami, I. (2015). Human papillomavirus molecular biology and disease association. *Reviews in Medical Virology, 25*(S1), 2–23. https://doi.org/10.1002/rmv.1822

Dowthwaite, S. A., Franklin, J. H., Palma, D. A., Fung, K., Yoo, J., & Nichols, A. C. (2012). The role of transoral robotic surgery in the management of oropharyngeal cancer: A review of the literature. *International Scholarly Research Notices, 2012*.

Dust, K., Carpenter, M., Chen, J. C., Grant, C., McCorrister, S., Westmacott, G. R., et al. (2022). Human papillomavirus 16 E6 and E7 oncoproteins alter the abundance of proteins associated with DNA damage response, immune signaling and epidermal differentiation. *Viruses, 14*(8), 1764. https://doi.org/10.3390/v14081764

El-Mofty, S. K., & Lu, D. W. (2003). Prevalence of human papillomavirus type 16 DNA in squamous cell carcinoma of the palatine tonsil, and not the oral cavity, in young patients: A distinct clinicopathologic and molecular disease entity. *The American Journal of Surgical Pathology, 27*, 1463–1470. [CrossRef] [PubMed].

El-Mofty, S. K., Zhang, M. Q., & Davila, R. M. (2008). Histologic identification of human papillomavirus (HPV)-related squamous cell carcinoma in cervical lymph nodes: A reliable predictor of the site of an occult head and neck primary carcinoma. *Head and Neck Pathology, 2*, 163–168.

Fakhry, C., Westra, W. H., Li, S., Cmelak, A., Ridge, J. A., Pinto, H., Forastiere, A., & Gillison, M. L. (2008). Improved survival of patients with human papillomavirus–positive head and neck squamous cell carcinoma in a prospective clinical trial. *Journal of the National Cancer Institute, 100*(4), 261–269.

Faraji, F., Eisele, D. W., & Fakhry, C. (2017). Emerging insights into recurrent and metastatic human papillomavirus-related oropharyngeal squamous cell carcinoma. *Laryngoscope Investigative Otolaryngology, 2*(1), 10–18.

Ferlay, J., Soerjomataram, I., Dikshit, R., Eser, S., Mathers, C., Rebelo, M., Parkin, D. M., Forman, D., & Bray, F. (2015). Cancer incidence and mortality worldwide: Sources, methods and major patterns in GLOBOCAN 2012. *International Journal of Cancer, 136*(5), E359–EE86.

Ferreira, C. C. (2023). The relation between human papillomavirus (HPV) and oropharyngeal cancer: A review. *PeerJ, 11*, e15568.

Fontan, C. T., Das, D., Bristol, M. L., James, C. D., Wang, X., Lohner, H., et al. (2020). Human papillomavirus 16 E2 repression of TWIST1 transcription is a potential mediator of HPV16 cancer outcomes. doi:https://doi.org/10.1101/2020.09.25.314484.

Gallus, R., Nauta, I. H., Marklund, L., Rizzo, D., Crescio, C., Mureddu, L., Tropiano, P., Delogu, G., & Bussu, F. (2023). Accuracy of p16 IHC in classifying HPV-driven OPSCC in different populations. *Cancers, 15*(3), 656.

Gao, G., Johnson, S. H., Kasperbauer, J. L., Eckloff, B. W., Tombers, N. M., Vasmatzis, G., & Smith, D. I. (2014). Mate pair sequencing of oropharyngeal squamous cell carcinomas reveals that HPV integration occurs much less frequently than in cervical cancer. *Journal of Clinical Virology, 59*(3), 195–200.

Ghosh, S., Shah, P. A., & Johnson, F. M. (2022). Novel systemic treatment modalities including immunotherapy and molecular targeted therapy for recurrent and metastatic head and neck squamous cell carcinoma. *International Journal of Molecular Sciences, 23*(14), 7889.

Gillison, M. L., Koch, W. M., Capone, R. B., Spafford, M., Westra, W. H., Wu, L., Zahurak, M. L., Daniel, R. W., Viglione, M., Symer, D. E., & Shah, K. V. (2000a). Evidence for a causal association between human papillomavirus and a subset of head and neck cancers. *Journal of the National Cancer Institute, 92*(9), 709–720.

Gillison, M. L., Koch, W. M., Capone, R. B., Spafford, M., Westra, W. H., Wu, L., et al. (2000b). Evidence for a causal association between human papillomavirus and a subset of head and neck cancers. *Journal of the National Cancer Institute, 92*(9), 709–720.

Goldenberg, D., Begum, S., Westra, W. H., Khan, Z., Sciubba, J., Pai, S. I., et al. (2008). Cystic lymph node metastasis in patients with head and neck cancer: An HPV-associated phenomenon. *Head Neck, 30*(7), 898–903.

Götz, C., Bischof, C., Wolff, K. D., & Kolk, A. (2019). Detection of HPV infection in head and neck cancers: Promise and pitfalls in the last ten years-a meta-analysis. *Molecular and Clinical Oncology, 10*(1), 17–28.

Graham, S. V. (2017). The human papillomavirus replication cycle, and its links to cancer progression: A comprehensive review. *Clinical Science, 131*(17), 2201–2221. https://doi.org/10.1042/cs20160786

Ha, P. K., & Califano, J. A. (2004). The role of human papillomavirus in oral carcinogenesis. *Critical Reviews in Oral Biology and Medicine, 15*(4), 188–196.

Hamada, K., Shirakawa, T., Gotoh, A., Roth, J. A., & Follen, M. (2006). Adenovirus-mediated transfer of human papillomavirus 16 E6/E7 antisense RNA and induction of apoptosis in cervical cancer. *Gynecologic Oncology, 103*(3), 820–830.

Harper, D. M., Franco, E. L., Wheeler, C., Ferris, D. G., Jenkins, D., Schuind, A., et al. (2004). Efficacy of a bivalent L1 virus-like particle vaccine in prevention of infection with human papillomavirus types 16 and 18 in young women: A randomised controlled trial. *Lancet, 364*(9447), 1757–1765.

Hasan, Y., Furtado, L., Tergas, A., Lee, N., Brooks, R., McCall, A., et al. (2020). A phase 1 trial assessing the safety and tolerability of a therapeutic DNA vaccination against HPV16 and HPV18 E6/E7 oncogenes after chemoradiation for cervical cancer. *International Journal of Radiation Oncology, Biology, Physics, 107*(3), 487–498.

Hashim, D., Genden, E., Posner, M., Hashibe, M., & Boffetta, P. (2019). Head and neck cancer prevention: From primary prevention to impact of clinicians on reducing burden. *Annals of Oncology, 30*(5), 744–756.

Hay, A., & Nixon, I. J. (2018). Recent advances in the understanding and management of oropharyngeal cancer. *F1000Research*, 7.

Hellner, K., & Münger, K. (2011). Human papillomaviruses as therapeutic targets in human cancer. *Journal of Clinical Oncology, 29*(13), 1785.

Herfs, M., Yamamoto, Y., Laury, A., Wang, X., Nucci, M. R., McLaughlin-Drubin, M. E., Münger, K., Feldman, S., McKeon, F. D., Xian, W., & Crum, C. P. (2012). A discrete population of squamocolumnar junction cells implicated in the pathogenesis of cervical cancer. *National Academy of Sciences of the United States of America, 109*(26), 10516–10521.

Herrero, R., Castellsagué, X., Pawlita, M., Lissowska, J., Kee, F., Balaram, P., Rajkumar, T., Sridhar, H., Rose, B., Pintos, J., & Fernández, L. (2003). Human papillomavirus and oral cancer: The International Agency for Research on Cancer multicenter study. *Journal of the National Cancer Institute, 95*(23), 1772–1783.

Hobbs, C. G., Sterne, J. A., Bailey, M., Heyderman, R. S., Birchall, M. A., & Thomas, S. J. (2006). Human papillomavirus and head and neck cancer: A systematic review and meta-analysis. *Clinical Otolaryngology, 31*(4), 259–266.

Holzinger, D., Wichmann, G., Baboci, L., Michel, A., Höfler, D., Wiesenfarth, M., Schroeder, L., Boscolo-Rizzo, P., Herold-Mende, C., Dyckhoff, G., & Boehm, A. (2017). Sensitivity and specificity of antibodies against HPV16 E6 and other early proteins for the detection of HPV16-driven oropharyngeal squamous cell carcinoma. *International Journal of Cancer, 140*(12), 2748–2757.

Hu, Z., Müller, S., Qian, G., Xu, J., Kim, S., Chen, Z., Jiang, N., Wang, D., Zhang, H., Saba, N. F., & Shin, D. M. (2015). Human papillomavirus 16 oncoprotein regulates the translocation of β-catenin via the activation of epidermal growth factor receptor. *Cancer, 121*(2), 214–225.

Huh, K., Zhou, X., Hayakawa, H., Cho, J.-Y., Libermann, T. A., Jin, J., et al. (2007). Human papillomavirus type 16 E7 oncoprotein associates with the cullin 2 ubiquitin ligase complex, which contributes to degradation of the retinoblastoma tumor suppressor. *Journal of Virology, 81*(18), 9737–9747.

Johnson, D. E., Burtness, B., Leemans, C. R., Lui, V. W., Bauman, J. E., & Grandis, J. R. (2020). Head and neck squamous cell carcinoma. *Nature Reviews Disease Primers, 6*(1), 92.

Jørgensen, K. R., & Jensen, J. B. (2020). Human papillomavirus and urinary bladder cancer revisited. *Amis, 128*(2), 72–79.

Jung, H. S., Rajasekaran, N., Song, S. Y., Kim, Y. D., Hong, S., Choi, H. J., et al. (2015). Human papillomavirus E6/E7-specific siRNA potentiates the effect of radiotherapy for cervical cancer in vitro and in vivo. *International Journal of Molecular Sciences, 16*(6), 12243–12260.

Junor, E., Kerr, G., Oniscu, A., Campbell, S., Kouzeli, I., Gourley, C., & Cuschieri, K. (2012). Benefit of chemotherapy as part of treatment for HPV DNA-positive but p16-negative squamous cell carcinoma of the oropharynx. *British Journal of Cancer, 106*(2), 358–365.

Kalof, A. N., & Cooper, K. (2007). Our approach to squamous intraepithelial lesions of the uterine cervix. *Journal of Clinical Pathology, 60*(5), 449–455.

Keck, M. K., Zuo, Z., Khattri, A., Stricker, T. P., Brown, C. D., Imanguli, M., Rieke, D., Endhardt, K., Fang, P., Brägelmann, J., & DeBoer, R. (2015). Integrative analysis of head and neck cancer identifies two biologically distinct HPV and three non-HPV subtypes. *Clinical Cancer Research, 21*(4), 870–881.

Khallouf, H., Grabowska, A. K., & Riemer, A. B. (2014). Therapeutic vaccine strategies against human papillomavirus. *Vaccine, 2*(2), 422–462.

Kim, S. H., Joung, J. Y., Chung, J., Park, W. S., Lee, K. H., & Seo, H. K. (2014). Detection of human papillomavirus infection and p16 immunohistochemistry expression in bladder cancer with squamous differentiation. *PLoS One, 9*(3), e93525.

Kobayashi, K., Hisamatsu, K., Suzui, N., Hara, A., Tomita, H., & Miyazaki, T. (2018). A review of HPV-related head and neck cancer. *Journal of Clinical Medicine, 7*(9), 241.

Kombe Kombe, A. J., Li, B., Zahid, A., Mengist, H. M., Bounda, G.-A., Zhou, Y., et al. (2021). Epidemiology and burden of human papillomavirus and related diseases, molecular pathogenesis, and vaccine evaluation. *Frontiers in Public Health*, 8. https://doi.org/10.3389/fpubh.2020.552028

Kreimer, A. R. (2014). Prospects for prevention of HPV-driven oropharynx cancer. *Oral Oncology, 50*(6), 555–559.

Kreimer, A. R., González, P., Katki, H. A., Porras, C., Schiffman, M., Rodriguez, A. C., Solomon, D., Jiménez, S., Schiller, J. T., Lowy, D. R., & van Doorn, L. J. (2011). Efficacy of a bivalent HPV 16/18 vaccine against anal HPV 16/18 infection among young women: A nested analysis within the Costa Rica vaccine trial. *The Lancet Oncology, 12*(9), 862–870.

Kuhs, K. A., Pawlita, M., Gibson, S. P., Schmitt, N. C., Trivedi, S., Argiris, A., Kreimer, A. R., Ferris, R. L., & Waterboer, T. (2016). Characterization of human papillomavirus antibodies in individuals with head and neck cancer. *Cancer Epidemiology, 42*, 46–52.

Lechner, M., Frampton, G. M., Fenton, T., Feber, A., Palmer, G., Jay, A., Pillay, N., Forster, M., Cronin, M. T., Lipson, D., & Miller, V. A. (2013). Targeted next-generation sequencing of head and neck squamous cell carcinoma identifies novel genetic alterations in HPV+ and HPV-tumors. *Genome Medicine, 5*, 1–2.

Leemans, C. R., Braakhuis, B. J. M., & Brakenhoff, R. H. (2011). The molecular biology of head and neck cancer. *Nature Reviews. Cancer, 11*(1), 9–22.

Leemans, C. R., Snijders, P. J., & Brakenhoff, R. H. (2018). The molecular landscape of head and neck cancer. *Nature Reviews Cancer, 18*(5), 269–282.

Lehtinen, M., Paavonen, J., Wheeler, C. M., Jaisamrarn, U., Garland, S. M., Castellsague, X., Skinner, S. R., et al. (2012). Overall efficacy of HPV-16/18 AS04-adjuvanted vaccine against grade 3 or greater cervical intraepithelial neoplasia: 4-year end-of-study analysis of the randomised, double-blind PATRICIA trial. *The lancet oncology 13*(1), 89–99.

Lewis, J. S., Jr., Khan, R. A., Masand, R. P., Chernock, R. D., Zhang, Q., Al-Naief, N. S., Muller, S., McHugh, J. B., Prasad, M. L., Brandwein-Gensler, M., & Perez-Ordonez, B. (2012). Recognition of nonkeratinizing morphology in oropharyngeal squamous cell carcinoma–a prospective cohort and interobserver variability study. *Histopathology, 60*(3), 427–436.

Li, S., Hong, X., Wei, Z., Xie, M., Li, W., Liu, G., et al. (2019). Ubiquitination of the HPV oncoprotein E6 is critical for E6/E6AP-mediated p53 degradation. *Frontiers in Microbiology, 10*, 2483.

Lim, X., Tan, S. H., Koh, W. L. C., Chau, R. M. W., Yan, K. S., Kuo, C. J., et al. (2013). Interfollicular epidermal stem cells self-renew via autocrine Wnt signaling. *Science (80-), 342*(6163), 1226–1230.

Lim, S. M., Cho, S. H., Hwang, I. G., Choi, J. W., Chang, H., Ahn, M. J., Park, K. U., Kim, J. W., Ko, Y. H., Ahn, H. K., & Cho, B. C. (2019). Investigating the feasibility of targeted next-generation sequencing to guide the treatment of head and neck squamous cell carcinoma. *Cancer Research and Treatment: Official Journal of Korean Cancer Association, 51*(1), 300–312.

Lindel, K., Beer, K. T., Laissue, J., Greiner, R. H., & Aebersold, D. M. (2001). Human papillomavirus positive squamous cell carcinoma of the oropharynx: A radiosensitive subgroup of head and neck carcinoma. *Cancer, 92*(4), 805–813.

Lindquist, D., Romanitan, M., Hammarstedt, L., Näsman, A., Dahlstrand, H., Lindholm, J., Onelöv, L., Ramqvist, T., Ye, W., Munck-Wikland, E., & Dalianis, T. (2007). Human papillomavirus is a favourable prognostic factor in tonsillar cancer and its oncogenic role is supported by the expression of E6 and E7. *Molecular Oncology, 1*(3), 350–355.

McBride, A. A. (2021). Human papillomaviruses: Diversity, infection and host interactions. *Nature Reviews Microbiology, 20*(2), 95–108. https://doi.org/10.1038/s41579-021-00617-5

McDermott, J. D., & Bowles, D. W. (2019). Epidemiology of head and neck squamous cell carcinomas: Impact on staging and prevention strategies. *Current Treatment Options in Oncology, 20*, 1–3.

McNeil, M. M., Weintraub, E. S., Duffy, J., Sukumaran, L., Jacobsen, S. J., Klein, N. P., et al. (2016). Risk of anaphylaxis after vaccination in children and adults. *The Journal of Allergy and Clinical Immunology, 137*(3), 868–878.

Milano, G., Guarducci, G., Nante, N., Montomoli, E., & Manini, I. (2023). Human papillomavirus epidemiology and prevention: Is there still a gender gap? *Vaccine, 11*(6), 1060.

Mirghani, H., Amen, F., Moreau, F., Guigay, J., Hartl, D. M., & Guily, J. L. (2014). Oropharyngeal cancers: Relationship between epidermal growth factor receptor alterations and human papillomavirus status. *European Journal of Cancer, 50*(6), 1100–1111.

Mirghani, H., Amen, F., Tao, Y., Deutsch, E., & Levy, A. (2015). Increased radiosensitivity of HPV-positive head and neck cancers: Molecular basis and therapeutic perspectives. *Cancer Treatment Reviews, 41*(10), 844–852.

Van Doorslaer, K., Li, Z., Xirasagar, S., Maes, P., Kaminsky, D., Liou, D., et al. (2016). The papillomavirus episteme: A major update to the papillomavirus sequence database. *Nucleic Acids Research, 45*, D1. https://doi.org/10.1093/nar/gkw879

Mirghani, H., Kuhs, K. A., & Waterboer, T. (2018). Biomarkers for early identification of recurrences in HPV-driven oropharyngeal cancer. *Oral Oncology, 82*, 108–114.

Moody, C. A. (2022). Regulation of the innate immune response during the human papillomavirus life cycle. *Viruses, 14*(8), 1797. https://doi.org/10.3390/v14081797

Moody, C. A., & Laimins, L. A. (2010). Human papillomavirus oncoproteins: Pathways to transformation. *Nature Reviews Cancer, 10*(8), 550–560. https://doi.org/10.1038/nrc2886

Morante, A. V., Baboolal, D. D., Simon, X., Pan, E. C.-Y., & Meneses, P. I. (2022). Human papillomavirus minor capsid protein L2 mediates intracellular trafficking into and passage beyond the endoplasmic reticulum. Microbiology. *Spectrum, 10*, 3. https://doi.org/10.1128/spectrum.01505-22

Morbini, P., & Benazzo, M. (2016). Human papillomavirus and head and neck carcinomas: Focus on evidence in the babel of published data. *Acta Otorhinolaryngologica Italica, 36*(4), 249.

Muñoz, N., Kjaer, S. K., Sigurdsson, K., Iversen, O. E., Hernandez-Avila, M., Wheeler, C. M., Perez, G., Brown, D. R., Koutsky, L. A., Tay, E. H., & Garcia, P. J. (2010). Impact of human papillomavirus (HPV)-6/11/16/18 vaccine on all HPV-associated genital diseases in young women. *Journal of the National Cancer Institute, 102*(5), 325–339.

Napier, S. S., & Speight, P. M. (2008). Natural history of potentially malignant oral lesions and conditions: An overview of the literature. *Journal of Oral Pathology & Medicine, 37*(1), 1–10.

Niedobitek, G., Pitteroff, S., Herbst, H., Shepherd, P., Finn, T., Anagnostopoulos, I., et al. (1990). Detection of human papillomavirus type 16 DNA in carcinomas of the palatine tonsil. *Journal of Clinical Pathology, 43*(11), 918–921.

Nouel, J., & White, E. A. (2022). ZER1 contributes to the carcinogenic activity of high-risk HPV E7 proteins. *MBio, 13*(6), e02033–e02022.

Olesen, T. B., Munk, C., Christensen, J., Andersen, K. K., & Kjaer, S. K. (2014). Human papillomavirus prevalence among men in sub-Saharan Africa: A systematic review and meta-analysis. *Sexually Transmitted Infections, 90*(6), 455–462.

Organization WH. (2017). Safety update of HPV vaccines. *Weekly Epidemiological Record, 92*(28), 398–401.

Palma, S., Gnambs, T., Crevenna, R., & Jordakieva, G. (2021). Airborne human papillomavirus (HPV) transmission risk during ablation procedures: A systematic review and meta-analysis. *Environmental Research, 192*, 110437.

Palmieri, B., Poddighe, D., Vadala, M., Laurino, C., Carnovale, C., & Clementi, E. (2017). Severe somatoform and dysautonomic syndromes after HPV vaccination: Case series and review of literature. *Immunologic Research, 65*, 106–116.

Pérez-Benavente, B. (2020). Alihamze Fathinajafabadi Nasresfahani, and Rosa Farràs. "Ubiquitin-regulated cell proliferation and cancer.". *Proteostasis and Disease: From Basic Mechanisms to Clinics*, 3–28.

Perrone, F., Suardi, S., Pastore, E., Casieri, P., Orsenigo, M., Caramuta, S., Dagrada, G., Losa, M., Licitra, L., Bossi, P., & Staurengo, S. (2006). Molecular and cytogenetic subgroups of oropharyngeal squamous cell carcinoma. *Clinical Cancer Research, 12*(22), 6643–6651.

Plummer, M., de Martel, C., Vignat, J., Ferlay, J., Bray, F., & Franceschi, S. (2016). Global burden of cancers attributable to infections in 2012: A synthetic analysis. *Lancet Global Health, 4*(9), e609–e616.

Pytynia, K. B., Dahlstrom, K. R., & Sturgis, E. M. (2014). Epidemiology of HPV-associated oropharyngeal cancer. *Oral Oncology, 50*(5), 380–386.

Rafael, T. S., Rotman, J., Brouwer, O. R., van der Poel, H. G., Mom, C. H., Kenter, G. G., de Gruijl, T. D., & Jordanova, E. S. (2022). Immunotherapeutic approaches for the treatment of HPV-associated (pre-) cancer of the cervix, vulva and penis. *Journal of Clinical Medicine, 11*(4), 1101.

Rahimi, S. (2020). HPV-related squamous cell carcinoma of oropharynx: A review. *Journal of Clinical Pathology, 73*(10), 624–629.

Reed, A. L., Califano, J., Cairns, P., Westra, W. H., Jones, R. M., Koch, W., Ahrendt, S., Eby, Y., Sewell, D., Nawroz, H., & Bartek, J. (1996). High frequency of p16 (CDKN2/MTS-1/INK4A) inactivation in head and neck squamous cell carcinoma. *Cancer Research, 56*(16), 3630–3633.

Reuschenbach, M., Karbach, J., Faulstich, F., Sauer, M., Kloor, M., Rafiyan, M. R., et al. (2013). Abstract A12: Vicoryx: A phase I/IIa clinical trial using a p16INK4a derived peptide as vaccine in patients with advanced human papillomavirus-associated cancer. *Cancer Prevention Research, 6*(11_Supplement), A12–A12.

Ringström, E., Peters, E., Hasegawa, M., Posner, M., Liu, M., & Kelsey, K. T. (2002). Human papillomavirus type 16 and squamous cell carcinoma of the head and neck. *Clinical Cancer Research, 8*(10), 3187–3192.

Roesler, R., & Schwartsmann, G. (2013). Failure of anti-EGFR therapy in p16-positive head and neck cancer. *The Lancet Oncology, 14*(11), e436–e4e7.

Romagosa, C., Simonetti, S., Lopez-Vicente, L., Mazo, A., Lleonart, M. E., Castellvi, J., & Ramon y Cajal S. (2011). p16Ink4a overexpression in cancer: A tumor suppressor gene associated with senescence and high-grade tumors. *Oncogene, 30*(18), 2087–2097.

Rumfield, C. S., Roller, N., Pellom, S. T., Schlom, J., & Jochems, C. (2020a). Therapeutic vaccines for HPV-associated malignancies. *ImmunoTargets and Therapy, 9*, 167.

Rumfield, C. S., Pellom, S. T., Morillon, Y. M., II, Schlom, J., & Jochems, C. (2020b). Immunomodulation to enhance the efficacy of an HPV therapeutic vaccine. *Journal for Immunotherapy of Cancer, 8*, 1.

Sabeena, S., Bhat, P., Kamath, V., & Arunkumar, G. (2017). Possible non-sexual modes of transmission of human papilloma virus. *The Journal of Obstetrics and Gynaecology Research, 43*(3), 429–435.

Saltz, L. B., Meropol, N. J., Loehrer, P. J., Sr., Needle, M. N., Kopit, J., & Mayer, R. J. (2004). Phase II trial of cetuximab in patients with refractory colorectal cancer that expresses the epidermal growth factor receptor. *Journal of Clinical Oncology, 22*(7), 1201–1208.

Satyanarayana, A., & Rudolph, K. L. (2004). p16 and ARF: Activation of teenage proteins in old age. *The Journal of Clinical Investigation, 114*(9), 1237–1240.

Scarth, J. A., Patterson, M. R., Morgan, E. L., & Macdonald, A. (2021). The human papillomavirus oncoproteins: A review of the host pathways targeted on the road to transformation. *Journal of General Virology, 102*, 3. https://doi.org/10.1099/jgv.0.001540

Schiller, J. T., Castellsagué, X., & Garland, S. M. (2012). A review of clinical trials of human papillomavirus prophylactic vaccines. *Vaccine, 30*, F123–FF38.

Schim van der Loeff, M. F., Mooij, S. H., Richel, O., de Vries, H. J. C., & Prins, J. M. (2014). HPV and anal cancer in HIV-infected individuals: A review. *Current HIV/AIDS Reports, 11*, 250–262.

Schmalbach, C. E., & Miller, F. R. (2007). Occult primary head and neck carcinoma. *Current Oncology Reports, 9*, 139–146.

Sisk, E. A., Soltys, S. G., Zhu, S., Fisher, S. G., Carey, T. E., & Bradford, C. R. (2002). Human papillomavirus and p53 mutational status as prognostic factors in head and neck carcinoma. *Head Neck, 24*(9), 841–849.

Snijders, P. J., Meijer, C. J., van den Brule, A. J., Schrijnemakers, H. F., Snow, G. B., & Walboomers, J. M. (1992). Human papillomavirus (HPV) type 16 and 33 E6/E7 region transcripts in tonsillar carcinomas can originate from integrated and episomal HPV DNA. *Journal of General Virology, 73*(8), 2059–2066.

Sung, H., Ferlay, J., Siegel, R. L., Laversanne, M., Soerjomataram, I., Jemal, A., & Bray, F. (2021). Global cancer statistics 2020: GLOBOCAN estimates of incidence and mortality worldwide for 36 cancers in 185 countries. *CA: A Cancer Journal for Clinicians, 71*(3), 209–249.

Suzuki, S., & Hosono, A. (2018). No association between HPV vaccine and reported post-vaccination symptoms in Japanese young women: Results of the Nagoya study. *Papillomavirus Research, 5*, 96–103.

Szymonowicz, K. A., & Chen, J. (2020). Biological and clinical aspects of HPV-related cancers. *Cancer Biology & Medicine, 17*(4), 864.

Tabor, M. P., Brakenhoff, R. H., van Houten, V. M. M., Kummer, J. A., Snel, M. H. J., Snijders, P. J. F., et al. (2001). Persistence of genetically altered fields in head and neck cancer patients: Biological and clinical implications. *Clinical Cancer Research, 7*(6), 1523–1532.

Tawk, B., Debus, J., & Abdollahi, A. (2022). Evolution of a paradigm switch in diagnosis and treatment of HPV-driven head and neck cancer—Striking the balance between toxicity and cure. *Frontiers in Pharmacology, 12*, 753387.

Van der Waal, I. (2009). Potentially malignant disorders of the oral and oropharyngeal mucosa; terminology, classification and present concepts of management. *Oral Oncology, 45*(4–5), 317–323.

Vardas, E., Giuliano, A. R., Goldstone, S., Palefsky, J. M., Moreira, E. D., Jr., Penny, M. E., et al. (2011). External genital human papillomavirus prevalence and associated factors among heterosexual men on 5 continents. *The Journal of Infectious Diseases, 203*(1), 58–65.

Villa, L. L. (2006). Prophylactic HPV vaccines: Reducing the burden of HPV-related diseases. *Vaccine, 24*, S23–SS8.

Waheed, M. Tahir, Nadja Thönes, Martin Müller, S. Waqas Hassan, N. Mona Razavi, Elke Lössl, Hans-Peter Kaul, and Andreas G. Lössl. (2011). Transplastomic expression of a modified human papillomavirus L1 protein leading to the assembly of capsomeres in tobacco: a step towards cost-effective second-generation vaccines. *Transgenic research 20*, 271–282.

Weinstein, G. S., O'Malley, B. W., Jr., Magnuson, J. S., Carroll, W. R., Olsen, K. D., Daio, L., Moore, E. J., & Holsinger, F. C. (2012). Transoral robotic surgery: A multicenter study to assess feasibility, safety, and surgical margins. *The Laryngoscope, 122*(8), 1701–1707.

Wichmann, G., Rudolph, J., Henger, S., Engel, C., Wirkner, K., Wenning, J. R., Zeynalova, S., Wiegand, S., Loeffler, M., Wald, T., & Dietz, A. (2023). Is high-risk sexual behavior a risk factor for oropharyngeal cancer? *Cancers, 15*(13), 3356.

Wilson, V. G. (2014). The role of ubiquitin and ubiquitin-like modification systems in papillomavirus biology. *Viruses, 6*(9), 3584–3611.

Wolf, J. M. (2023). Human papillomavirus infection: Epidemiology, biology, host interactions, cancer development, prevention, and therapeutics; doi:https://doi.org/10.22541/au.168670985.56660105/v1.

Wong, R. J. (2008). Current status of FDG-PET for head and neck cancer. *Journal of Surgical Oncology, 97*(8), 649–652.

Yanofsky, V. R., Patel, R. V., & Goldenberg, G. (2012). Genital warts: A comprehensive review. *The Journal of Clinical and Aesthetic Dermatology, 5*(6), 25.

Zhang, M. Q., El-Mofty, S. K., & Dávila, R. M. (2008). Detection of human papillomavirus-related squamous cell carcinoma cytologically and by in situ hybridization in fine-needle aspiration biopsies of cervical metastasis: A tool for identifying the site of an occult head and neck primary. *Cancer, 114*(2), 118–123.

Zhang, L., Poh, C. F., Williams, M., Laronde, D. M., Berean, K., Gardner, P. J., et al. (2012). Loss of heterozygosity (LOH) profiles—Validated risk predictors for progression to oral cancer. *Cancer Prevention Research, 5*(9), 1081–1089.

Zhen, S., Lu, J., Liu, Y.-H., Chen, W., & Li, X. (2020). Synergistic antitumor effect on cervical cancer by rational combination of PD1 blockade and CRISPR-Cas9-mediated HPV knockout. *Cancer Gene Therapy, 27*(3–4), 168–178.

Kaposi's Sarcoma Associated Herpes Virus

8

Monaza Bibi and Nosheen Masood

Abstract

Kaposi's sarcoma-associated herpesvirus is another name for human herpesvirus 8. This virus is an oncogenic virus that is a member of the Herpesviridae family. There are two phases to the viral cycle: a brief lytic phase and a latency phase that causes a permanent infection in target cells and the production of a few genes, such as v-FLIP, v-cyclin, and LANA-1. The seroprevalence and risk factors of infection vary globally, and saliva seems to play a major role in the transmission of viruses. Kaposi's sarcoma in all of its epidemiological forms, classic, endemic, iatrogenic, epidemic, and non-epidemic, contains KSHV. KSHV primarily exists in a dormant state in a Kaposi's sarcoma lesion, but less than 5% of virus particles are replicative and have been linked to the possible growth of nearby cells, indicating that they play important roles in the carcinogenesis process. Kaposi's sarcoma is caused by KSHV-encoded oncogenic proteins that can alter cellular pathways to produce features common to all cancers, such as immune escape, angiogenesis, inflammation, and stimulation of cell proliferation. These proteins are also involved in the development of apoptosis. There are still significant questions, nevertheless, particularly regarding the precise pathways of transmission, variations in KSHV seroprevalence, and the global frequency of Kaposi's sarcoma. We summarized here the most recent data on the genome, replication, seroprevalence, structure, diversification, and lytic and latent oncogenesis proteins of the KSHV virus that cause Kaposi's sarcoma. Finally, we discussed the viral, immunological, and environmental variables that

M. Bibi (✉)
Microbiology and Biotechnology Research Lab, Department of Biotechnology, Fatima Jinnah Women University, Rawalpindi, Pakistan

N. Masood
Department of Biotechnology, Fatima Jinnah Women University, Rawalpindi, Pakistan

N. Masood, A. Yasmin (eds.), *The Microbiome and Cancer*,
https://doi.org/10.1007/978-981-97-6948-3_8

may be linked to KSHV transmission and that may also influence the course of Kaposi's sarcoma.

Keywords

Oncogenic viruses · Kaposi's sarcoma · HHV-8 · KSHV · Latency · Oncogenic viral proteins · and LANA-1

8.1 Introduction

Kaposi's sarcoma-associated herpesvirus KSHV, commonly known as *human herpesvirus 8 (HHV-8)*, is an oncogenic virus identified in 1994 in the United States (Chang et al., 1994). For the first time, two viral fragments about 330 and 631 base pairs (bp) were found in skin biopsies taken from AIDS patients who had developed Kaposi's sarcoma (KS). These fragments, discovered using a representational difference analysis assay, were phylogenetically similar to the Epstein-Barr virus (EBV) and the saimiri herpesvirus (*Saimiriine herpesvirus 2, SaHV-2).* The following year, KSHV was found in two hemopathies: primary effusion lymphoma (PEL) (Cesarman et al., 1995), a rare kind of non-Hodgkin's malignant lymphoma (LMNH), and multicentric Castleman's disease (MCD), a lymphoproliferative condition (Soulier et al., 1995). The International Agency for Research on Cancer (IARC) has classified KSHV as one of the ten carcinogenic infectious agents in humans (https://monographs.iarc.who.int/list-of-classifications retrieved on December 4, 2021). This decision was made after epidemiological and molecular studies supported the link between KSHV and Kaposi's sarcoma.

8.2 Classification

Phylogenic research indicates that KSHV is the first and only *Human rhinovirus* discovered to date and that it is a member of the *Herpesviridae* family and the *Gammaherpesvirinae* subfamily (Russo et al., 1996; Moore & Chang, 2001). KSHV shares similarities with two other *rhadinoviruses*: *Herpesvirus saimiri,* which was discovered in 1968 in squirrel monkeys, and *Herpesvirus ateles*, which was discovered in 1972 in spider monkeys (Melendez et al., 1968, 1972). Since then, a number of studies have reported that macaques, African green monkeys, monkeys, and chimpanzees are infected with a *rhadinovirus* that is linked to KSHV (Desrosiers et al., 1997; Bosch et al., 1998; Alexander et al., 2000; Greensill et al., 2000; Schultz et al., 2000).

8.3 Structure

8.3.1 Viral Particle

KSHV is a large double-stranded DNA virus that is roughly 165 to 170 kilo base pairs in size. It is linked to a typical herpesvirus icosahedral capsid, which is made up of four structural proteins, SCIP (small capsomer-interacting protein), MCP (major capsid protein), TRI-1 and TRI-2 (triplex component 1 and 2), and CSAF (scaffolding or assembly protein) (Nealon et al., 2001; Zhu et al., 2005) that, although not a component of the capsid, is necessary for its assembly. Tegument, which is described as an electron-dense substance, envelops the capsid and is made up of inner and outer layers linked to the eight proteins that the ORF codes for: 11, 21, 33, 45, 52, 63, 64, and 75 (Sathish et al., 2012; Yan et al., 2019). They could have a role in both the initial stages of viral replication and the genome's entrance during the initial infection. Ultimately, an envelope derived from the lipid bilayer of a KSHV-infected cell surrounds the tegument (Dollery, 2019). This layer contains eight glycoproteins that are involved in the contact with host cells: gB, K8.1A, K8.1B, gH, gL, gM, and gN. The diameter of a KSHV virus particle is between 110 and 150 nm (Fig. 8.1) (Akula et al., 2002; Spear & Longnecker, 2003; Zhu et al., 2005; Muniraju et al., 2019; Yan et al., 2019; Birkmann et al., 2001).

8.3.2 Viral Genome

The genome of KSHV was initially sequenced from the BC-1 PEL cell line. The 801 pb long terminal repeat sequences that are very GC-rich flank the center long unique region (LUR), which is approximately 137 kilo bp in length (Russo et al., 1996). The 13 pre-microRNAs (miRNAs) encoding for 25 miRNAs and the 90 open reading frames (ORFs) are included in the LUR. Fifteen ORFs are unique to the *Rhadinovirus* (K1 to K15), but the majority of ORFs (from ORF4 to ORF75) are shared by the human herpesvirus (Zhu et al., 2005) (Fig. 8.2). Many of these ORFs encode different proteins that are part of both latent and lytic infection programs. Cellular homologous genes, such as v-IL-6 (viral interleukin 6), v-BCL-2, v-FLIP (viral Fas-associated protein with death domain-like interleukin-1β-converting enzyme/caspase-8-inhibitory protein), and v-Cyclin, are among the viral genes encoded by KSHV (Cesarman et al., 2019). Moreover, in line with its capacity for transformation, KSHV encodes a variety of proteins that exhibit proliferative, anti-apoptotic, angiogenic, immune-evading, and inflammatory characteristics (Mesri et al., 2014). Circular RNAs, or circRNAs, have been discovered more recently to be produced by KSHV from a number of KSHV genes. These include K7.3, polyadenylated nuclear (PAN) RNA, and K10 (vIRF4, viral interferon regulatory factor 4). Each and every KSHV circRNA is integrated into KSHV virions and has the ability to be produced as early products in freshly infected cells (Abere et al., 2020). The KSHV genome is not integrated into the host cell's genome during the latency phase, when it expresses as an episome, also known as a circular mini chromosome. To replicate and express the several proteins that are intrinsic in the lytic phase, the KSHV genome adopts a linear shape.

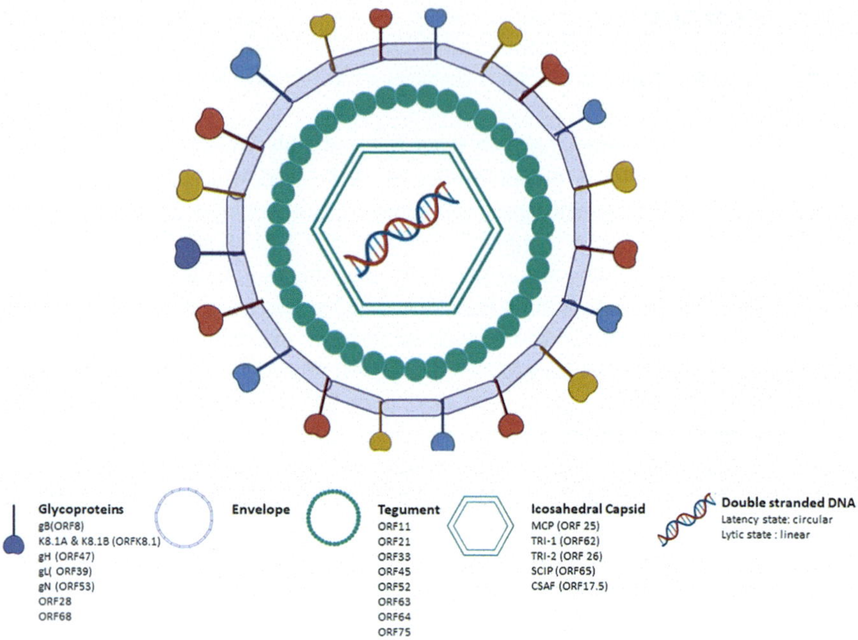

Fig. 8.1 Structure of the herpesvirus associated with *Kaposi's sarcoma virus*. The KSHV virion consists of the following four morphologically different components, just like the other human herpesviruses: The components of a virus include its double-stranded DNA genome; its icosahedral capsid, which encloses its DNA; a lipid envelope made from cell membranes; and tegument, the electron-dense substance that lies between the envelope and the capsid. Multiple viral proteins are included in each layer. *gX* glycoprotein, *MCP* major capsid protein, *ORF* open reading frame, *SCIP* small capsomer-interacting protein, *CSAF* scaffolding or assembly protein, DNA acid desoxyribonucleic. Triplex components 1 and 2 are referred to as TRI 1 and 2. Created using BioRender.com

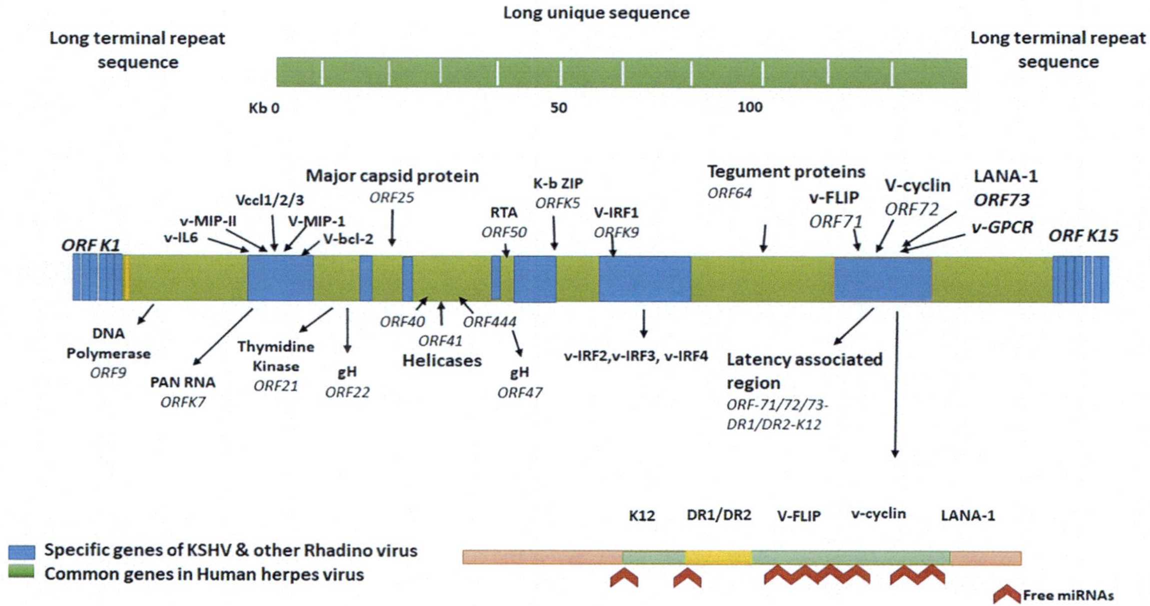

Fig. 8.2 Kaposi's sarcoma herpesvirus genomic structure. The long, unique center region, measuring around 137 kilo base pairs, is surrounded by extremely GC-rich 801 pb-long terminal repeat sequences. There are 90 open reading frames (ORFs) and 13 pre-miRNAs (miRNAs) in the genome, which together encode 25 miRNAs. Fifteen ORFs are unique to the Rhadinovirus (K1 to K15, region in red), while the majority of ORFs are shared with the human herpesvirus (region in orange). As an oncovirus, KSHV also encoded proteins with proliferative, antiapoptotic, angiogenic, immuno-evading, and inflammatory characteristics, as well as those implicated in carcinogenesis (shown in blue). The latency-associated region, situated at the right end of the genome, is home to the majority of the genes that encode the proteins involved in the latency state, such as LANA-1, v-FLIP, v-cyclin, mtRNAs, and Kaposin. DNA, deoxyribonucleic acid; *gX* protein glycoprotein, *Kb* kilobase, *K-bZIP* K-basic leucine zipper protein, *ORF* open reading frame, *MCP* major capsid protein, *LANA-1* latency-associated nuclear antigen 1). The terms PAN RNA, RTA, TRI 1 and 2, v-IL6, v-MIP, v-encoded chemokines, v-IRF, v-cyclin, and viral macrophage inflammatory protein are acronyms for transduction and transcription activator, viral interleukin, and viral macrophage inflammatory protein. Two instances of pathogenic proteins include the viral G-protein coupled receptor (v-GPCR) and the viral Fas-associated protein with death domain-like interleukin-1β-converting enzyme/caspase-8-inhibitory protein (v-FLIP)

8.4 Viral Replication

Similar to other herpesviruses, KSHV can experience two distinct infection programs: latent infection, which is its default pathway, and episodes of lytic reactivation, which result in the production of infectious particles and the death of the host cell. These programs are distinguished by intricate yet distinctive gene expression patterns.

8.4.1 Entry in the Cell

KSHV entrance is a multistep, sequential procedure (Chakraborty et al., 2012). Initially, KSHV's surface glycoproteins connect to their target cell in an unspecific manner. Its attachment is facilitated by many interactions with cell surface proteoglycans, which are mainly mediated by gB, gHgL, and K8.1A (Akula et al., 2001a, b; Birkmann et al., 2001). These proteoglycans are not necessary, but they improve KSHV entrance by concentrating the virions on the target cell's surface (Akula et al., 2001a, b; Birkmann et al., 2001; Dollery et al., 2019). Then depending on the target cell, the viral glycoproteins can interact with their specific cellular receptors, such as heparan sulphate, α3β1 integrin, xCT (cystine-glutamate transporter), or DC-SIGN (Dendritic Cell-Specific Intercellular adhesion molecule-3-Grabbing Non-integrin), and promote various endocytosis routes (mostly fusion or endocytosis), which allow the virus particle to enter (Akula et al., 2003; Inoue et al., 2003; Kerur et al., 2010; Kumar & Chandran, 2016).

8.4.2 KSHV Latent Cycle

During a cell division, the viral circular episome, which is in the latent stage, attaches itself to the host cell's chromosome and replicates in accordance with the replication machinery of each daughter cell. This kind of replication is nonproductive; it doesn't cause the host cell to lyse or produce infectious virions. A chronic latent infection is created when only a tiny portion of the viral genes are expressed, which permits the KSHV to evade the immune system and preserve the genome in proliferating cells. The latent phase's primary protein is the LANA-1 (latency-associated nuclear antigen 1) protein, which is encoded by ORF-73 (Uppal et al., 2014). During cell division, the KSHV genome is passed to daughter cells by the interaction of LANA-1 with the host DNA's histones through MeCP2 (methyl CpG binding protein 2) and DEK. This process contributes to the maintenance of the viral genome in the nucleus. Additionally, LANA-1 functions as a transcriptional regulator, namely, by inhibiting the expression of the protein known as RTA (replication and transcription activator), which is necessary for the virus to enter the multiplication phase. The other two major proteins expressed are homologues of human cellular proteins, specifically the ORF-71/K13-encoded v-FLIP and the ORF-72-encoded v-cyclin. Additionally, these three latency proteins have the capacity to regulate a number of oncogenic processes, including cell survival, differentiation, and proliferation, which permits immortalization and the transformation of the cell and, in turn, induces the different phenotypic traits linked to KSHV-associated illnesses (Yan et al., 2019).

In addition, a few other proteins are expressed during the latent cycle: 25 miRNAs (micro RNAs), which are noncoding single-stranded RNAs of roughly 19–23 nucleotides and are encoded by 13 pre-miRNAs; these proteins are known as Kaposin proteins and are encoded by ORF-K12 (Gottwein et al., 2011; Guo et al., 2017; Qin et al., 2017). These latter play a dual role in controlling the viral cycle (e.g., miRNA-K7-5p and miRNA-K9-5p) and repress the posttranscriptional expression of the viral protein RTA (Lin et al., 2011). In addition, they regulate the cell cycle and interact with the host cell to support the virus's persistence and the emergence of associated diseases (Samols et al., 2007; Gallaher et al., 2013; Guo et al., 2017; Qin et al., 2017; Hussein et al., 2019).

8.4.3 KSHV Lytic Cycle

Tetradecanoyl phorbol myristyl acetate (TPA) and n-butyrate (Yu et al., 1999) are chemical substances that can cause KSHV to reactivate and enter the replicative phase. Other stimuli that can affect the virus include oxidative stress, hypoxia, and reactive oxygen species (Davis et al., 2001; Ye et al., 2011). New infectious virus particles are created during this replicative phase, which also causes the infected cells to lyse. KSHV exhibits temporal and sequential gene expression patterns that are separated into three phases: immediate-early (IE), early (E), and late (L). These patterns are similar to those of other herpesviruses. Transactivators of the E and L genes are encoded by the IE genes; one such gene is ORF-50, which codes for the RTA protein, which is the leader for the start of the lytic phase (Gradoville et al., 2000; Guito & Lukac, 2012). The RTA protein does this by forming an association with the cellular transcription factor RBP-Jk, which enables it to activate its own promoter and sustain itself at a level high enough to continue the lytic cycle (Aneja & Yuan, 2017; DeCotiis & Lukac, 2017). Alongside ORF50, other genes are also expressed simultaneously, including ORF K8 and ORF57. Recent reports have also suggested that PAN RNA, a noncoding viral RNA produced by ORF-K7, may contribute to viral reactivation by binding to the LANA-1 protein and removing its inhibitory effect on the expression of lytic genes (Conrad, 2016; Rossetto et al., 2013). Following the interferon-expressed phase (IE phase), the E genes produced an encoded viral protein that was mainly needed for gene expression and DNA replication, specifically the viral thymidine kinase (ORF21), viral phosphotransferase (ORF36, homologue of the UL97 protein kinase encoded by human *cytomegalovirus*), viral DNA polymerase (ORF9) involved in the replication of the viral genome, bZIP, vIRF-1 (viral interferon regulatory factor), v-IL-6, v-CCLs (viral-encoded chemokines), and v-GPCR (viral G-protein coupled receptor) (Jacobs & Damania, 2011). Eventually, the L genes that code for the maturation and structural proteins of the viral particles start to express around 24 h after the lytic phase begins. The nucleus is where the new viral particle is assembled once all three steps have been completed. In the nucleus of freshly synthesized capsids, the KSHV genome is integrated. Teguments are then acquired in the cytoplasm, and the virus buds through host membranes to produce envelopes. Ultimately, the host cell releases the viral progeny (Aneja & Yuan, 2017) (Fig. 8.3).

Importantly, the distinction between the latency and lytic phases is not as evident as previously assumed. So far, the concept of interactions between the two phases of the cycle, as well as the expression of some viral transcripts overlapping on both phases, has been broadly accepted (Wang et al., 2006; Lei et al., 2010; Mesri et al., 2010; Abere et al., 2017).

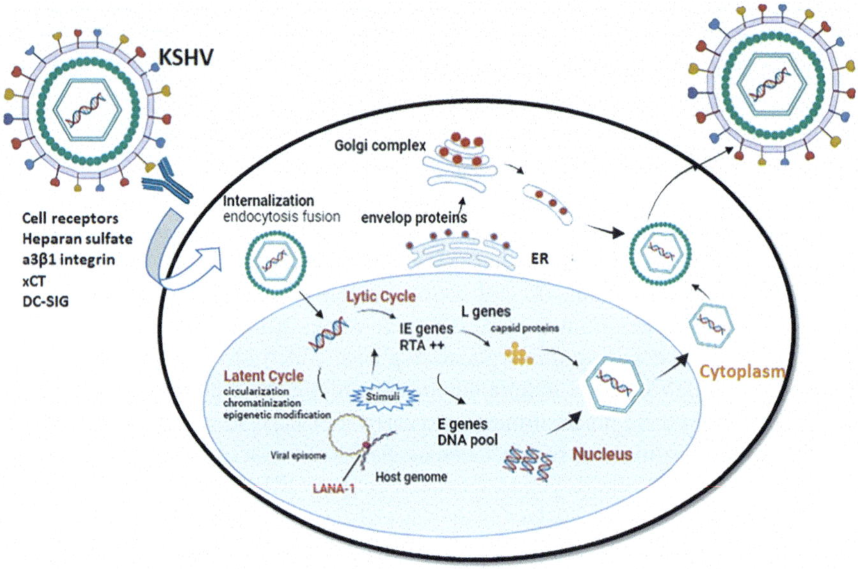

Fig. 8.3 Kaposi's sarcoma-associated herpesvirus viral cycles. The precise cellular receptors that the KSHV virus's glycoproteins bind to vary depending on the target cell. This interaction triggers distinct endocytosis pathways that allow the viral particle to enter. The nucleocapsid is transferred to the cell nucleus where only the viral genome is released after it has entered the cytoplasm. The following two viral cycles can occur in the nucleus: (1) the lytic cycle, which results in the lysis of host cells and the creation of fresh infectious particles, and (2) the latent cycle that keeps the viral episome in the nucleus and allows it to attach to the host cell's DNA via LANA-1. The lytic genes are expressed in three phases, immediate-early (IE), early (E), and late (L), in response to various stimuli (chemical, physiological, etc.). This temporal and sequential expression pattern is responsible for the reactivation of the KSHV. xCT stands for cystine-glutamate transporter; DC-SIGN stands for DC-Specific Intercellular adhesion molecule-3-Grabbing Non-integrin. Created with BioRender.com

8.5 Epidemiology

8.5.1 KSHV Seroprevalence and Transmission

In contrast to other human herpesviruses, KSHV seroprevalence has a north-south gradient globally. Indeed, more than half of the population in sub-Saharan Africa is infected with KSHV, whereas KSHV infection is anecdotic in Western Europe, North America, and Asia, with less than 10% of the population having encountered the virus. KSHV seroprevalence ranges from 10 to 30% across the Mediterranean basin and Eastern Europe (Mesri et al., 2010) (Table 8.1). These discrepancies' causes are still unknown. However, it seems that environmental factors like infectious agents (malaria and co-infection with other parasites) (Ikoma et al., 2018; Nalwoga et al., 2018; Newton et al., 2018) or the existence of metal-rich soils (iron, silica, or aluminum) in some regions of the world may raise the risk of developing Kaposi's sarcoma in addition to the risk of KSHV transmission (Ziegler, 1993; Simonart et al., 1998; Simonart, 2006; Pelser et al., 2009).

8.5.1.1 Transmission in Countries with Low Seroprevalence

In non-endemic areas, men having sex with men (MSM) are the primary carriers of KSHV, with a seroprevalence that varies from 30 to 60% based on the presence of HIV (Casper et al., 2006, 2007; Giuliani et al., 2007; Sullivan et al., 2010; Lee et al., 2018). No matter their HIV status, MSM have an overall seroprevalence of 33% (Liu et al., 2017). Numerous studies have demonstrated a correlation between risky sexual behavior and the transmission of KSHV. These risky behaviors include the number of different sexual partners (Martin et al., 1998; Melbye et al., 1998; Smith et al., 1999; Rohner et al., 2016), the length of sexual activity, HIV infection, a history of other sexually transmitted infections (STIs), oral or inhaled poppers, and oral-penile or oral-anal contacts. It's unclear how important saliva appears to be in the transmission process (Dukers et al., 2000; Pauk et al., 2000; Casper et al., 2002). In fact, KSHV-DNA is commonly and sporadically found in saliva in MSM infected with the virus (Smith et al., 1999) while its detection in seminal fluid varies

Table 8.1 Kaposi's sarcoma-associated herpesvirus seroprevalence and transmission risk factors

Geographical area	Seroprevalence	Risk factors	Reference
Sub-Saharan Africa	> 50%	Infectious agents, co-infection with other parasites, soils rich in metals	Pelser et al. (2009), Mesri et al. (2010), Newton et al. (2018)
Western Europe, North America, and Asia	< 10%	Infectious agents, co-infection with other parasites, soils rich in metals	Mesri et al. (2010), Ikoma et al. (2018), Nalwoga et al. (2018)
Around the Mediterranean basin and Eastern Europe	10–30%	Infectious agents, co-infection with other parasites, soils rich in metals	Ziegler (1993), Simonart et al. (1998) Ikoma et al. (2018)

according to the HIV status and the population (Corbellino et al., 1996; Monini et al., 1996; Pellett et al., 1999; Kaspersen & Höllsberg, 2013). However, KSHV is not often detected in the urine or anal area (Cannon et al., 2003). The use of other intimate touch practices, particularly during oral-anal sex, is the basis for one of the transmission theories in this population (Casper et al., 2006; Cannon et al., 2003). Although the findings of many research are contradictory, this concept is further supported by the fact that there is no evidence of sexual transmission in the heterosexual population (Eltom et al., 2002; Janier et al., 2002; Pauk et al., 2000; Engels et al., 2007; Malope et al., 2008). Blood and transplants are the two additional potential routes through which KSHV can spread in these areas. While individuals with drug use histories, hemophiliacs, and many transfusion recipients have been found to have a seroprevalence akin to blood donors, blood giving itself does not offer proof of transmission (Lennette et al., 1996; Simpson et al., 1996; Shimizu et al., 2001; Renwick et al., 2002; Crum et al., 2003; Chiereghin et al., 2017). This minimal transmission risk is likely a result of the safety of blood products due to systematic leukocyte depletion. Conversely, it is important to take into account the possibility of KSHV transmission through solid organ transplants. Regardless of the kind of graft, seroconversion is reported to occur in 14–31% of instances (Frances et al., 2009; Lebbe et al., 2013; Riva et al., 2013).

8.5.1.2 Transmission in Endemic Regions

The epidemiology and modalities of KSHV transmission seem to differ in highly endemic countries, where the seroprevalence is frequently already high in childhood and rises with age until a plateau is achieved at the end of puberty (Gessain et al., 1999; Plancoulaine et al., 2000). Numerous investigations have revealed KSHV seropositivity in familial situations, indicating possible early mother-to-child or sibling transmission (Mayama et al., 1998; Mbulaiteye et al., 2006). Saliva is also believed to be a major factor in the transmission (Dedicoat et al., 2004), and the following theories have been put forth: sharing toothbrushes (Romano et al., 2010; Butler et al., 2011), pre-chewing food, and applying maternal saliva to insect bites. Maternal-fetal, perinatal, and human milk transfer is uncommon (Brayfield et al., 2004). Since the significance of co-infection with HIV-1, Kaposi's sarcoma is quite common in specific regions of sub-Saharan Africa. Nonetheless, it is still unclear if KSHV seropositivity and HIV infection are related, particularly in heterosexual populations (Campbell et al., 2009; Shebl et al., 2011).

8.5.2 Molecular Epidemiology

Three viral subtypes A, B, and C were identified from lesions of AIDS patients with Kaposi's sarcoma as a result of the initial sequencing of the two genes that were first revealed for the study of KSHV genetic variation. However, less than 3% of these genes change between subtypes, which hindered their capacity to fully characterize the genetic diversity of KSHV (Hladik et al., 2006).

8.5.2.1 ORF-K1

Similar to the genes encoding STP (saimiri transformation protein) in SaHV and LMP-1 (latent membrane protein) in EBV, a gene ORF-K1, near the end left of the KSHV genome, piqued the interest of multiple groups toward the end of the 1990s (Cook et al., 1999; Zong et al., 1999). The glycosylated transmembrane protein with 289 amino acids that this ORF encodes is also present in the immunoglobulin receptor family. Its extracellular domain has two hypervariable regions (VR1 and VR2) and two conserved regions (C1 and C2), while its C-terminal intracytoplasmic domain has a conserved ITAM motif that is implicated in the oncogenic process and the activation of transduction pathways (Zong et al., 1999, 2002; Lee et al., 2005; Bhatt & Damania, 2012). The seven subtypes (A, B, C, D, E, F, and Z) that are currently identified have up to 44% of their amino acid sequence variability in common. This diversity is primarily found in the VR1 (54–93 AA) and VR2 (191–228 AA) domains of the gene. Subtypes B and D differ by around 30 and 24%, respectively, from subtypes A and C, whereas subtypes A and C differ by approximately 15% from one another. When the amino acid sequences of these subtypes varied by roughly 10%, they were further separated into genotypic variants (e.g., A1, A2, A3) (Zong et al., 1999). Subtypes A and C are present around the world but are more common in North America, Western Europe, the Mediterranean region, and Asia (Ötvös et al., 2014; Ouyang et al., 2014; Varmazyar et al., 2017). For now, the variety of KSHV is primarily related to the patient's place of origin. As with the subtype B, the variant A5 was first and is primarily reported from Africa (Isaacs et al., 2016; Mamimandjiami et al., 2021). A few individuals in Uganda (variant F1) and Caucasian MSM living in France (variant F2) are known to belong to subtype D, while subtype E is found in Native Americans in Brazil. A small cohort of children in Zambia is known to belong to subtype Z (Fig. 8.4) (Kasolo et al., 1998; Whitby et al., 2004; Kajumbula et al., 2006; Jary et al., 2020).

8.5.2.2 Other Genes

The second most variable gene in KSHV, based on a different molecular classification, is ORF-K15, which is situated at the right end of the LUR. A transcript encoding a 45 kilo Dalton K15 protein with 12 membrane domains is produced by this gene when all 8 of its alternatively spliced exons are included (Brinkmann et al., 2003). The K15 gene has at least three divergent variants that are classified as P (predominant), M (minor), and N (Poole et al., 1999; Hayward et al., 2007). These alleles share less than 33% of the same amino acid sequence. Other ORFs, such as ORF-73 or ORF-26, whose nine discovered subtypes (A, B, C, D, E, J, K, Q, and R) have a geographical distribution parallel to that of the subtypes obtained with ORF-K1 (Tornesello et al., 2010). These ORFs have also been used to explain KSHV diversity. The full KSHV genome can now be understood and sequencing capacities have increased due to next-generation sequencing (NGS). Four additional genes were found to differ between the various KSHV strains by whole genome sequencing: K4.2, K8.1, K11/vIRF2, and K12/Kaposin. This suggests that the full genome should be taken into account to accurately characterize each strain (Olp et al., 2015). Furthermore, Sallah et al. (2018) indicated that large-scale genome sequencing is

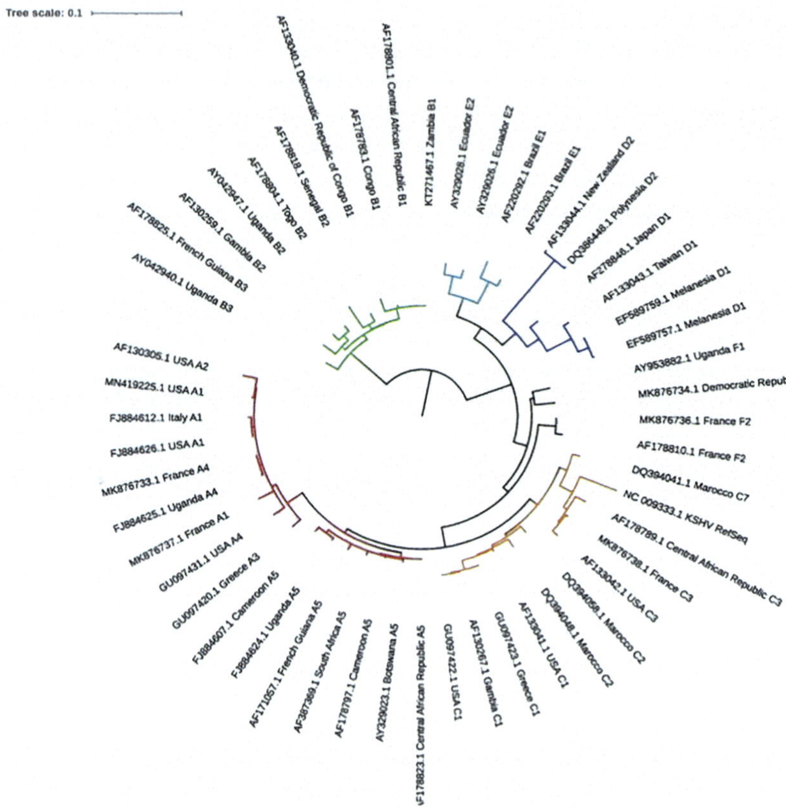

Fig. 8.4 The variety of Kaposi's sarcoma-associated herpesvirus worldwide is described via a phylogenetic tree built using 54 nucleotide sequences of ORF-K1 and released from the GenBank database. Using the GTR model, four rate categories of the gamma shape parameter, and 1000 bootstrap resampling, phylogenic analysis was carried out using PhyML (version 3.0). The Interactive Tree of Life, or iTOL v6., was an online program used to manage and visualize the tree. The tree has middle roots, and the branches of the following subtypes are distinguished by distinct colors: subtype A is red, subtype B is green, subtype C is orange, subtype D is dark blue, subtype E is light blue, and subtype F is black. The following is how label names are put together: subtype/variant, country of origin, and NCBI identification number. Retrieved from (Jary et al., 2020)

also required to fully capture genetic variation, including KSHV recombination, and they presented data supporting the need to revise the nomenclature for KSHV genotypes.

8.5.2.3 Molecular Diversity and Pathogenicity

Various KSHV subtypes may exhibit distinct tumorigenic and pathogenic characteristics, which could lead to varying clinical presentation severity and course (Mancuso et al., 2008; Perez and Tous, 2017; Bellocchi et al., 2020). The findings presented in the literature, however, are not always consistent. White et al. (2008), for instance, did not report any variation in the clinical presentation between categories A and B in patients with the epidemic KS form in Zimbabwe, in contrast to the findings of Kadyrova et al. (2003) who examined patients with classic, post-transplant, and epidemic forms in Russia. However, when a subtype A was included, Mancuso et al. (2008) reported that the classic form of KS evolved more quickly and linked it to increased blood KSHV viral levels. It was also observed by Jary (2020) that in MSM with the epidemic KS form, the KSHV viral load in the blood compartment was larger for subtype A than for subtype C, irrespective of the immunovirological state. Isaacs et al. (2016) reported that variant A5 in the epidemic KS form in Africa was linked to more than 10 KS lesions at diagnosis, while variants A1 and A4 were linked to a decreased risk of lower limb involvement and an extensive form, respectively. However, an in vitro study revealed that the K1 gene in the epidemic form had a stronger transforming activity than those in the classic form. Additionally, there was a correlation seen between this increased activity and the clinical presentation's severity (Nakasone et al., 2019).

8.6 KSHV Oncogenesis and Kaposi's Sarcoma

8.6.1 KSHV, Cells Tropism, and Risk Factors

KSHV can infect a variety of cell types in laboratory cell culture, including dendritic cells, epithelial cells, fibroblasts, and mesenchymal stem cells, after primo-infection. In vivo, KSHV typically infects B cells, macrophages, and endothelial cells (Bechtel et al., 2003). KSHV then becomes dormant, most notably in monocytes and B lymphocytes. Numerous research studies have indicated the possible contribution of certain environmental factors, like the ingestion of plant-based chemical molecules (Whitby et al., 2007), the existence of a malaria parasite (Wakeham et al., 2011; Nalwoga et al., 2015), or the uptake of iron from volcanic soils (Simonart et al., 2004), to the promotion of KSHV pathogenesis and transmission through KSHV reactivation. Traditionally, KHSV diseases are strongly influenced by iatrogenic immunosuppression following transplantation or HIV infection ((Simonart et al., 2004), Piselli et al., 2009). Although the idea of a synergy between HIV and KSHV boosting cell transformation has been proposed, more research is still required. It is true that several HIV secretory proteins, like Tat and Nef, may be released into the bloodstream to encourage the development of Kaposi's sarcoma (Ensoli et al., 1990; Zhu et al., 2014). Specifically, in vitro research has demonstrated that the Tat protein, in conjunction with the protein K1, reactivates KSHV to stimulate the proliferation of KSHV-infected endothelium cells (Ensoli et al., 1994; Rusnati and Presta, 2002; Wood and Feller, 2008; Aoki and Tosato, 2007; Zeng et al., 2007). According to the COHERE Study in EuroCoord (2016), one of the risk factors for developing Kaposi's sarcoma is the ability of an HIV-1 RNA viral load to be detectable in vivo even after antiretroviral treatment and successful immune system reconstitution. Importantly, KSHV has been shown in experiments to be required to start and sustain tumor growth (Grossmann et al., 2006; Alkharsah et al., 2011). It is present in the lesions of all epidemiological forms of Kaposi's sarcoma (Boshoff et al., 2001). Per infected cell, the viral genome is present in one to two copies on average (Boshoff and Chang, 2001; Asahi-Ozaki 2006). The KSHV-infected cell is able to multiply and become immortal due to the oncogenic viral proteins' manipulation of multiple cellular pathways. This also enables the cell to escape from the immune system, induce angiogenesis, and suppress apoptosis features shared by all forms of cancer (Jha et al., 2016). The spindle cells (typical KSHV-infected cells) in Kaposi's sarcoma are thought to originate from the endothelial cell line, while it is yet unclear if these cells are vascular or lymphatic. In fact, the spindle cells exhibit the phenotypic traits of both vascular and lymphatic endothelial cells and express markers of both types of cells (VEGF-3, LYVE-1, and podoplanin, or CD34, CD31, and CD36, respectively) (Hong et al., 2004; Wang et al., 2004; Gasperini et al., 2012). However, neither of these two endothelium lineages is adequately represented by their genetic expression (Cancian et al., 2013). It is clear that viral latent proteins play a crucial role in the development of Kaposi's sarcoma since the KSHV virus is present in most of the disease's cells in a constant state of latency. But a tiny percentage of spindle cells experience spontaneous lytic viral reactivation, which results in the production of virions and the expression of lytic proteins (<5% according to the literature) (Decker et al., 1996; Katano et al., 2000). These virions may promote the growth of nearby cells by inducing the release of pro-inflammatory and pro-angiogenic factors, which suggests their critical roles in the tumorigenesis process (Katano et al., 2000; Sun et al., 1999; Parravicini et al., 2000; Grundhoff and Ganem, 2004).

8.6.2 KSHV Latency Proteins and Oncogenesis

The oncogenesis process and a latent infection establishment are significantly aided by latency proteins encoded in genes. Certain proteins, like PI3K-AKT-mTOR, MAPK, and NF-kB, have the capacity to constitutively and/or transiently alter cellular signaling pathways that are critical for tumor transformation and the infected cells survival (Hunte et al., 2018; Kim et al., 2020) (Table 8.2). They also take part

Table 8.2 Proteins with gene codes that are essential for the latency/lytic cycles and carcinogenesis of the herpesvirus linked to Kaposi's sarcoma

Protein	Cycle	Function	Reference
LANA-1	Latency	Promotes tumorigenesis, inhibits p53, promotes cell proliferation, contributes to angiogenesis, increases telomerase expression	Fuentes-Gonzalez et al., (2013), Uppal et al., (2015), Broussard and Damania, (2020)
v-cyclin	Latency	Accelerates G1/S phase transition	Ojala et al., (1999)
v-FLIP	Latency	Activates NF-kB pathway, induces inflammatory cytokine production	Chaudhary et al., (1999), Matta and Chaudhary, (2004); Ballon et al., (2015)
KSHV-encoded miRNAs	Latency	Maintains latency, inhibits apoptosis, promotes immune escape, modulates angiogenesis, cell migration, and adhesion	O Hara et al., (2009); Guo et al., (2017), Qin et al., (2017), Hussein et al., (2019)

in the synthesis of signals that are pro-inflammatory and pro-angiogenic, which are important in the etiology of KSHV-related malignancies. LANA-1, among other things, plays a key role in driving carcinogenesis and is essential for the formation of KSHV latency. Initially, during primo-infection, LANA-1 engages numerous elements of the host epigenetic machinery to facilitate the production of latent KSHV episomes during the early establishment of the latent state (Uppal et al., 2015; Broussard and Damania, 2020).

Furthermore, during cell division, LANA-1 is necessary for segregation, replication, and episomal maintenance (Uppal et al., 2015). LANA is involved in the stimulation of the pathways leading to cell proliferation and transformation and can impede the advancement of the cell cycle and apoptosis, namely, through the inhibition of the transcription factor p53. LANA-1 also has a role in angiogenesis and cell immortalization (Verma et al., 2004; Fuentes-Gonzalez et al., 2013), extending the life of the infected cell by upregulating telomerase expression. The few other expressed genes, Kaposin, v-FLIP, v-cyclin, and miRNAs, can likewise control many host-signaling pathways. Specifically, miRNAs produced by KSHV contribute to the preservation of viral dormancy and influence carcinogenesis through stifling the expression of numerous host and viral genes. While some of them influence the suppression of apoptosis, others encourage immune escape by controlling the host reaction (Guo et al., 2017; Qin et al., 2017; Hussein et al., 2019). Additionally, several miRNAs have the capacity to regulate angiogenesis, adhesion, migration, and transformation of endothelial cells all of which are critical for the spread of KSHV infection and the pathogenicity of the virus (O Hara et al., 2009). On the other hand, v-FLIP can trigger the production of inflammatory cytokines and chemokines as well as lead to cell survival and proliferation during latency (Chaudhary et al., 1999; Matta and Chaudhary, 2004; Ballon et al., 2015) by activating the NF-kB pathway, a crucial cellular pathway. Additionally, KSHV encodes v-cyclin, a virus that binds to cellular cyclin-dependent kinase 6 (CDK6) and is a homolog of cellular cyclin. Cell cycle inhibitor proteins like p21 and p27 are guaranteed to be phosphorylated and inactivated by the v-cyclin-CDK6 complex, which accelerates the G1/S phase of the cell cycle (Ojala et al., 1999).

8.6.3 Lytic Proteins and Oncogenesis

Only a tiny percentage of tumor cells produce lytic proteins, whereas all infected cells display latency proteins. On the other hand, their existence is necessary for KSHV carcinogenesis (Manners et al., 2018). In reality, by generating fresh infectious particles, they contribute to the persistence of the viral infection as well as the immune system's evasion and the synthesis of growth factors and cytokines. As a result, they have the ability to affect neighboring cells by creating an angiogenic and inflammatory milieu that is conducive to tumor growth. Of them, ORF-K1, v-IL-6, v-BCL2, v-GPCR, v-IRFs 1–4, vCCCL-1, vCCCL-2, vCCCL-3, and v-IAPs are the primary lytic proteins implicated in the pathogenicity of the KSHV; the majority of these proteins are homologs of cellular proteins. For instance, the v-GPCR promotes angiogenesis by inducing the secretion of pro-inflammatory and pro-angiogenic factors, such as VEGF, IL-6, and IL-8, which causes tumor progression in a paracrine manner (Martin et al., 2008; Azzi et al., 2014; Cousins and Nicholas, 2014; Gramolelli and Schulz, 2015). By activating the PI3-AKT, MAPK-ERK, and JAK-STAT pathways through gp130, on the other hand, v-IL-6 can promote cell survival and proliferation. In addition, pro-inflammatory cytokines like human IL-6 can be excreted by viruses through IL-6, which makes the environment of KS more pro-inflammatory and pro-angiogenic (Gramolelli and Schulz, 2015). Conversely, through VEGF induction, the v-CCL proteins drive angiogenesis and aid in immune escape in Kaposi's sarcoma (Bala et al., 2012). These viral chemokines can also contribute paracrinely to the survival of uninfected or latently infected cells, which means they have a role in the pathophysiology of KSHV (Choi and Nicholas, 2008). Lastly, by targeting cellular IRFs, the lytic proteins known as v-IRFs can suppress the synthesis and signaling of type I interferons (INFs I) and consequently lower immunological response (Golas et al., 2020).

8.7 Rationale and Feasibility of KSHV Vaccine

To prevent KSHV infection or cure disorders linked to it, there is currently no KSHV vaccination available. Even though HIV-1 infection may now be controlled immunovirologically, Kaposi's sarcoma is still one of the most prevalent malignancies in Africa and can still develop in HIV-1-positive individuals (Yanik et al., 2016; Palich et al., 2020). Furthermore, there aren't many treatment options for Kaposi's sarcoma because some of them have a minimal hazardous dose (Reid et al., 2019) and are difficult to get in resource-constrained nations, particularly those in sub-Saharan Africa. Accordingly, the creation of a KSHV vaccine may have a significant effect on global public health, particularly in endemic regions and for immunocompromised or at-risk individuals. The following several immunological pathways are part of the immune response to KSHV that is necessary to stop the development of Kaposi's sarcoma: (1) natural killer cells through the downregulation of MHC type 1 molecules on KSHV-infected tumor cells (Dupuy et al., 2012; Beldi-Ferchiou et al., 2016) and (2) response of KSHV-specific CD8 T-cells, which is lower in HIV-positive patients with Kaposi's sarcoma than in KSHV-positive patients with no symptoms (Guihot et al., 2006; Lambert et al., 2006).

Two KSHV proteins, LANA and K12/Kaposin, as well as early and late lytic proteins, have been reported to be targeted by KSHV-specific CD8 T cells (Wu et al., 2012). (3) KSHV infection also induces the production of neutralizing antibodies (nAbs), which may be the most promising approach to developing KSHV prophylactic vaccines (Dialyna et al., 2004; Kumar et al., 2013). The gH/gL complex is the main antigenic determinant of KSHV-specific nAbs. This is based on the analysis of the antigenic targets of KSHV-specific nAbs (Mortazavi et al. 2020).

8.8 Conclusions

KSHV is present in all epidemiological variants of Kaposi's sarcoma and is one of the seven oncogenic viruses that are currently recognized. Oncogenesis and transmission mechanisms underlying Kaposi's sarcoma, however, are intricate. In fact, there are regional differences in the global distribution of KSHV infection, which may indicate the influence of unidentified cofactors. Other than that, KSHV infection in the general population seldom results in Kaposi's sarcoma, despite the fact that KSHV encodes multiple oncogenes that may possibly produce a tumor phenotype. This suggests the significance of cofactors, such as immunological deficiency. To date, further research is needed to advance our understanding of the molecular traits, oncogenesis, and transmission of KSHV.

References

Abere, B., Mamo, T. M., Hartmann, S., Samarina, N., Hage, E., Rückert, J., Hotop, S.-K., Büsche, G., & Schulz, T. F. (2017). The Kaposi's sarcoma-associated herpesvirus (KSHV) non-structural membrane protein K15 is required for viral lytic replication and may represent a therapeutic target. *PLoS Pathogens, 13*(7), e1006639.

Abere, B., Li, J., Zhou, H., Toptan, T., Moore, P. S., & Chang, Y. (2020). Kaposi's sarcoma-associated herpesvirus-encoded CircRNAs are expressed in infected tumor tissues and are incorporated into Virions. *MBio, 11*(1), e03027–e03019.

Akula, S. M., Wang, F. Z., Vieira, J., & Chandran, B. (2001a). Human herpesvirus 8 interaction with target cells involves Heparan sulfate. *Virology, 282*(2), 245–255.

Akula, S. M., Pramod, N. P., Wang, F. Z., & Chandran, B. (2001b). Human herpesvirus 8 envelope-associated glycoprotein B interacts with Heparan sulfate-like moieties. *Virology, 284*(2), 235–249.

Akula, S. M., Pramod, N. P., Wang, F. Z., & Chandran, B. (2002). Integrin Alpha3beta1 (CD 49c/29) is a cellular receptor for Kaposi's sarcoma-associated herpesvirus (*KSHV/HHV-8*) entry into the target cells. *Cell, 108*(3), 407–419.

Akula, S. M., Naranatt, P. P., Walia, N. S., Wang, F. Z., Fegley, B., & Chandran, B. (2003). Kaposi's sarcoma-associated herpesvirus (human herpesvirus 8) infection of human fibroblast cells occurs through endocytosis. *Journal of Virology, 77*(13), 7978–7990.

Alexander, L., Denekamp, L., Knapp, A., Auerbach, M. R., Damania, B., & Desrosiers, R. C. (2000). The primary sequence of rhesus monkey *Rhadinovirus* isolate 26-95: Sequence similarities to Kaposi's sarcoma-associated herpesvirus and rhesus monkey *Rhadinovirus isolate 17577*. *Journal of Virology, 74*(7), 3388–3398.

Alkharsah, K. R., Singh, V. V., Bosco, R., Santag, S., Grundhoff, A., Konrad, A., Sturzi, M., Wirth, D., Dittrich-Breiholz, O., Kracht, M., & Schulz, T. F. (2011). Deletion of Kaposi's sarcoma-associated herpesvirus FLICE inhibitory protein, VFLIP, from the viral genome compromises the activation of STAT1-responsive cellular genes and spindle cell formation in endothelial cells. *Journal of Virology, 85*(20), 10375–10388.

Aneja, K. K., & Yuan, Y. (2017). Reactivation and lytic replication of Kaposi's sarcoma-associated herpesvirus: An update. *Frontiers in Microbiology, 8*, 613.

Aoki, Y., & Tosato, G. (2007). Interactions between HIV-1 tat and KSHV. *Current Topics in Microbiology and Immunology, 312*, 309–326.

Asahi-Ozaki, Y., Sato, Y., Kanno, T., Sata, T., & Katano, H. (2006). Quantitative analysis of Kaposi sarcoma-associated herpesvirus (KSHV) in KSHV-associated diseases. *Journal of Infectious Diseases, 193*(6), 773–782.

Azzi, S., Smith, S. S., Dwyer, J., Leclair, H. M., Alexia, C., Hebda, J. K., Dupin, N., Bidere, N., & Gavard, J. (2014). YGLF motif in the Kaposi sarcoma herpes virus G-protein-coupled receptor adjusts NF-KB activation and paracrine actions. *Oncogene, 33*(44), 5609–5618.

Bala, K., Bosco, R., Gramolelli, S., Haas, D. A., Kati, S., Pietrek, M., Havemeier, A., Yakushko, Y., Singh, V. V., Dittrich-Breiholz, O., Kracht, M., & Schulz, T. F. (2012). Kaposi's sarcoma herpesvirus K15 protein contributes to virus-induced angiogenesis by recruiting PLCγ1 and activating NFAT1-dependent RCAN1 expression. *PLoS Pathogens, 8*(3), e1002927.

Ballon, G., Akar, G., & Cesarman, E. (2015). Systemic expression of Kaposi sarcoma herpesvirus (KSHV) Vflip in endothelial cells leads to a profound Proinflammatory phenotype and myeloid lineage remodeling in vivo. *PLoS Pathogens, 11*(4), e1004581.

Bechtel, J. T., Liang, Y., Hvidding, J., & Ganem, D. (2003). Host range of Kaposi's sarcoma-associated herpesvirus in cultured cells. *Journal of Virology, 77*(12), 6474–6481.

Beldi-Ferchiou, A., Lambert, M., Dogniaux, S., Vély, F., Vivier, E., Olive, D., Dupuy, S., Levasseur, F., Zucman, D., Lebbe, C., Sene, D., Hivroz, C., & Caillat-Zucman, S. (2016). PD-1 mediates functional exhaustion of activated NK cells in patients with Kaposi sarcoma. *Oncotarget, 7*(45), 72961–72977.

Bellocchi, M. C., Svicher, V., & Ceccherini-Silberstein, F. (2020). HHV-8 genetic diversification and its impact on severe clinical presentation of associated diseases. *Journal of Infectious Diseases, 222*(8), 1250–1253.

Bhatt, A. P., & Damania, B. (2012). AKTivation of PI3K/AKT/MTOR signaling pathway by KSHV. *Frontiers in Immunology, 3*, 401.

Birkmann, A., Mahr, K., Ensser, A., Yağuboğlu, S., Titgemeyer, F., Fleckenstein, B., & Neipel, F. (2001). Cell surface Heparan Sulfate Is a Receptor for Human Herpesvirus 8 and Interacts with Envelope Glycoprotein K8.1. *Journal of Virology, 75*(23), 11583–11593.

Bosch, M. L., Strand, K. B., & Rose, T. M. (1998). Gammaherpesvirus Sequence Comparisons. *Journal of Virology, 72*(11), 8458–8459.

Boshoff, C., & Chang, Y. (2001). Kaposi's sarcoma-associated herpesvirus: A new DNA tumor virus. *Annual Review of Medicine, 52*, 453–470.

Brayfield, B. P., Kankasa, C., West, J. T., Muyanga, J., Bhat, G., Klaskala, W., Mitchell, C. D., & Wood, C. (2004). Distribution of Kaposi sarcoma-associated herpesvirus/human herpesvirus 8 in maternal saliva and breast Milk in Zambia: Implications for transmission. *Journal of Infectious Diseases, 189*(12), 2260–2270.

Brinkmann, M. M., Glenn, M., Rainbow, L., Kieser, A., Henke-Gendo, C., & Schulz, T. F. (2003). Activation of mitogen-activated protein kinase and NF-KappaB pathways by a Kaposi's sarcoma-associated herpesvirus K15 membrane protein. *Journal of Virology, 77*, 9346–9358.

Broussard, G., & Damania, B. (2020). Regulation of KSHV latency and lytic reactivation. *Viruses, 12*(9), 1034.

Butler, L. M., Were, W. A., Balinandi, S., Downing, R., Dollard, S., Neilands, T. B., Gupta, S., Rutherford, G. W., & Mermin, J. (2011). Human herpesvirus 8 infection in children and adults in a population-based study in rural Uganda. *Journal of Infectious Diseases, 203*(5), 625–634.

Campbell, T. B., Borok, M., Ndemera, B., Fiorillo, S., White, I. E., Zhang, X., Machekano, R. N., Katzenstein, D., & Gwanzura, L. (2009). Lack of evidence for frequent heterosexual transmission of human herpesvirus 8 in Zimbabwe. *Clinical Infectious Diseases, 48*(12), 1601–1608.

Cancer Project Working Group for the Collaboration of Observational HIV Epidemiological Research Europe (COHERE) Study in EuroCoord. (2016). Changing incidence and risk factors for Kaposi sarcoma by time since starting antiretroviral therapy: Collaborative analysis of 21 European cohort studies. *Clinical Infectious Diseases, 63*(10), 1373–1379.

Cancian, L., Hansen, A., & Boshoff, C. (2013). Cellular origin of Kaposi's sarcoma and Kaposi's sarcoma-associated herpesvirus-induced cell reprogramming. *Trends in Cell Biology, 23*(8), 421–432.

Cannon, M. J., Dollard, S. C., Black, J. B., Edlin, B. R., Hannah, C., Hogan, S. E., Patel, M. M., Jaffe, H. W., Offermann, M. K., Spira, T. J., et al. (2003). Risk factors for Kaposi's Sarcoma in Men Seropositive for Both Human Herpesvirus 8 and Human Immunodeficiency Virus. *AIDS, 17*(2), 215–222.

Casper, C., Wald, A., Pauk, J., Tabet, S. R., Corey, L., & Celum, C. L. (2002). Correlates of prevalent and incident Kaposi's sarcoma-associated herpesvirus infection in men who have sex with men. *Journal of Infectious Diseases, 185*(7), 990–993.

Casper, C., Carrell, D., Miller, K. G., Judson, F. D., Meier, A. S., Pauk, J. S., Morrow, R. A., Corey, L., Wald, A., & Celum, C. (2006). HIV Serodiscordant sex partners and the prevalence of human herpesvirus 8 infection among HIV negative men who have sex with men: Baseline data from the EXPLORE study. *Sexually Transmitted Infections, 82*(3), 229–235.

Casper, C., Krantz, E., Selke, S., Kuntz, S. R., Wang, J., Huang, M. L., Pauk, J. S., Corey, L., & Wald, A. (2007). Frequent and asymptomatic oropharyngeal shedding of human herpesvirus 8 among immunocompetent men. *Journal of Infectious Diseases, 195*(1), 30–36.

Cesarman, E., Chang, Y., Moore, P. S., Said, J. W., & Knowles, D. M. (1995). Kaposi's sarcoma-associated herpesvirus-like DNA sequences in AIDS-related body-cavity-based lymphomas. *New England Journal of Medicine, 332*(17), 1186–1191.

Cesarman, E., Damania, B., Krown, S. E., Martin, J., Bower, M., & Whitby, D. (2019). Kaposi Sarcoma. *Nature Reviews Disease Primers, 5*, 9.

Chakraborty, S., Veettil, M. V., & Chandran, B. (2012). Kaposi's sarcoma associated herpesvirus entry into target cells. *Frontiers in Microbiology, 3*, 6.

Chang, Y., Cesarman, E., Pessin, M. S., Lee, F., Culpepper, J., Knowles, D. M., & Moore, P. S. (1994). Identification of herpesvirus-like DNA sequences in AIDS-associated Kaposi's sarcoma. *Science, 266*(5192), 1865–1869.

Chaudhary, P. M., Jasmin, A., Eby, M. T., & Hood, L. (1999). Modulation of the NF-kappa B pathway by virally encoded death effector domains-containing proteins. *Oncogene, 18*(41), 5738–5746.

Chiereghin, A., Barozzi, P., Petrisli, E., Piccirilli, G., Gabrielli, L., Riva, G., Potenza, L., Cappelli, G., De Ruvo, N., Libri, I., Maggiore, U., Morelli, M. C., Potena, L., Todeschini, P., Gibertoni, D., Labanti, M., Sangiorgi, G., La Manna, G., Pinna, A. D., ... & Lazzarotto, T. (2017). Multicenter prospective study for laboratory diagnosis of HHV8 infection in solid organ donors and transplant recipients and evaluation of the clinical impact after transplantation. *Transplantation, 101*(8), 1935–1944.

Choi, Y. B., & Nicholas, J. (2008). Autocrine and paracrine promotion of cell survival and virus replication by human herpesvirus 8 chemokines. *Journal of Virology, 82*(13), 6501–6513.

Conrad, N. K. (2016). New insights into the expression and functions of the Kaposi's sarcoma-associated herpesvirus long noncoding PAN RNA. *Virus Research, 212*, 53–63.

Cook, P. M., Whitby, D., Calabro, M. L., Luppi, M., Kakoola, D. N., Hjalgrim, H., Ariyoshi, K., Ensoli, B., Davison, A. J., & Schulz, T. F. (1999). Variability and evolution of Kaposi's sarcoma-associated herpesvirus in Europe and Africa. *International Collaborative Group. AIDS, 13*(9), 1165–1176.

Corbellino, M., Bestetti, G., Galli, M., & Parravicini, C. (1996). Absence of HHV-8 in prostate and semen. *New England Journal of Medicine, 335*(16), 1237. author reply 1238–1239.

Cousins, E., & Nicholas, J. (2014). Molecular biology of human herpesvirus 8: Novel functions and virus-host interactions implicated in viral pathogenesis and replication. *Viruses and Human Cancer, 193*, 227–268.

Crum, N. F., Wallace, M. R., Stephan, K., Blazes, D. L., Aronson, N., Tasker, S. A., Thomas, A. G., Wegner, S., Casper, C., Wald, A., Corey, L., & Brodine, S. K. (2003). Correlates of human Herpesvirus-8 Seropositivity among U.S. military members recently infected with human immunodeficiency virus. *Sexually Transmitted Diseases, 30*(9), 713–718.

Davis, D. A., Rinderknecht, A. S., Zoeteweij, J. P., Aoki, Y., Read-Connole, E. L., Tosato, G., Blauvelt, A., & Yarchoan, R. (2001). Hypoxia induces lytic replication of Kaposi sarcoma-associated herpesvirus. *Blood, 97*(10), 3244–3250.

Decker, L. L., Shankar, P., Khan, G., Freeman, R. B., Dezube, B. J., Lieberman, J., & Thorley-Lawson, D. A. (1996). The Kaposi sarcoma-associated herpesvirus (KSHV) is present as an intact latent genome in KS tissue but replicates in the peripheral blood mononuclear cells of KS patients. *Journal of Experimental Medicine, 184*(1), 283–288.

DeCotiis, J. L., & Lukac, D. M. (2017). KSHV and the role of notch receptor dysregulation in disease progression. *Pathogens, 6*(3), 34.

Dedicoat, M., Newton, R., Alkharsah, K. R., Sheldon, J., Szabados, I., Ndlovu, B., Page, T., Casabonne, D., Gilks, C. F., & Cassol, S. A. (2004). Mother-to-child transmission of human Herpesvirus-8 in South Africa. *Journal of Infectious Diseases, 190*(6), 1068–1075.

Desrosiers, R. C., Sasseville, V. G., Czajak, S. C., Zhang, X., Mansfield, K. G., Kaur, A., Johnson, R. P., Lackner, A. A., & Jung, J. U. (1997). A herpesvirus of rhesus monkeys related to the human Kaposi's sarcoma-associated herpesvirus. *Journal of Virology, 71*(12), 9764–9769.

Dialyna, I. A., Graham, D., Rezaee, R., Blue, C. E., Stavrianeas, N. G., Neisters, H. G. M., Spandidos, D. A., & Blackbourn, D. J. (2004). Anti-HHV-8/KSHV antibodies in infected individuals inhibit infection in vitro. *AIDS, 18*(9), 1263–1270.

Dollery, S. J. (2019). Towards understanding KSHV fusion and entry. *Viruses, 11*(11), 1073.

Dollery, S. J., Santiago-Crespo, R. J., Chatterjee, D., & Berger, E. A. (2019). Glycoprotein K8.1A of Kaposi's sarcoma-associated herpesvirus is a critical B cell tropism determinant independent of its Heparan sulfate binding activity. *Journal of Virology, 93*(14), e01876–e01818.

Dukers, N. H., Renwick, N., Prins, M., Geskus, R. B., Schulz, T. F., Weverling, G. J., Coutinho, R. A., & Goudsmit, J. (2000). Risk factors for human herpesvirus 8 Seropositivity and seroconversion in a cohort of homosexual men. *American Journal of Epidemiology, 151*(3), 213–224.

Dupuy, S., Lambert, M., Zucman, D., Choukem, S.-P., Tognarelli, S., Pages, C., Lebbe, C., & Caillat-Zucman, S. (2012). Human herpesvirus 8 (HHV8) sequentially shapes the NK cell repertoire during the course of asymptomatic infection and Kaposi sarcoma. *PLoS Pathogens, 8*(3), e1002486.

Eltom, M. A., Mbulaiteye, S. M., Dada, A. J., Whitby, D., & Biggar, R. J. (2002). Transmission of human herpesvirus 8 by sexual activity among adults in Lagos, Nigeria. *AIDS, 16*(18), 2473–2478.

Engels, E. A., Atkinson, J. O., Graubard, B. I., McQuillan, G. M., Gamache, C., Mbisa, G., et al. (2007). Risk factors for human herpesvirus 8 infection among adults in the United States and evidence for sexual transmission. *Journal of Infectious Diseases, 196*(2), 199–207.

Ensoli, B., Barillari, G., Salahuddin, S. Z., Gallo, R. C., & Wong-Staal, F. (1990). Tat protein of HIV-1 stimulates growth of cells derived from Kaposi's sarcoma lesions of AIDS patients. *Nature, 345*(6270), 84–86.

Ensoli, B., Gendelman, R., Markham, P., Fiorelli, V., Colombini, S., Raffeld, M., Cafaro, A., Chang, H. K., Brady, J. N., & Gallo, R. C. (1994). Synergy between basic fibroblast growth factor and HIV-1 tat protein in induction of Kaposi's sarcoma. *Nature, 371*(6499), 674–680.

Frances, C., Marcelin, A. G., Legendre, C., Chevret, S., Dussaix, E., Lejeune, J., Euvrard, S., Bigorie, A., Schulz, T. F., Agbalika, F., & Lebbe, C. (2009). The impact of preexisting or acquired Kaposi sarcoma herpesvirus infection in kidney transplant recipients on morbidity and survival. *American Journal of Transplantation, 9*, 2580–2586.

Fuentes-González, A. M., Contreras-Paredes, A., Manzo-Merino, J., & Lizano, M. (2013). The modulation of apoptosis by oncogenic viruses. *Virology Journal, 10*, 182.

Gallaher, A. M., Das, S., Xiao, Z., Andresson, T., Kieffer-Kwon, P., Happel, C., & Ziegelbauer, J. (2013). Proteomic screening of human targets of viral MicroRNAs reveals functions associated with immune evasion and angiogenesis. *PLoS Pathogens, 9*(9), e1003584.

Gasperini, P., Espigol-Frigole, G., McCormick, P. J., Salvucci, O., Maric, D., Uldrick, T. S., Polizzotto, M. N., Yarchoan, R., & Tosato, G. (2012). Kaposi sarcoma herpesvirus promotes endothelial-to-mesenchymal transition through notch-dependent signaling. *Cancer Research, 72*(5), 1157–1169.

Gessain, A., Mauclère, P., van Beveren, M., Plancoulaine, S., Ayouba, A., Essame-Oyono, J. L., Martin, P. M. V., & de Thé, G. (1999). Human herpesvirus 8 primary infection occurs during childhood in Cameroon, Central Africa. *International Journal of Cancer, 81*, 189–192.

Giuliani, M., Cordiali-Fei, P., Castilletti, C., Di Carlo, A., Palamara, G., Boros, S., & Rezza, G. (2007). Incidence of human herpesvirus 8 (HHV-8) infection among HIV-uninfected individuals at high risk for sexually transmitted infections. *BMC Infectious Diseases, 7*(1), 143.

Golas, G., Jang, S. J., Naik, N. G., Alonso, J. D., Papp, B., & Toth, Z. (2020). Comparative analysis of the viral interferon regulatory factors of KSHV for their requisite for virus production and inhibition of the type I interferon pathway. *Virology, 541*, 160–173.

Gottwein, E., Corcoran, D. L., Mukherjee, N., Skalsky, R. L., Hafner, M., Nusbaum, J. D., Shamulailatpam, P., Love, C. L., Dave, S. S., Tuschl, T., et al. (2011). Viral MicroRNA Targetome of KSHV-infected primary effusion lymphoma cell lines. *Cell Host & Microbe, 10*(5), 515–526.

Gradoville, L., Gerlach, J., Grogan, E., Shedd, D., Nikiforow, S., Metroka, C., & Miller, G. (2000). Kaposi's sarcoma-associated herpesvirus open Reading frame 50/Rta protein activates the entire viral lytic cycle in the HH-B2 primary effusion lymphoma cell line. *Journal of Virology, 74*(13), 6207–6212.

Gramolelli, S., & Schulz, T. F. (2015). The role of Kaposi sarcoma-associated herpesvirus in the pathogenesis of Kaposi sarcoma. *Journal of Pathology, 235*(3), 368–380.

Greensill, J., Sheldon, J. A., Renwick, N. M., Beer, B. E., Norley, S., Goudsmit, J., & Schulz, T. F. (2000). Two distinct Gamma-2 herpesviruses in African green monkeys: A second Gamma-2 herpesvirus lineage among Old World primates? *Journal of Virology, 74*(4), 1572–1577.

Grossmann, C., Podgrabinska, S., Skobe, M., & Ganem, D. (2006). Activation of NF-KappaB by the latent VFLIP gene of Kaposi's sarcoma-associated herpesvirus is required for the spindle shape of virus-infected endothelial cells and contributes to their Proinflammatory phenotype. *Journal of Virology, 80*(15), 7179–7185.

Grundhoff, A., & Ganem, D. (2004). Inefficient establishment of KSHV latency suggests an additional role for continued lytic replication in Kaposi sarcoma pathogenesis. *Journal of Clinical Investigation, 113*(1), 124–136.

Guihot, A., Dupin, N., Marcelin, A.-G., Gorin, I., Bedin, A.-S., Bossi, P., Galicier, L., Oksenhendler, E., Autran, B., & Carcelain, G. (2006). Low T cell responses to human herpesvirus 8 in patients with AIDS-related and classic Kaposi sarcoma. *Journal of Infectious Diseases, 194*(8), 1078–1088.

Guito, J., & Lukac, D. M. (2012). KSHV Rta promoter specification and viral reactivation. *Frontiers in Microbiology, 3*, 30.

Guo, Y., Li, W., Qin, J., Lu, C., & Fan, W. (2017). Kaposi's sarcoma-associated herpesvirus (KSHV)-encoded MicroRNAs promote matrix metalloproteinases (MMPs) expression and pro-Angiogenic cytokine secretion in endothelial cells. *Journal of Medical Virology, 89*(7), 1274–1280.

Hayward, G. S., & Zong, J. C. (2007). Modern evolutionary history of the human KSHV genome. *Current Topics in Microbiology and Immunology, 312*, 1–42.

Hladik, W., Dollard, S. C., Mermin, J., Fowlkes, A. L., Downing, R., Amin, M. M., Banage, F., Nzaro, E., Kataaha, P., Dondero, T. J., Pellett, P. E., & Lachritz, E. M. (2006). Transmission of human herpesvirus 8 by blood transfusion. *New England Journal of Medicine, 355*(13), 1331–1338.

Hong, Y.-K., Foreman, K., Shin, J. W., Hirakawa, S., Curry, C. L., Sage, D. R., Libermann, T., Dezube, B. J., Fingeroth, J. D., & Detmar, M. (2004). Lymphatic reprogramming of blood vascular endothelium by Kaposi sarcoma-associated herpesvirus. *Nature Genetics, 36*(7), 683–685.

Hunte, R., Alonso, P., Thomas, R., Bazile, C. A., Ramos, J. C., van der Weyden, L., Dominguez-Bendala, J., Khan, W. N., & Shembade, N. (2018). CADM1 is essential for KSHV-encoded VGPCR-and VFLIP-mediated chronic NF-KB activation. *PLoS Pathogens, 14*(8), e1006968.

Hussein, H. A. M., Alfhili, M. A., Pakala, P., Simon, S., Hussain, J., McCubrey, J. A., & Akula, S. M. (2019). MiRNAs and their roles in KSHV pathogenesis. *Virus Research, 266*, 15–24.

Ikoma, M., Gantt, S., Casper, C., Ogata, Y., Zhang, Q., Basom, R., Dyen, M. R., Rose, T. M., & Barcy, S. (2018). KSHV Oral shedding and plasma viremia result in significant changes in the extracellular tumorigenic MiRNA expression profile in individuals infected with the malaria parasite. *PLoS One, 13*(2), e0192659.

Inoue, N., Winter, J., Lal, R. B., Offermann, M. K., & Koyano, S. (2003). Characterization of entry mechanisms of human herpesvirus 8 by using an Rta-dependent reporter cell line. *Journal of Virology, 77*(15), 8147–8152.

Isaacs, T., Abera, A. B., Muloiwa, R., Katz, A. A., & Todd, G. (2016). Genetic Diversity of HHV8 Subtypes in South Africa: A5 Subtype is Associated with Extensive Disease in AIDS-KS. *Journal of Medical Virology, 88*(2), 292–303.

Jacobs, S. R., & Damania, B. (2011). The viral interferon regulatory factors of KSHV: Immunosuppressors or oncogenes? *Frontiers in Immunology, 2*, 19.

Janier, M., Agbalika, F., de La Salmonière, P., Lassau, F., Lagrange, P., & Morel, P. (2002). Human herpesvirus 8 Seroprevalence in an STD Clinic in Paris: A study of 512 patients. *Sexually Transmitted Diseases, 29*(12), 698–702.

Jary, A., Leducq, V., Desire, N., Petit, H., Palich, R., Joly, V., Canestri, A., Gothland, A., Lambert-Niclot, S., Surgers, L., Amiel, C., Descamps, D., Spano, J. P., Katlama, C., Calvez, V., & Marcelin, A. G. (2020). New Kaposi's sarcoma-associated herpesvirus variant in men who have sex with men associated with severe pathologies. *Journal of Infectious Diseases, 222*(8), 1320–1328.

Jha, H. C., Banerjee, S., & Robertson, E. S. (2016). The role of Gammaherpesviruses in cancer pathogenesis. *Pathogens, 5*(1), 18.

Kadyrova, E., Lacoste, V., Duprez, R., Pozharissky, K., Molochkov, V., Huerre, M., & Gessain, A. (2003). Molecular epidemiology of Kaposi's sarcoma-associated herpesvirus/human herpesvirus 8 strains from Russian patients with classic, Posttransplant, and AIDS-associated Kaposi's sarcoma. *Journal of Medical Virology, 71*(4), 548–556.

Kajumbula, H., Wallace, R. G., Zong, J.-C., Hokello, J., Sussman, N., Simms, S., Rockwell, R. F., Pozos, R., Hayward, G. S., & Boto, W. (2006). Ugandan Kaposi's sarcoma-associated herpesvirus phylogeny: Evidence for cross-ethnic transmission of viral subtypes. *Intervirology, 49*(3), 133–143.

Kasolo, F. C., Monze, M., Obel, N., Anderson, R. A., French, C., & Gompels, U. A. (1998). Sequence analyses of human Herpesvirus-8 strains from both African human immunodeficiency virus-negative and -positive childhood endemic Kaposi's sarcoma show a close relationship with strains identified in febrile children and high variation in the K1 glycoprotein. *Journal of General Virology, 79*(12), 3055–3065.

Kaspersen, M. D., & Höllsberg, P. (2013). Seminal shedding of human herpesviruses. *Virology Journal, 10*(1), 226.

Katano, H., Sato, Y., Kurata, T., Mori, S., & Sata, T. (2000). Expression and localization of human herpesvirus 8-encoded proteins in primary effusion lymphoma, Kaposi's sarcoma, and multicentric Castleman's disease. *Virology, 269*(2), 335–344. [Google Scholar] [CrossRef] [Green Version].

Kerur, N., Veettil, M. V., Sharma-Walia, N., Sadagopan, S., Bottero, V., Paul, A. G., & Chandran, B. (2010). Characterization of entry and infection of Monocytic THP-1 cells by Kaposi's sarcoma-associated herpesvirus (KSHV): Role of Heparan sulfate, DC-SIGN, Integrins and signaling. *Virology, 406*, 103–116.

Kim, H., Jang, J. H., Song, Y. E., & Seo, T. (2020). Kaposi's sarcoma-associated herpesvirus viral protein kinase phosphorylates extracellular signal-regulated kinase and activates MAPK/ERK signaling pathway. *Biochemical and Biophysical Research Communications, 521*(4), 1083–1088.

Kumar, B., & Chandran, B. (2016). KSHV entry and trafficking in target cells-hijacking of cell signal pathways, actin and membrane dynamics. *Viruses, 8*(11), 305.

Kumar, P., Kuwa, N. Y., Minhas, V., Marimo, C., Shea, D. M., Kankasa, C., & Wood, C. (2013). Higher levels of neutralizing antibodies against KSHV in KS patients compared to asymptomatic individuals from Zambia. *PLoS One, 8*(8), e71254.

Lambert, M., Gannagé, M., Karras, A., Abel, M., Legendre, C., Kerob, D., Agbalika, F., Girard, P. M., Lebbe, C., & Caillat-Zucman, S. (2006). Differences in the frequency and function of HHV8-specific CD8 T cells between asymptomatic HHV8 infection and Kaposi sarcoma. *Blood, 108*(12), 3871–3880.

Lebbe, C., Porcher, R., Marcelin, A. G., Agbalika, F., Dussaix, E., Samuel, D., Varnous, S., Euvrard, S., Bigorie, A., Creusvaux, H., Legendre, C., & Frances, C. (2013). Human herpesvirus 8 (HHV8) transmission and related morbidity in organ recipients. *American Journal of Transplantation, 13*, 207–213.

Lee, B.-S., Lee, S.-H., Feng, P., Chang, H., Cho, N.-H., & Jung, J. U. (2005). Characterization of the Kaposi's sarcoma-associated herpesvirus K1 signalosome. *Journal of Virology, 79*(19), 12173–12184.

Lee, Y.-M., Hung, P.-S., & Lin, C.-W. (2018). Seroepidemiology and phylogenetic analysis of human herpesvirus type 8 in injection drug users and men who have sex with men in northern Taiwan. *Journal of International Medical Research, 48*(7), 300060518764747.

Lei, X., Bai, Z., Ye, F., Xie, J., Kim, C.-G., Huang, Y., & Gao, S.-J. (2010). Regulation of NF-KappaB inhibitor IkappaBalpha and viral replication by a KSHV MicroRNA. *Nature Cell Biology, 12*(2), 193–199.

Lennette, E. T., Blackbourn, D. J., & Levy, J. A. (1996). Antibodies to human herpesvirus type 8 in the general population and in Kaposi's sarcoma patients. *The Lancet, 348*(9034), 858–861.

Lin, X., Liang, D., He, Z., Deng, Q., Robertson, E. S., & Lan, K. (2011). MiR-K12-7-5p encoded by Kaposi's sarcoma-associated herpesvirus stabilizes the latent state by targeting viral ORF50/RTA. *PLoS One, 6*(6), e16224.

Liu, Z., Fang, Q., Zuo, J., Chen, Y., Minhas, V., Wood, C., & Zhang, T. (2017). Global epidemiology of human herpesvirus 8 in men who have sex with men: A systematic review and meta-analysis. *Journal of Medical Virology, 90*(4), 582–591.

Malope, B. I., MacPhail, P., Mbisa, G., MacPhail, C., Stein, L., Ratshikhopha, E. M., Ndhlovu, L., Sitas, F., & Whitby, D. (2008). No evidence of sexual transmission of Kaposi's sarcoma herpes virus in a heterosexual south African population. *AIDS, 22*(5), 519–526.

Mamimandjiami, A. I., Mouinga-Ondémé, A., Ramassamy, J.-L., Djuicy, D. D., Afonso, P. V., Mahé, A., et al. (2021). Epidemiology and genetic variability of HHV-8/KSHV among rural populations and Kaposi's sarcoma patients in Gabon, Central Africa. Review of the geographical distribution of HHV-8 K1 genotypes in Africa. *Viruses, 13*(2), 175.

Mancuso, R., Biffi, R., Valli, M., Bellinvia, M., Athanasia, T., Ferrucci, S., Brambilla, L., Delbue, S., Ferrante, P., Tinelli, C., & Clerici, M. (2008). HHV8 a subtype is associated with rapidly evolving classic Kaposi's sarcoma. *Journal of Medical Virology, 80*, 2153–2160.

Manners, O., Murphy, J. C., Coleman, A., Hughes, D. J., & Whitehouse, A. (2018). Contribution of the KSHV and EBV lytic cycles to Tumourigenesis. *Current Opinion in Virology, 32*, 60–70.

Martin, J. N., Ganem, D. E., Osmond, D. H., Page-Shafer, K. A., Macrae, D., & Kedes, D. H. (1998). Sexual transmission and the natural history of human herpesvirus 8 infection. *New England Journal of Medicine, 338*(14), 948–954.

Martin, D., Galisteo, R., Ji, Y., Montaner, S., & Gutkind, J. S. (2008). An NF-KappaB gene expression signature contributes to Kaposi's sarcoma virus VGPCR-induced direct and paracrine neoplasia. *Oncogene, 27*(13), 1844–1852.

Matta, H., & Chaudhary, P. M. (2004). Activation of alternative NF-kappa B pathway by human herpes virus 8-encoded Fas-associated death domain-like IL-1 Beta-converting enzyme inhibitory protein (VFLIP). *Proceedings of the National Academy of Sciences USA, 101*(26), 9399–9404.

Mayama, S., Cuevas, L. E., Sheldon, J., Omar, O. H., Smith, D. H., Okong, P., Silvel, B., Hart, C. A., & Schulz, T. F. (1998). Prevalence and transmission of Kaposi's sarcoma-associated herpesvirus (human herpesvirus 8) in Ugandan children and adolescents. *International Journal of Cancer, 77*(6), 817–820.

Mbulaiteye, S., Marshall, V., Bagni, R. K., Wang, C. D., Mbisa, G., Bakaki, P. M., Owor, A. M., Ndugwa, C. M., Engels, E. A., Katongole-Mbidde, E., Biggar, R. J., & Whitby, D. (2006). Molecular evidence for mother-to-child transmission of Kaposi sarcoma-associated herpesvirus in Uganda and K1 gene evolution within the host. *Journal of Infectious Diseases, 193*(9), 1250–1257.

Melbye, M., Cook, P. M., Hjalgrim, H., Begtrup, K., Simpson, G. R., Biggar, R. J., & Schulz, T. F. (1998). Risk factors for Kaposi's-sarcoma-associated herpesvirus (KSHV/HHV-8)

Seropositivity in a cohort of homosexual men, 1981–1996. *International Journal of Cancer, 77*(4), 543–548.
Melendez, L. V., Daniel, M. D., Hunt, R. D., & Garcia, F. G. (1968). An apparently new herpesvirus from primary kidney cultures of the squirrel monkey (*Saimiri Sciureus*). *Laboratory Animal Care, 18*(4), 374–381.
Melendez, L. V., Hunt, R. D., King, N. W., Barahona, H. H., Daniel, M. D., Fraser, C. E., & Garcia, F. G. (1972). Herpesvirus Ateles, a new lymphoma virus of monkeys. *Nature New Biology, 235*(54), 182–184.
Mesri, E. A., Cesarman, E., & Boshoff, C. (2010). Kaposi's sarcoma and its associated herpesvirus. *Nature Reviews Cancer, 10*(10), 707–719.
Mesri, E. A., Feitelson, M. A., & Munger, K. (2014). Human viral oncogenesis: A cancer hallmarks analysis. *Cell Host & Microbe, 15*(3), 266–282.
Monini, P., de Lellis, L., Fabris, M., Rigolin, F., & Cassai, E. (1996). Kaposi's sarcoma-associated herpesvirus DNA sequences in prostate tissue and human semen. *New England Journal of Medicine, 334*(17), 1168–1172.
Moore, P. S., & Chang, Y. (2001). Molecular virology of Kaposi's sarcoma-associated herpesvirus. *Philosophical Transactions of the Royal Society of London. Series B: Biological Sciences, 356*(1408), 499–516.
Mortazavi, Y., Lidenge, S. J., Tran, T., West, J. T., Wood, C., & Tso, F. Y. (2020). The Kaposi's sarcoma-associated herpesvirus (KSHV) GH/GL complex is the predominant neutralizing antigenic determinant in KSHV-infected individuals. *Viruses, 12*(3), 256.
Muniraju, M., Mutsvunguma, L. Z., Foley, J., Escalante, G. M., Rodriguez, E., Nabiee, R., Totonchy, J., Mulama, D. H., Nyagol, J., Wussow, F., et al. (2019). Kaposi sarcoma-associated herpesvirus glycoprotein H is indispensable for infection of epithelial, endothelial, and fibroblast cell types. *Journal of Virology, 93*(16), e00630–e00619.
Nalwoga, A., Cose, S., Wakeham, K., Miley, W., Ndibazza, J., Drakeley, C., Elliott, A., Whitby, D., & Newton, R. (2015). Association between malaria exposure and Kaposi's sarcoma-associated herpes virus Seropositivity in Uganda. *Tropical Medicine & International Health, 20*(6), 665–672.
Nalwoga, A., Cose, S., Nash, S., Miley, W., Asiki, G., Kusemererwa, S., Yarchoan, R., Labo, N., Whitby, D., & Newton, R. (2018). Relationship between anemia, malaria coinfection, and Kaposi sarcoma-associated herpesvirus Seropositivity in a population-based study in rural Uganda. *Journal of Infectious Diseases, 218*(7), 1061–1065.
Nealon, K., Newcomb, W. W., Pray, T. R., Craik, C. S., Brown, J. C., & Kedes, D. H. (2001). Lytic replication of Kaposi's sarcoma-associated herpesvirus results in the formation of multiple capsid species: Isolation and molecular characterization of a, B, and C capsids from a Gammaherpesvirus. *Journal of Virology, 75*(6), 2866–2878.
Newton, R., Labo, N., Wakeham, K., Marshall, V., Roshan, R., Nalwoga, A., Sebina, I., Muhangi, L., Webb, E. L., Miley, W., et al. (2018). Determinants of Gammaherpesvirus shedding in saliva among Ugandan children and their mothers. *Journal of Infectious Diseases, 218*(6), 892–900.
O'Hara, A. J., Chugh, P., Wang, L., Netto, E. M., Luz, E., Harrington, W. J., Dezube, B. J., Damania, B., & Dittmer, D. P. (2009). Pre-micro RNA signatures delineate stages of endothelial cell transformation in Kaposi sarcoma. *PLoS Pathogens, 5*(8), e1000389.
Ojala, P. M., Tiainen, M., Salven, P., Veikkola, T., Castaños-Vélez, E., Sarid, R., Biberfeld, P., & Mäkelä, T. P. (1999). Kaposi's sarcoma-associated herpesvirus-encoded v-cyclin triggers apoptosis in cells with high levels of cyclin-dependent kinase 6. *Cancer Research, 59*(20), 4984–4989.
Olp, L. N., Jeanniard, A., Marimo, C., West, J. T., & Wood, C. (2015). Whole-genome sequencing of Kaposi's sarcoma-associated herpesvirus from Zambian Kaposi's sarcoma biopsy specimens reveals unique viral diversity. *Journal of Virology, 89*(24), 12299–12308.
Ötvös, R., Juhasz, A., Szalai, E., Ujvari, D., Ötvös, K., Szabo, K., Remenyik, E., Szekely, L., Gergely, L., & Konya, J. (2014). Molecular typing of human herpesvirus 8 isolates from patients with Kaposi's sarcoma in Hungary. *Anticancer Research, 34*(2), 893–898.

Ouyang, X., Zeng, Y., Fu, B., Wang, X., Chen, W., Fang, Y., Luo, M., & Wang, L. (2014). Genotypic analysis of Kaposi's sarcoma-associated herpesvirus from patients with Kaposi's sarcoma in Xinjiang, China. *Viruses, 6*(12), 4800–4810.

Palich, R., Veyri, M., Valantin, M.-A., Marcelin, A.-G., Guihot, A., Pourcher, V., Jary, A., Solas, C., Makinson, A., & Poizot-Martin, I. (2020). Recurrence and occurrence of Kaposi's sarcoma in patients living with human immunodeficiency virus (HIV) and on antiretroviral therapy, despite suppressed HIV viremia. *Clinical Infectious Diseases, 70*(11), 2435–2438.

Parravicini, C., Chandran, B., Corbellino, M., Berti, E., Paulli, M., Moore, P. S., & Chang, Y. (2000). Differential viral protein expression in Kaposi's sarcoma-associated herpesvirus-infected diseases: Kaposi's sarcoma, primary effusion lymphoma, and multicentric Castleman's disease. *American Journal of Pathology, 156*(3), 743–749.

Pauk, J., Huang, M. L., Brodie, S. J., Wald, A., Koelle, D. M., Schacker, T., Celum, C., Selke, S., & Corey, L. (2000). Mucosal Shedding of Human Herpesvirus 8 in Men. *New England Journal of Medicine, 343*(19), 1369–1377.

Pellett, P. E., Spira, T. J., Bagasra, O., Boshoff, C., Corey, L., de Lellis, L., Huang, M. L., Lin, J. C., Matthews, S., Monini, P., Rimessi, P., Sosa, C., Wood, C., & Stewart, J. A. (1999). Multicenter comparison of PCR assays for detection of human herpesvirus 8 DNA in semen. *Journal of Clinical Microbiology, 37*(5), 1298–1301.

Pelser, C., Dazzi, C., Graubard, B. I., Lauria, C., Vitale, F., & Goedert, J. J. (2009). Risk of classic Kaposi sarcoma with residential exposure to volcanic and related soils in Sicily. *Annals of Epidemiology, 19*(8), 597–601.

Pérez, C. L., & Tous, M. I. (2017). Diversity of human herpesvirus 8 genotypes in patients with AIDS and non-AIDS associated Kaposi's sarcoma, Castleman's disease and primary effusion lymphoma in Argentina. *Journal of Medical Virology, 89*(11), 2020–2028.

Piselli, P., Busnach, G., Citterio, F., Frigerio, M., Arbustini, E., Burra, P., Pinna, A. D., Bresadola, V., Ettorre, G. M., Baccarani, U., Buda, A., Lauro, A., Zanus, G., Cimaglia, C., Spagnoletti, G., Lenardon, A., Agozzino, M., Gambato, M., Zanfi, C., ... & Serraino, D. (2009). Risk of Kaposi sarcoma after solid-organ transplantation: Multicenter study in 4767 recipients in Italy, 1970–2006. *Transplantation Proceedings, 41*(4), 1227–1230.

Plancoulaine, S., Abel, L., van Beveren, M., Trégouët, D. A., Joubert, M., Tortevoye, P., de The, G., & Gessain, A. (2000). Human herpesvirus 8 transmission from mother to child and between siblings in an endemic population. *The Lancet, 356*, 1062–1065.

Poole, L. J., Zong, J. C., Ciufo, D. M., Alcendor, D. J., Cannon, J. S., Ambinder, R., Orenstein, J. M., Reitz, M. S., & Hayward, G. S. (1999). Comparison of genetic variability at multiple loci across the genomes of the major subtypes of Kaposi's sarcoma-associated herpesvirus reveals evidence for recombination and for two distinct types of open Reading frame K15 alleles at the right-hand end. *Journal of Virology, 73*(8), 6646–6660.

Qin, J., Li, W., Gao, S.-J., & Lu, C. (2017). KSHV MicroRNAs: Tricks of the devil. *Trends in Microbiology, 25*(8), 648–661.

Reid, E., Suneja, G., Ambinder, R. F., Ard, K., Baiocchi, R., Barta, S. K., Carchman, E., Cohen, A., Crysler, O. V., Gupta, N., Gustafson, C., Hall, A., Johung, K. L., Klopp, A., LaCasce, A. S., Lin, C., Mehta, A., Menon, M. P., Morgan, D., ... & Freedman-Cass, D. A. (2019). AIDS-related Kaposi sarcoma, version 2.2019, NCCN clinical practice guidelines in oncology. *Journal of the National Comprehensive Cancer Network, 17*(2), 171–189.

Renwick, N., Dukers, N. H. T. M., Weverling, G. J., Sheldon, J. A., Schulz, T. F., Prins, M., Coutinho, R. A., & Goudsmit, J. (2002). Risk factors for human herpesvirus 8 infection in a cohort of drug users in The Netherlands, 1985–1996. *Journal of Infectious Diseases, 185*(12), 1808–1812.

Riva, G., Barozzi, P., Quadrelli, C., Vallerini, D., Zanetti, E., Forghieri, F., Chiereghin, A., Libri, I., Maggiore, U., Buzio, C., Lazzarotto, T., Narni, F., Luppi, M., & Potenza, L. (2013). Human herpesvirus 8 (HHV8) infection and related diseases in Italian transplant cohorts. *American Journal of Transplantation, 13*, 1619–1620.

Rohner, E., Wyss, N., Heg, Z., Faralli, Z., Mbulaiteye, S. M., Novak, U., Zwahlen, M., Egger, M., & Bohlius, J. (2016). HIV and human herpesvirus 8 co-infection across the globe: Systematic review and meta-analysis. *International Journal of Cancer, 138*(1), 45–54.

Romano, R., Tabacchi, F., Paganotti, G. M., Russo, G., Gramolelli, S., Marinucci, F., Nelli, L. C., Coluzzi, B., & Mario, C. (2010). Evaluation of bloodsucking arthropod bite as a possible risk co-factor in human Herpesvirus-8 transmission route. *Parassitologia, 52*(3), 405–410.

Rossetto, C. C., Tarrant-Elorza, M., Verma, S., Purushothaman, P., & Pari, G. S. (2013). Regulation of viral and cellular gene expression by Kaposi's sarcoma-associated herpesvirus Polyadenylated nuclear RNA. *Journal of Virology, 87*(10), 5540–5553.

Rusnati, M., & Presta, M. (2002). HIV-1 tat protein and endothelium: From protein/cell interaction to AIDS-associated pathologies. *Angiogenesis, 5*(3), 141–151.

Russo, J. J., Bohenzky, R. A., Chien, M. C., Chen, J., Yan, M., Maddalena, D., Parry, J. P., Peruzzi, D., Edelman, I. S., Chang, Y., et al. (1996). Nucleotide sequence of the Kaposi sarcoma-associated herpesvirus (HHV8). *Proceedings of the National Academy of Sciences of the United States of America, 93*(25), 14862–14867.

Sallah, N., Palser, A. L., Watson, S. J., Labo, N., Asiki, G., Marshall, V., Newton, R., Whitby, D., Kellam, P., & Barroso, I. (2018). Genome-wide sequence analysis of Kaposi sarcoma-associated herpesvirus shows diversification driven by recombination. *Journal of Infectious Diseases, 218*(10), 1700–1710.

Samols, M. A., Skalsky, R. L., Maldonado, A. M., Riva, A., Lopez, M. C., Baker, H. V., & Renne, R. (2007). Identification of cellular genes targeted by KSHV-encoded MicroRNAs. *PLoS Pathogens, 3*(5), e65.

Sathish, N., Wang, X., & Yuan, Y. (2012). Tegument proteins of Kaposi's sarcoma-associated herpesvirus and related gamma-herpesviruses. *Frontiers in Microbiology, 3*, 98.

Schultz, E. R., Rankin, G. W., Blanc, M. P., Raden, B. W., Tsai, C. C., & Rose, T. M. (2000). Characterization of two divergent lineages of *macaque Rhadinoviruses* related to Kaposi's sarcoma-associated herpesvirus. *Journal of Virology, 74*(11), 4919–4928.

Shebl, F. M., Dollard, S. C., Pfeiffer, R. M., Biryahwaho, B., Amin, M. M., Munuo, S. S., Hladik, W., Parsons, R., Graubard, B. I., & Mbulaiteye, S. M. (2011). Human herpesvirus 8 Seropositivity among sexually active adults in Uganda. *PLoS One, 6*, e21286.

Shimizu, S., Katano, H., Sata, T., Chen, K. R., Tagami, H., Hanabusa, H., & Shimizu, H. (2001). Absence of anti-human herpesvirus 8 antibody in 32 Japanese hemophiliacs with advanced HIV infection. *Archives of Dermatological Research, 293*(8), 380–381.

Simonart, T. (2004). Iron: A target for the Management of Kaposi's sarcoma? *BMC Cancer, 4*(1), 1.

Simonart, T. (2006). Role of environmental factors in the pathogenesis of classic and African-endemic Kaposi sarcoma. *Cancer Letters, 244*(1), 1–7.

Simonart, T., Noel, J. C., Andrei, G., Parent, D., Van Vooren, J. P., Hermans, P., Lunardi-Yskandar, Y., Lambert, C., Dieye, T., Farber, C. M., Liesnard, C., Snoeck, R., Heenen, M., & Boelaert, J. R. (1998). Iron as a potential co-factor in the pathogenesis of Kaposi's sarcoma? *International Journal of Cancer, 78*(6), 720–726.

Simpson, G. R., Schulz, T. F., Whitby, D., Cook, P. M., Boshoff, C., Rainbow, L., Howard, M. R., Gao, S. J., Bohenzky, R. A., Simmonds, P., Lee, C., de Ruiter, A., Hatzakis, A., Tedder, R. S., Weller, L. V. D., Weiss, R. A., & Moore, P. S. (1996). Prevalence of Kaposi's sarcoma associated herpesvirus infection measured by antibodies to recombinant capsid protein and latent immunofluorescence antigen. *The Lancet, 348*(9039), 1133–1138.

Smith, N. A., Sabin, C. A., Gopal, R., Bourboulia, D., Labbet, W., Boshoff, C., Barlow, D., Band, B., Peters, B. S., Ruiter, A. D., Brown, D. W. G., Weiss, R. A., Best, J. M., & Whitby, D. (1999). Serologic evidence of human herpesvirus 8 transmission by homosexual but not heterosexual sex. *Journal of Infectious Diseases, 180*(2), 600–606.

Soulier, J., Grollet, L., Oksenhendler, E., Cacoub, P., Cazals-Hatem, D., Babinet, P., d'Agay, M. F., Clauvel, J. P., Raphael, M., & Degos, L. (1995). Kaposi's sarcoma-associated herpesvirus-like DNA sequences in multicentric Castleman's disease. *Blood, 86*(4), 1276–1280.

Spear, P. G., & Longnecker, R. (2003). Herpesvirus entry: An update. *Journal of Virology, 77*(19), 10179–10185.

Sullivan, S. G., Hirsch, H. H., Franceschi, S., Steffen, I., Amari, E. B. E., Mueller, N. J., Magouras, I., Biggar, R. J., Rickenbach, M., & Clifford, G. M. (2010). Kaposi sarcoma herpes virus antibody response and viremia following highly active antiretroviral therapy in the Swiss HIV cohort study. *AIDS, 24*(14), 2245–2252.

Sun, R., Lin, S. F., Staskus, K., Gradoville, L., Grogan, E., Haase, A., & Miller, G. (1999). Kinetics of Kaposi's sarcoma-associated herpesvirus gene expression. *Journal of Virology, 73*(3), 2232–2242.

Tamanaha-Nakasone, A., Uehara, K., Tanabe, Y., Ishikawa, H., Yamakawa, N., Toyoda, Z., Kurima, K., Kina, S., Tsuneki, M., Okubo, Y., Yamaguchi, S., Utsumi, D., Takahashi, K., Arakawa, H., Arasaki, A., & Kinjo, T. (2019). K1 gene transformation activities in AIDS-related and classic type Kaposi's sarcoma: Correlation with clinical presentation. *Scientific Reports, 9*, 6416.

Tornesello, M. L., Biryahwaho, B., Downing, R., Hatzakis, A., Alessi, E., Cusini, M., Ruocco, V., Katongole-Mbidde, E., Loquercio, G., Buonaguro, L., & Buonaguro, F. M. (2010). Human herpesvirus type 8 variants circulating in Europe, Africa and North America in classic, endemic and epidemic Kaposi's sarcoma lesions during pre-AIDS and AIDS era. *Virology, 398*, 280–289.

Uppal, T., Banerjee, S., Sun, Z., Verma, S. C., & Robertson, E. S. (2014). KSHV LANA—The master regulator of KSHV latency. *Viruses, 6*(12), 4961–4998.

Uppal, T., Jha, H. C., Verma, S. C., & Robertson, E. S. (2015). Chromatinization of the KSHV genome during the KSHV life cycle. *Cancers, 7*(1), 112–142.

Varmazyar, S., Shoja, Z., Kakavand-Ghalehnoei, R., Shahmahmoodi, S., Marashi, S. M., & Jalilvand, S. (2017). Molecular typing of human herpesvirus 8 among HIV positive in comparison to HIV-negative individuals in Iran. *Journal of Medical Virology, 89*(4), 703–709.

Verma, S. C., Borah, S., & Robertson, E. S. (2004). Latency-associated nuclear antigen of Kaposi's sarcoma-associated herpesvirus up-regulates transcription of human telomerase reverse transcriptase promoter through interaction with transcription factor Sp1. *Journal of Virology, 78*(19), 10348–10359.

Wakeham, K., Webb, E. L., Sebina, I., Muhangi, L., Miley, W., Johnson, W. T., Ndibazza, J., Elliott, A. M., Whitby, D., & Newton, R. (2011). Parasite infection is associated with Kaposi's sarcoma associated herpesvirus (KSHV) in Ugandan women. *Infectious Agents and Cancer, 6*(1), 15.

Wang, H.-W., Trotter, M. W. B., Lagos, D., Bourboulia, D., Henderson, S., Mäkinen, T., Elliman, S., Flanagan, A. M., Alitalo, K., & Boshoff, C. (2004). Kaposi sarcoma herpesvirus-induced cellular reprogramming contributes to the lymphatic endothelial gene expression in Kaposi sarcoma. *Nature Genetics, 36*(7), 687–693.

Wang, L., Dittmer, D. P., Tomlinson, C. C., Fakhari, F. D., & Damania, B. (2006). Immortalization of primary endothelial cells by the K1 protein of Kaposi's sarcoma-associated herpesvirus. *Cancer Research, 66*(7), 3658–3666.

Whitby, D., Marshall, V. A., Bagni, R. K., Wang, C. D., Gamache, C. J., Guzman, J. R., Kron, M., Ebbesen, P., & Biggar, R. J. (2004). Genotypic characterization of Kaposi's sarcoma-associated herpesvirus in asymptomatic infected subjects from isolated populations. *Journal of General Virology, 85*(1), 155–163.

Whitby, D., Marshall, V. A., Bagni, R. K., Miley, W. J., McCloud, T. G., Hines-Boykin, R., Goedert, J. J., Conde, B. A., Nagashima, K., Mikovits, J., Dittmer, D. P., & Newman, D. J. (2007). Reactivation of Kaposi's sarcoma-associated herpesvirus by natural products from Kaposi's sarcoma endemic regions. *International Journal of Cancer, 120*(2), 321–328.

White, T., Hagen, M., Gudza, I., White, I. E., Ndemera, B., Gwanzura, L., Borok, M., & Campbell, T. B. (2008). Genetic diversity of the Kaposi's sarcoma herpesvirus K1 protein in AIDS-KS in Zimbabwe. *Journal of Clinical Virology, 42*, 165–171.

Wood, N. H., & Feller, L. (2008). The malignant potential of HIV-associated Kaposi sarcoma. *Cancer Cell International, 8*(1), 14.

Wu, T.-T., Qian, J., Ang, J., & Sun, R. (2012). Vaccine Prospect of Kaposi sarcoma-associated herpesvirus. *Current Opinion in Virology, 2*(4), 482–488.

Yan, L., Majerciak, V., Zheng, Z.-M., & Lan, K. (2019). Towards better understanding of KSHV life cycle: From transcription and posttranscriptional regulations to pathogenesis. *Virologica Sinica, 34*(2), 135–161.

Yanik, E. L., Achenbach, C. J., Gopal, S., Coghill, A. E., Cole, S. R., Eron, J. J., Moore, R. D., Mathews, W. C., Drozd, D. R., Hamdan, A., Ballestas, M. E., & Engels, E. A. (2016). Changes in clinical context for Kaposi's sarcoma and non-Hodgkin lymphoma among people with HIV infection in the United States. *Journal of Clinical Oncology, 34*(28), 3276–3283.

Ye, F., Zhou, F., Bedolla, R. G., Jones, T., Lei, X., Kang, T., Guadalupe, M., & Gao, S.-J. (2011). Reactive oxygen species hydrogen peroxide mediates Kaposi's sarcoma-associated herpesvirus reactivation from latency. *PLoS Pathogens, 7*(5), e1002054.

Yu, Y., Black, J. B., Goldsmith, C. S., Browning, P. J., Bhalla, K., & Offermann, M. K. (1999). Induction of human Herpesvirus-8 DNA replication and transcription by butyrate and TPA in BCBL-1 cells. *Journal of General Virology, 80*(Pt 1), 83–90.

Zeng, Y., Zhang, X., Huang, Z., Cheng, L., Yao, S., Qin, D., & Zhang, L. (2007). Intracellular tat of human immunodeficiency virus type 1 activates lytic cycle replication of Kaposi's sarcoma-associated herpesvirus: Role of JAK/STAT signaling. *Journal of Virology, 81*(5), 2401–2417.

Zhu, F. X., Chong, J. M., Wu, L., & Yuan, Y. (2005). Virion proteins of Kaposi's sarcoma-associated herpesvirus. *Journal of Virology, 79*(2), 800–811.

Zhu, X., Guo, Y., Yao, S., Yan, Q., Xue, M., Hao, T., & Lu, C. (2014). Synergy between Kaposi's sarcoma-associated herpesvirus (KSHV) VIL-6 and HIV-1 Nef protein in promotion of angiogenesis and oncogenesis: Role of the AKT signaling pathway. *Oncogene, 33*(16), 1986–1996.

Ziegler, J. L. (1993). Endemic Kaposi's sarcoma in Africa and local volcanic soils. *The Lancet, 342*(8879), 1348–1351.

Zong, J. C., Ciufo, D. M., Alcendor, D. J., Wan, X., Nicholas, J., Browning, P. J., Rady, P. L., Tyring, S. K., Orenstein, J. M., Rabkin, C. S., Su, I. J., Powell, K. F., Croxson, M., Foreman, K. E., Nickoloff, B. J., Alkan, S., & Hayward, G. S. (1999). High-level variability in the ORF-K1 membrane protein gene at the left end of the Kaposi's sarcoma-associated herpesvirus genome defines four major virus subtypes and multiple variants or clades in different human populations. *Journal of Virology, 73*(5), 4156–4170.

Zong, J., Ciufo, D. M., Viscidi, R., Alagiozoglou, L., Tyring, S., Rady, P., Orenstein, J., Boto, W., Kalumbuja, H., Romano, N., Melbye, M., Kang, G. H., Boshoff, C., & Hayward, G. S. (2002). Genotypic analysis at multiple loci across Kaposi's sarcoma herpesvirus (KSHV) DNA molecules: Clustering patterns, novel variants and Chimerism. *Journal of Clinical Virology, 23*(2), 119–148.

Cancer and Fungal Infections

9

Sumeyye Ucar, Ab Qayoom Naik, and Tabassum Zafar

Abstract

This chapter explores the specific issues experienced by cancer patients due to their vulnerability to numerous fungal pathogens, as well as the complex link between fungal infections and cancer. Aspergillosis, blastomycosis, histoplasmosis, candidiasis, coccidioidomycosis, cryptococcosis, Malassezia, paracoccidioidomycosis, sporotrichosis, and zygomycosis are just a few of the specific fungal infections that are covered in detail in this review, along with their prevalence and effects on patients receiving cancer treatment. Examined is the epidemiology of invasive fungal infections (IFIs), with an emphasis on finding patterns and risk factors. The necessity of fast and precise detection is emphasized while discussing diagnostic techniques for IFIs, particularly in pediatric cancer patients who may have particular difficulties. The chapter delves into the development of therapeutic treatments aimed at reducing the risks of fungal infections in the cancer population, including preventive measures and antifungal therapy options. With the goal of better understanding the intricate relationship between fungal infections and cancer, this thorough review will be of great use to researchers, physicians, and other healthcare professionals who work with cancer patients.

Keywords

Fungal infections · Cancer patients · Invasive fungal infections (IFIs) · Diagnostic techniques · Antifungal therapy

S. Ucar
Faculty of Medicine, Department of Anatomy, Erciyes University, Melikgazi/Kayseri, Turkey

A. Q. Naik (✉) · T. Zafar
Laboratory of Endocrinology, Department of Biosciences, Barkatullah University, Bhopal, MP, India

N. Masood, A. Yasmin (eds.), *The Microbiome and Cancer*,
https://doi.org/10.1007/978-981-97-6948-3_9

Abbreviations

AML	Acute myelogenous leukemia
BMI	Body mass index
BMT	Bone marrow transplant
CMCD	Chronic mucocutaneous candidiasis disease
EBV	Epstein-Barr virus
HEPA	High-efficiency particulate air
HIV	Human immunodeficiency virus
HPV	Human papillomavirus
HSCT	Hematopoietic stem cell transplantation
IA	Invasive aspergillosis
IFDs	Invasive fungal diseases
IFIs	Invasive fungal infections
PCR	Polymerase chain reaction
RES	Reticuloendothelial system

9.1 Introduction

Cancer incidence and mortality rates are escalating globally at a rapid pace. This trend is driven by a multitude of factors, including population growth and aging, alongside shifts in the prevalence and distribution of key cancer risk factors, many of which are linked to socioeconomic progress. As the world's population expands and ages, cancer is increasingly becoming a primary cause of death, partly due to notable declines in mortality rates for stroke and coronary heart disease compared to cancer in numerous countries (Bray et al., 2018).

Colorectal cancer is a significant problem, with approximately 150,000 new cases being diagnosed in the United States each year. As a result, research efforts have been aimed at understanding the risk factors for different types of cancer in order to reduce their incidence. Each cancer has its own specific risk factors, such as a history of inflammatory bowel disease, type 2 diabetes, and colorectal cancer, which are known to increase the risk of developing colorectal cancer (Amersi et al., 2005; Bray et al., 2018).

Risk factors for cancer can vary by region and country, in line with cancer incidence and death rates. Factors such as age, smoking, diet high in saturated fats, body weight, alcohol, pollutants, radiation, genetic background, and infection by bacterial and fungal species have been linked to the risk of developing cancer (Fig. 9.1). Studies have shown that microbial infections are responsible for initiating approximately 2.2 million new cancer incidents. Currently, it is approximate 16% of all cancer cases are associated with microbial infections and toxicities (Eyvazi et al., 2020; Plummer et al., 2016).

There is increasing interest in the role of infectious agents in the development of cancer. Presently, numerous infectious agents have been identified that cause or contribute to certain cancers in humans (Cogliano et al., 2011; O'Mahony &

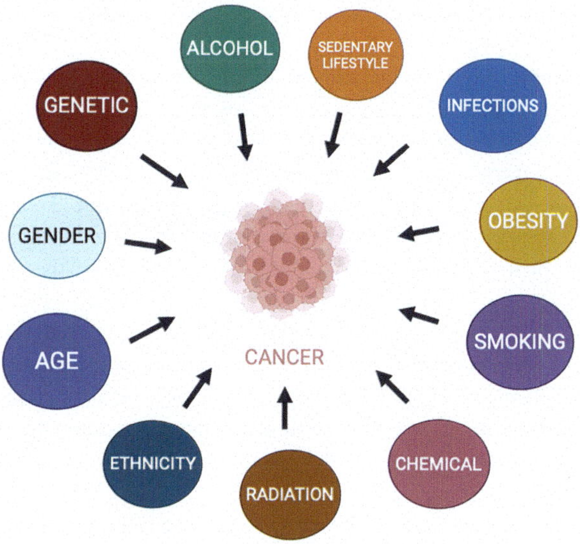

Fig. 9.1 Risk factors of carcinogenesis (figure was created in BioRender)

Hegarty, 2009; "Personal habits and indoor burns," 2012). Epidemiological data demonstrated that Helicobacter pylori infections are associated with gastric cancer, Chlamydia pneumoniae is associated with lung cancer, and Chlamydia trachomatis has been shown to be a predisposing factor for the cervical cancer. It has been determined that some human viruses also display oncogenic properties. Epstein-Barr virus (EBV), which causes nasopharyngeal carcinoma and Burkitt lymphoma, has been detected in tumor cells. Studies show that EBV plays a role in the etiopathogenesis of stomach cancer, Hodgkin's lymphoma, as well as palatal tonsil and tongue cancer (Chang and Parsonnet, 2010; Neves et al., 2017).

Cervical cancer is caused by human papillomavirus (HPV) infection in women. Other viruses such as human Merkel cell polyomavirus, human T-lymphotropic retrovirus type 1 (HTLV-1), and human immunodeficiency virus (HIV) types 1 and 2 are important in leukemia, lung cancer, and lymphomas. Parasitic infections have also been shown to contribute to tumorigenesis; Schistosoma haematobium is a primary factor in gastric cancer and liver flukes in bladder cancer (Delbue et al., 2017; Jemal et al., 2011; Maginnis and Atwood, 2009; Yang et al., 2023).

Invasive fungal infections (IFIs) remain a significant defiance in cancer development, often associated with death and death rates. Despite advancements in diagnostics and therapeutics for fungal infections over the last 20 years, IFI mortality has continued to rise, with a growth rate of 35.7%. This suggests a potential impact of IFIs on cancer progression, necessitating further investigation. Common fungal genera and species implicated in cancers include Candida albicans, C. glabrata, C. tropicalis, Aspergillus flavus, A. parasiticus, Fusarium verticillioides, and F. proliferatum. However, ongoing efforts are crucial to better understand the relationship among them fungal infections and cancers, as the battle against cancer is unlikely to be swiftly won.

IFI is a significant cause of morbidity and mortality in cancer patients, especially those undergoing hematopoietic stem cell transplantation (HSCT). However, these infections increasingly appear to be significant complications in cancer patients receiving chemotherapy alone and can be difficult to diagnose. Invasive procedures such as tissue biopsy are often required to confirm infection. Treatment remains difficult due to variable susceptibility of these organisms to antifungal agents, often necessitating empiric therapy while awaiting identification of the causative agent (Angarone, 2014).

When cleanliness is not given sufficient importance, fungal infections often occur in patients with suppressed immune systems, for example, Aspergillus spp. They are airborne fungi and spread through ventilation systems. The risk of infection can be greatly reduced by using high-efficiency particulate air (HEPA) filtration. Transmission of candida from the hands of medical professionals is also a significant cause of nosocomial infections. Therefore, handwashing is vital, but infections are often encountered because the cleaning required to reduce infection is not always implemented (Garber, 2001).

9.2 Cancer Patients and Fungal Infections

9.2.1 Fungal Infections

Fungi are common in our immediate living environments, and many can produce abundant airborne spores. Therefore, humans are exposed daily to various environmental fungal pathogens such as bioaerosols. While most fungi do not cause disease in healthy individuals, they can cause a range of diseases in the immunocompromised, including superficial, allergic, chronic, and life-threatening invasive fungal diseases (IFDs) (Fisher et al., 2022). The risk of IFD includes older people; those with weakened immune systems due to HIV, cancer chemotherapy, or immunosuppression therapy requiring organ transplantation; as well as people with serious viral infections such as influenza virus (H1N1) and COVID-19 (Armstrong-James et al., 2020; Gargand et al. al., 2021; Wauters et al., 2012). In the last group of patients, mainly Aspergillus spp., Candida spp. including C. auris., and India, there have been increases in infection caused by fungal groups, particularly Mucoromycota species, which exhibit strong intrinsic and acquired resistance to antifungal treatments (Arastehfar et al., 2020; Janssen et al., 2021; Singh et al., 2021).

Fungi of medical importance are classified according to their main growth patterns, such as yeast and mold. Yeasts are single-celled fungi that reproduce by budding (Candida spp.) or division (Schizosaccharomyces pombe). Molds are multicellular fungi that reproduce with filamentous cell protrusions or hyphae. Medically important fungi transform from multicellular hyphae in the environment to single-celled yeast forms in the tissue, and these pathogens are called dimorphic fungi (such as Blastomyces dermatitidis, Histoplasma capsulatum) (Garber, 2001).

9.2.2 Aspergillosis

Aspergillus fungi are hyphal organisms widely distributed in soil and decaying plant material. The genus comprises 185 species, with Aspergillus fumigatus, Aspergillus niger, Aspergillus flavus, and Aspergillus terreus being the most common human pathogens. A. fumigatus is responsible for about 90% of cases. Accurate identification of Aspergillus species is clinically important; for instance, A. terreus is resistant to amphotericin B. Aerosolized Aspergillus conidia, the reproductive spores, can remain suspended in the air and be carried long distances by dust and other particles. Although inhaled conidia are typically neutralized by the body's immune system in healthy individuals, they can cause invasive disease in those with compromised immunity. Macrophages and neutrophils play crucial roles in defending against Aspergillus conidia. The key risk factors for aspergillus infections include neutropenia, high conidia exposure surpassing macrophage clearance capacity, and the use of corticosteroids, which can suppress macrophage function. Aspergillosis, particularly the invasive form, has a high mortality rate and its incidence is rising. Invasive aspergillosis is now a leading cause of mortality, largely due to successful prophylactic antifungal treatments and advanced diagnostic techniques for candida infections. Infection usually occurs when spores colonize the airways in the upper and lower respiratory tract. Cutaneous infections can also occur, often resulting from wound contamination, intradermal transplantation, or hematogenous spread. Immunocompromised individuals are particularly vulnerable to hematogenous spread of the infection (Lockhart & Guarner, 2019; Sutcu & Salman, 2015).

9.2.3 Blastomycosis

Blastomyces dermatitidis is the fungus responsible for causing blastomycosis and belongs to the ascomycete group, which demonstrates thermal dimorphism. Other fungi in this group include Histoplasma capsulatum, Coccidioides immitis, Coccidioides posadasii, Paracoccidioides brasiliensis, Sporothrix schenckii, and Penicillium marneffei. The infectious conidia of B. dermatitidis are abundant in soil at temperatures ranging from 25 °C to 22 °C and can also be found in bird and animal feces. Aerosolized conidia are inhaled into the host's lungs and transform into a pathogenic yeast form at body temperature (37 °C). In its yeast form, the fungus disrupts the host's cytokine response, leading to disease.

Blastomycosis can affect both immunocompromised and immunocompetent individuals. Although most cases are sporadic, outbreaks can occur, particularly affecting children. Blastomycosis is uncommon in children, accounting for only 2–13% of cases. The primary route of entry into the body is through the lungs, where spores are inhaled. Rarely, transmission can occur at birth through inhalation of infected vaginal secretions or through inoculation into the skin. The incubation period for blastomycosis varies from 3 weeks to several months (Mazi et al., 2021).

9.2.4 Histoplasmosis

Histoplasma capsulatum is a fungus that exhibits thermal dimorphism, existing as a mold at temperatures of 25–30 °C and as a yeast at 37 °C. It thrives in nitrate-rich soils heavily contaminated with bird droppings or decaying wood. Infection with Histoplasma typically occurs in environments such as caves or areas frequented by bats.

In humans, infection begins with the inhalation of microconidia. The initial infection manifests as bronchopneumonia, leading to the development of pulmonary lesions. Giant cells form, followed by granuloma formation and necrosis. Yeast cells then disseminate to the reticuloendothelial system (RES) via the lymph nodes. Spleen involvement indicates dissemination and typically occurs after primary lung involvement.

The first pulmonary lesion usually resolves within 2–4 months, leaving behind calcifications that can be mistaken for tuberculosis. Lead-like calcifications may also appear in the lungs and spleen. There is a risk of reactivation in these areas, leading to recurrent infection (Araúz & Papineni, 2021).

9.2.5 Candidiasis

Infections caused by Candida species are known as candidiasis or moniliasis. Candida naturally resides in the gastrointestinal and genitourinary tracts of healthy individuals. Species that commonly infect humans include Candida albicans, Candida parapsilosis, Candida tropicalis, Candida krusei, Candida lusitaniae, and Candida glabrata. Candida dubliniensis, previously classified as C. albicans, is now recognized as a distinct species responsible for invasive Candida infections.

Candida can exist in three forms: blastospores, yeast cells, and chlamydospores. Chlamydospores, which are double-walled structures located at the terminal end of pseudohyphae, form pseudomycelium, representing the tissue stage of candida. While most candida species produce pseudohyphae, only C. albicans and C. dubliniensis produce true hyphae.

Candida is a leading cause of fungal infections in intensive care, premature, chronically ill, and immunocompromised patients, as well as in healthcare settings. Mortality rates due to candidemia in children range from 10% to 26%, with lung involvement being the most common, followed by liver, kidney, brain, heart, spleen, and eye involvement.

Risk factors for candidemia include malignancy, primary and secondary immune deficiencies, central venous catheters, arterial catheters, hyperalimentation, and antifungal and broad-spectrum antibiotic use, particularly vancomycin and anaerobic spectrum antibiotics. Among Candida species, C. albicans is the most common cause of invasive fungal infections due to its presence in normal human mucosa.

C. parapsilosis has become the second most isolated species in children with invasive candidiasis. Approximately 20% of C. lusitaniae strains are resistant to amphotericin B. Fluconazole is effective against most candida infections but is ineffective against all C. krusei strains and 20% of C. glabrata strains.

C. albicans is considered the most pathogenic species, with virulence factors including adhesion, biofilm formation, secretion of hydrolytic enzymes, and morphological changes. Both innate and adaptive immune responses play crucial roles in the defense against candida infections, primarily manifesting as mucosal or disseminated infections in immunosuppressed patients. T-cell response is particularly important in combating candida infections (Garber, 2001; McCarty et al., 2021; Pfaller & Diekema, 2007).

9.2.6 Coccidioidomycosis

Coccidioidomycosis is a fungal infection caused by Coccidioides immitis, a dimorphic fungus belonging to the genus Coccidioides. This infection primarily affects the lungs and is typically self-limiting, but it can progress to a severe and potentially fatal disease. C. immitis exists in a mycelial form in soil and in a spherule form in human tissue. Although the disease is commonly observed in long-term residents of arid regions, recovery from the infection usually provides lifelong immunity.

The mycelial form of C. immitis enters the host through inhalation or via damaged skin, where it transforms into spherules. Within these spherules, endospores develop and disseminate throughout the host. Human-to-human transmission is rare but can occur through organ transplantation from an infected individual. Maternal-fetal and maternal-neonatal infections have also been reported.

The arthroconidia produced by C. immitis in soil and on culture surfaces are highly infectious and can reproduce within 4–5 days in surface cultures. Patient isolation is unnecessary, but covering open lesions can help prevent transmission. It is believed that most infections result from exposure to a single spore (Johnson et al., 2021; Stockamp & Thompson 3rd., 2016).

9.2.7 Cryptococcosis

Cryptococcal infections are invasive fungal diseases caused by Cryptococcus neoformans, a monomorphic, encapsulated yeast. The virulence factors of this fungus include its capsule, melanization ability, capacity to reproduce at 37 °C, and production of urease and phospholipase. C. neoformans has five serotypes (A, B, C, D, and AD) based on reactions with anticapsular serum. Serotypes B and C are referred to as Cryptococcus gattii.

There are notable differences between C. neoformans and C. gattii. While C. neoformans primarily affects immunocompromised individuals, including those with HIV/AIDS, cases of pediatric cryptococcosis occur with equal frequency in immunocompromised and immunocompetent patients. The fungus is commonly found in temperate climates and can be present in soil contaminated with bird droppings, on vegetables and fruits, and can be carried by cockroaches. Bird keepers and laboratory workers are at the highest risk of infection (Garber, 2001; Zavala & Baddley, 2020).

9.2.8 Malassezia

It is the causative agent of tinea versicolor, which is a fungal dermatosis. It may rarely cause fungemia in patients with catheters. Malassezia commensal is a lipophilic yeast and causes disease in skin areas rich in sebaceous glands. The transformation of yeast into hyphal forms facilitates the formation of invasive disease. Thick-walled blastospores and hyphae cluster together, creating the characteristic "spaghetti and meat balls" appearance. The most common species is Malassezia furfur. Additionally, Malassezia pachydermatis has caused outbreaks in neonatal intensive care units. The major cause of tinea versicolor is Malassezia globosa. Additionally, Malassezia species may cause acne and seborrheic dermatitis in newborns. It can be found in sebum-rich skin in asymptomatic individuals (Theelen et al., 2018).

9.2.9 Paracoccidioidomycosis

The causative agent of paracoccidioidomycosis is fungi of the genus Paracoccidioides, which show temperature dimorphism. As a result of phylogenetic studies, two species were determined: Paracoccidioides brasiliensis (S1, PS2, PS3, and phylogenetic species) and Paracoccidioides lutzii (such as pb01). While it exists in mycelial form at room temperature, it is detected as yeast in human tissues. Infection occurs as a result of inhalation of conidia. Acute disease is common in immunosuppressed cases and children. Fever, malaise, lymphadenopathy, and hepatosplenomegaly findings due to reticuloendothelial system involvement are prominent. Mortality in acute disease is 25%. Chronic disease is seen in people with intense exposure to soil. Pulmonary involvement with progressive dyspnea, chest pain, and hemoptysis is most common. Histological identification of the fungus in tissue samples is necessary for definitive diagnosis. The first treatment of choice is itraconazole and the average treatment duration is 6 months. Amphotericin B is used when the disease is widespread and unresponsive to treatment (Marques, 2012; Sutcu & Salman, 2015).

9.2.10 Sporotrichosis

It is a common fungal infection caused by Sporothrix schenckii, which begins as an ulcerated nodule at a local trauma site and spreads along regional lymphatic channels. It is a rare problem in children and occurs as a result of frequent exposure to soil and plants. S. schenckii shows temperature dimorphism and is found as a mold at 25–30 °C and as a yeast in human tissue. It is seen both sporadically and in epidemics all over the world. In humans, the agent enters the skin where its integrity is disrupted. Farmers, gardeners, veterinarians, and laboratory workers are at risk. Besides, inhalation of numerous spores, pulmonary infection may occur as a result. T-cell-mediated immune response is important in limiting infection, and widespread infection is detected in immunosuppressed patients (Queiroz-Telles et al., 2019).

9.2.11 Zygomycosis

Zygomycosis is a group of opportunistic fungal infections characterized by vascular invasion, thrombosis, and necrosis. The agents are Absidia mucor, Rhizopus, and Rhizomucor genera from the Zygomycetes class. These saprophytic and parasitic fungi are abundant in soil, rotten plants, and animals. Zygomycosis infections are seen in people with impaired immunity such as diabetes mellitus, hematological malignancies, chronic acidosis, iron accumulation, corticosteroid use, organ transplantation, and AIDS patients. The disease occurs as a result of inhalation of spores in the environment. More rarely, percutaneous transmission may cause cutaneous or subcutaneous zygomycosis. Consuming contaminated foods and beverages is associated with gastrointestinal disease. The main cells in the immune response against the organism are neutrophils, and zygomycosis infections are more common in patients with neutropenia and neutrophil dysfunction. Zygomycosis causes various clinical conditions such as sinusitis, pulmonary, gastrointestinal, cutaneous, and disseminated disease. The most common clinical form is rhinocerebral and pulmonary disease. In rhinocerebral disease following the symptoms of headache, retro-orbital pain, fever, and runny nose, which are compatible with sinusitis, orbital findings such as periorbital edema, proptosis, ptosis, and ophthalmoplegia are observed. Nasal discharge is mostly black and thick, and the nasal mucosa contains black and necrotic areas. While brain abscesses occur in the frontal and frontotemporal lobes due to the direct spread of infection in the nasal cavity and sinuses, they may occur in the occipital lobe and brain stem as a result of disseminated infection (Gonzalez et al., 2002; Roden et al., 2005).

9.3 Epidemiology

The epidemiology of IFI in cancer patients has evolved over time. In the 1980s, Candida species were the predominant cause of IFI in cancer patients, with candidiasis being more common than mold infections. The introduction of azole antifungals marked a significant shift in fungal pathogens. Azoles were preferred over polyene antifungals due to their improved tolerability and became widely used for prophylaxis, leading to a decline in candidal infections. As a result of azole use, azole-resistant Candida species such as Candida krusei and C. Glabrata have emerged. These resistant strains account for more than half of the candida isolates identified in cancer patients (Horn et al., 2009; Marr et al., 2000; Pfaller et al., 2010).

The use of new era antifungal agents, such as echinocandins, has been effective against fungi, especially in the treatment of azole-resistant candidal isolates. At the same time, as the use of echinocandins has increased, the incidence of echinocandin-resistant organisms such as Candida parapsilosis has increased. More research is needed to clarify the mechanisms and to manage and prevent the development of antifungal resistance (Sipsas et al., 2009).

The true incidence of IFIs in patients with cancer is difficult to assess, primarily because data are available from single center or regional retrospective studies. Reported incidence rates range from 5% to 30%. However, in the last two decades, there has been a significant shift in the causative agents of IFE, with an increase in infections caused by molds and mucormycosis agents such as Aspergillus spp. and Fusarium spp.

Autopsy studies conducted at MD Anderson Cancer Center from 1989 to 2003 provide insight into the prevalence of IFI. The most significant finding was that the incidence of invasive mold infections increased from 60% to 76%, while candidal infections decreased from 40% to 26%. Significant increases have been observed, especially in infections caused by Aspergillus spp., endemic fungi, and mucormycoses (Chamilos et al., 2005; Husain et al., 2003; Leventakos et al., 2010; Marr et al., 2002).

Fungal infections may occur with esophageal carcinoma. The most common fungus is Candida spp., an opportunistic yeast that settles in the genitourinary and gastrointestinal tract mucosa of healthy people (approximately 30–50% of people) without causing infection (Hosseini et al., 2022). The first line of defense against fungal infections is the anatomical barrier provided by structures such as skin and mucosa. These barriers prevent microorganisms from entering the body and produce antimicrobial secretions that help destroy fungal organisms. In immunocompetent individuals, an intact immune system and anatomical barriers effectively prevent these fungal elements.

In cancer patients, these barriers can be breached by mucosal damage caused by invasive procedures such as central venous catheters and chemotherapy, allowing fungal organisms to penetrate tissues and enter the bloodstream. Immunosuppression increases the susceptibility of cancer patients to fungal infections (Bow, 2008; Guiot et al., 1994; Maertens et al., 2001).

9.4 Diagnosis of IFI

Diagnosis of fungal infections is difficult due to their variety, including persistent fever, sepsis, fungemia, and organ-invasive diseases. Isolating and identifying the causative fungus often requires invasive procedures, and many molds are difficult to grow in the laboratory. To identify the fungus that is the cause of the infection, it is necessary to culture tissue samples and perform histopathological examination, fungal antigen analyses, and molecular tests such as PCR for fungal-specific DNA (Baddley et al., 2010).

The gold standard for diagnosing fungal infections is histopathological identification in specimens if the patient is suitable for biopsy and isolation of fungal organisms from blood or other clinical specimens in culture.

A variety of non-culture-based tests are now available that can help diagnose fungal infections using serum or urine samples. These analyses include serological analyses and fungal antigen detection tests. Antigen detection assays have also proven effective in diagnosing infections caused by cryptococcosis, histoplasmosis, and blastomycosis (Pappagianis and Zimmer, 1990; Connolly et al., 2012).

Polymerase chain reaction (PCR) is a powerful diagnostic technique that has the potential to provide accurate and definitive diagnosis through noninvasive testing. PCR assays are known for their high specificity, which can range from 92% to 100% depending on the gene amplified (Kawazu et al., 2004; Loeffler et al., 2000; Pham et al., 2003; White et al., 2006). The practicality of the PCR test has not yet been proven, but with the development of commercial or standardized assays, this testing method may become more widely used and acceptable in the future (Angarone, 2014; Connolly 2012).

9.5 Fungal Infections in Pediatric Patients with Cancer

In neonates and pediatric populations, IFIs pose significant challenges. High doses of chemotherapy used in conditions like acute myelogenous leukemia (AML) can cause profound immunosuppression. This, combined with the intensity and duration of chemotherapy, often leads to prolonged neutropenia, increasing the risk of IFIs.

To address these challenges, large studies and cooperative group clinical trials are being conducted to better understand IFIs in pediatric cancer patients. Researchers are exploring various predictors of infection outcomes, such as body mass index (BMI), age, ethnicity, leukemia cytogenetics, and the size of the treating institution. Data from these trials are crucial in determining the incidence and types of infections associated with pediatric cancer patients, which may lead to the development of alternative strategies to improve outcomes (Lange et al., 2005; Mor et al., 2011; Sung, 2010; Sung et al., 2007).

The issue of invasive fungal infections (IFI) in adult patients with leukemia is increasingly recognized. Studies are underway to determine the incidence of infection, the specific fungal organisms involved, and the factors that increase the risk of infection in this population. As more data accumulates and awareness grows, clinicians can consider more effective treatments for these vulnerable patients.

This knowledge is now being applied to IFI studies in pediatric populations, particularly in children with leukemia. Collaborative group studies are ongoing to evaluate the most effective chemotherapy regimens, which also allow researchers to determine the incidence of IFIs and associated deaths. In immunocompromised pediatric patients undergoing aggressive chemotherapy, the added risk of IFIs is significant, necessitating careful evaluation of antifungal therapy.

While aspergillus and Candida are the primary fungal pathogens causing infection-related death in pediatric cancer patients, emerging pathogens are becoming a concern. Continued research focusing on these invasive fungal pathogens is crucial to successfully prevent and treat IFIs in pediatric patients with leukemia (Sung, 2010).

9.6 Prophylaxis

When considering prophylaxis for invasive fungal diseases (IFD), it's important to carefully balance the likelihood of IFD and the types of fungi that are likely to cause IFD with the risks and benefits of antifungal agents. Studies have demonstrated that fluconazole prophylaxis is safe and effective for patients undergoing bone marrow transplant (BMT) and those with leukemia, consistently reducing the incidence of IFDs, which are mostly caused by Candida species (Marr et al., 2000).

In patient populations exposed to extensive fluconazole use, such as those undergoing BMT, there has been a proportional increase in invasive fungal diseases (IFDs) caused by non-albicans Candida species. Surprisingly, the emergence of IFDs due to fluconazole-resistant non-albicans Candida species has been less problematic for BMT patients, despite over 20 years of azole prophylaxis.

In vitro antifungal susceptibility testing remains useful in epidemiological studies to guide treatment and identify the emergence of resistant strains in these patients. With invasive aspergillosis (IA) emerging as the primary cause of IFD in cancer patients, efforts have focused on identifying patients at greatest risk for IA and determining optimal management strategies.

Previously, IA was considered a contraindication for BMT. However, it is now accepted that "secondary" prophylaxis with voriconazole can enable the safe performance of BMT in patients with prior IA (Cordonnier et al., 2010).

In a trial involving patients with acute myelogenous leukemia (AML) or myelodysplastic syndrome (MDS), the use of posaconazole effectively reduced invasive fungal diseases (IFDs), including invasive aspergillosis (IA). Additionally, the trial noted a survival benefit associated with posaconazole treatment (Cornely et al., 2007).

It's crucial to recognize that participants in certain prophylaxis studies had varying risks for invasive fungal disease (IFD). The benefits of prophylaxis might have been more pronounced if only patients at higher risk were enrolled. For instance, in one study, patients undergoing bone marrow transplant (BMT) for acute myelogenous leukemia (AML), who were at the highest risk, showed fewer IFDs with voriconazole prophylaxis in that subgroup (Wingard et al., 2010).

The breadth of the fungal family also indicates its endless diversity, which easily adapts when exposed to antifungal chemicals. The ecological breadth of fungal species, combined with their unique and diverse ecological diet in rapidly changing environments, means that they will always be intertwined with human health.

At the same time, the need to control fungal diseases in agricultural environments needs to be carried out in conjunction with the clinic. Today, the fight against mantras must be addressed by networks, infrastructures, research funding, greater coordination between policymakers, funding agencies and researchers, and by involving antifungal manufacturers and users (Fisher et al., 2022).

9.7 Antifungal Treatment

A wide range of agents are available with different mechanisms of action, spectrums of activity, and tolerability. The primary classes include polyenes, echinocandins, and azoles. Treatment selection is based on factors such as the specific fungal pathogen, infection site, drug tolerability, and toxicity profile. Polyenes, including amphotericin B deoxycholate and lipid-associated preparations, work by binding to ergosterol in the fungal cell membrane. This binding creates ion channels in the membrane, leading to its physical disruption. Polyenes exhibit broad-spectrum activity against most fungi. However, a significant drawback of amphotericin B deoxycholate is its nephrotoxicity, which can result in renal failure necessitating dialysis (Bates et al., 2001; Wingard et al., 1999).

Echinocandins, such as caspofungin, micafungin, and anidulafungin, are lipopeptides that hinder the synthesis of 1,3-β-glucan, a polysaccharide crucial for fortifying the fungal cell wall. This inhibition alters the osmotic balance within the fungal cell, ultimately causing cell damage. Echinocandins have also demonstrated effectiveness in treating refractory aspergillosis when used alongside voriconazole. Common adverse effects of echinocandins include elevated liver aminotransferases (especially with caspofungin), gastrointestinal disturbances, and headaches (Arathoon et al., 2002; Marr et al., 2004).

The azoles are a class of antifungal agents with a similar mechanism of action but varying activity spectra. They work by inhibiting lanosterol 14-α-demethylase, a key enzyme in ergosterol biosynthesis, leading to alterations in the fungal cell membrane. Four commonly used azoles in patients with malignancies are fluconazole, itraconazole, voriconazole, and posaconazole.

Fluconazole is a narrow-spectrum azole mainly used for treating candidal infections. It is well-tolerated and available in both oral and intravenous forms. Itraconazole, on the other hand, has a broader spectrum, effective against Candida spp., Aspergillus spp., and endemic fungi. However, its oral capsule form can have erratic gastrointestinal absorption, potentially causing upset.

New era zoles like voriconazole and posaconazole offer broader spectrums of activity and improved tolerability. Voriconazole, similar in structure to fluconazole, also targets Aspergillus spp. and is often the preferred choice for proven or suspected invasive aspergillosis based on available data (Herbrecht et al., 2002; Walsh et al., 2008).

9.8 Summary

The present topic "Cancer and Fungal Infections" explores the complex interplay between different fungal infections and cancer patients. The importance of comprehending how fungi can cause infections in cancer patients is emphasized in the beginning. Providing insights into their manifestations and implications for cancer patients, the review thoroughly covers a number of specific infections, including aspergillosis, blastomycosis, histoplasmosis, candidiasis, coccidioidomycosis, cryptococcosis, Malassezia, paracoccidioidomycosis, sporotrichosis, and zygomycosis. The incidence and risk factors related to these illnesses are clarified by the epidemiological aspects. Invasive fungal infections (IFI) and their diagnostic problems are addressed, with a focus on the significance of prompt and precise detection. The review broadens its scope to include juvenile cancer patients, clarifying special factors to take into account when treating fungal infections in this population. A thorough review of preventative methods and antifungal medications is provided, presenting a full picture of approaches to fungal infections in cancer patients.

References

Amersi, F., Agustin, M., & Ko, C. Y. (2005). Colorectal cancer: Epidemiology, risk factors, and health services. *Clinics in Colon and Rectal Surgery, 18*(3), 133–140. https://doi.org/10.1055/s-2005-916274

Angarone, M. (2014). Fungal infections in cancer patients. *Cancer Treatment and Research, 161*, 129–155. https://doi.org/10.1007/978-3-319-04220-6_4

Arastehfar, A., Carvalho, A., Nguyen, M. H., Hedayati, M. T., Netea, M. G., Perlin, D. S., & Hoenigl, M. (2020). COVID-19-associated candidiasis (CAC): An underestimated complication in the absence of immunological predispositions? *J Fungi (Basel), 6*(4), 211. https://doi.org/10.3390/jof6040211

Arathoon, E. G., Gotuzzo, E., Noriega, L. M., Berman, R. S., DiNubile, M. J., & Sable, C. A. (2002). Randomized, double-blind, multicenter study of caspofungin versus amphotericin B for treatment of oropharyngeal and esophageal candidiases. *Antimicrobial Agents and Chemotherapy, 46*(2), 451–457. https://doi.org/10.1128/aac.46.2.451-457.2002

Araúz, A. B., & Papineni, P. (2021). Histoplasmosis. *Infectious Disease Clinics of North America, 35*(2), 471–491. https://doi.org/10.1016/j.idc.2021.03.011

Armstrong-James, D., Youngs, J., Bicanic, T., Abdolrasouli, A., Denning, D. W., Johnson, E., Mehra, V., Pagliuca, T., Patel, B., Rhodes, J., Schelenz, S., Shah, A., van de Veerdonk, F. L., Verweij, P. E., White, P. L., & Fisher, M. C. (2020). Confronting and mitigating the risk of COVID-19 associated pulmonary aspergillosis. *The European Respiratory Journal, 56*(4), 2002554. https://doi.org/10.1183/13993003.02554-2020

Baddley, J. W., Andes, D. R., Marr, K. A., Kontoyiannis, D. P., Alexander, B. D., Kauffman, C. A., Oster, R. A., Anaissie, E. J., Walsh, T. J., Schuster, M. G., Wingard, J. R., Patterson, T. F., Ito, J. I., Williams, O. D., Chiller, T., & Pappas, P. G. (2010). Factors associated with mortality in transplant patients with invasive aspergillosis. *Clinical Infectious Diseases : An Official Publication of the Infectious Diseases Society of America, 50*(12), 1559–1567. https://doi.org/10.1086/652768

Bates, D. W., Su, L., Yu, D. T., Chertow, G. M., Seger, D. L., Gomes, D. R., Dasbach, E. J., & Platt, R. (2001). Mortality and costs of acute renal failure associated with amphotericin B therapy. *Clinical Infectious Diseases, 32*(5), 686–693. https://doi.org/10.1086/319211

Bray, F., Ferlay, J., Soerjomataram, I., Siegel, R. L., Torre, L. A., & Jemal, A. (2018). Global cancer statistics 2018: GLOBOCAN estimates of incidence and mortality worldwide for 36 cancers in 185 countries. *CA: A Cancer Journal for Clinicians, 68*(6), 394–424. https://doi.org/10.3322/caac.21492

Chamilos, G., Marom, E. M., Lewis, R. E., Lionakis, M. S., & Kontoyiannis, D. P. (2005). Predictors of pulmonary zygomycosis versus invasive pulmonary aspergillosis in patients with cancer. *Clinical Infectious Diseases : An Official Publication of the Infectious Diseases Society of America, 41*(1), 60–66. https://doi.org/10.1086/430710

Chang, A. H., & Parsonnet, J. (2010). Role of bacteria in oncogenesis. *Clinical Microbiology Reviews, 23*(4), 837–857.

Cogliano, V. J., Baan, R., Straif, K., Grosse, Y., Lauby-Secretan, B., El Ghissassi, F., Bouvard, V., Benbrahim-Tallaa, L., Guha, N., Freeman, C., Galichet, L., & Wild, C. P. (2011). Preventable exposures associated with human cancers. *Journal of the National Cancer Institute, 103*(24), 1827–1839.

Connolly, P., Hage, C. A., Bariola, J. R., Bensadoun, E., Rodgers, M., Bradsher, R. W., Jr., & Wheat, L. J. (2012). Blastomyces dermatitidis antigen detection by quantitative enzyme immunoassay. *Clinical and vaccine immunology: CVI, 19*(1), 53–56. https://doi.org/10.1128/CVI.05248-11

Cordonnier, C., Rovira, M., Maertens, J., Olavarria, E., Faucher, C., Bilger, K., Pigneux, A., Cornely, O. A., Ullmann, A. J., Bofarull, R. M., de la Cámara, R., Weisser, M., Liakopoulou, E., Abecasis, M., Heussel, C. P., Pineau, M., Ljungman, P., & Einsele, H. (2010). Voriconazole for secondary prophylaxis of invasive fungal infections in allogeneic stem cell transplant recipients: Results of the VOSIFI study. *Haematologica, 95*(10), 1762–1768. https://doi.org/10.3324/haematol.2009.020073

Cornely, O. A., Maertens, J., Winston, D. J., Perfect, J., Ullmann, A. J., Walsh, T. J., Helfgott, D., Holowiecki, J., Stockelberg, D., Goh, Y. T., Petrini, M., Hardalo, C., Suresh, R., & Angulo-Gonzalez, D. (2007). Posaconazole vs. fluconazole or itraconazole prophylaxis in patients with neutropenia. *The New England Journal of Medicine, 356*(4), 348–359. https://doi.org/10.1056/NEJMoa061094

Delbue, S., Comar, M., & Ferrante, P. (2017). Review on the role of the human Polyomavirus JC in the development of tumors. *Infectious agents and cancer, 12*, 10.

Eyvazi, S., Vostakolaei, M. A., Dilmaghani, A., Borumandi, O., Hejazi, M. S., Kahroba, H., & Tarhriz, V. (2020). The oncogenic roles of bacterial infections in development of cancer. *Microbial Pathogenesis, 141*, 104019. https://doi.org/10.1016/j.micpath.2020.104019

Fisher, M. C., Alastruey-Izquierdo, A., Berman, J., Bicanic, T., Bignell, E. M., Bowyer, P., Bromley, M., Brüggemann, R., Garber, G., Cornely, O. A., Gurr, S. J., Harrison, T. S., Kuijper, E., Rhodes, J., Sheppard, D. C., Warris, A., White, P. L., Xu, J., Zwaan, B., & Verweij, P. E. (2022). Tackling the emerging threat of antifungal resistance to human health. *Nature Reviews. Microbiology, 20*(9), 557–571. https://doi.org/10.1038/s41579-022-00720-1

Garber, G. (2001). An overview of fungal infections. *Drugs, 61*(Suppl 1), 1–12. https://doi.org/10.2165/00003495-200161001-00001

Gonzalez, C. E., Rinaldi, M. G., & Sugar, A. M. (2002). Zygomycosis. *Infectious Disease Clinics of North America, 16*(4), 895–914, vi. https://doi.org/10.1016/s0891-5520(02)00037-5

Herbrecht, R., Denning, D. W., Patterson, T. F., Bennett, J. E., Greene, R. E., Oestmann, J. W., Kern, W. V., Marr, K. A., Ribaud, P., Lortholary, O., Sylvester, R., Rubin, R. H., Wingard, J. R., Stark, P., Durand, C., Caillot, D., Thiel, E., Chandrasekar, P. H., Hodges, M. R., et al. (2002). Voriconazole versus amphotericin B for primary therapy of invasive Aspergillosis. *The New England Journal of Medicine, 347*(6), 408–415. https://doi.org/10.1056/NEJMoa020191

Horn, D. L., Neofytos, D., Anaissie, E. J., Fishman, J. A., Steinbach, W. J., Olyaei, A. J., Marr, K. A., Pfaller, M. A., Chang, C. H., & Webster, K. M. (2009). Epidemiology and outcomes of candidemia in 2019 patients: Data from the prospective antifungal therapy alliance registry. *Clinical Infectious Diseases : An Official Publication of the Infectious Diseases Society of America, 48*(12), 1695–1703. https://doi.org/10.1086/599039

Hosseini, K., Ahangari, H., Chapeland-Leclerc, F., Ruprich-Robert, G., Tarhriz, V., & Dilmaghani, A. (2022). Role of fungal infections in carcinogenesis and cancer development: A literature review. *Adv Pharm Bull, 12*(4), 747–756. https://doi.org/10.34172/apb.2022.076

Husain, F. A., Martin, M. J., Mullenix, P. S., Steele, S. R., & Elliott, D. C. (2003). Serum lactate and base deficit as predictors of mortality and morbidity. *American Journal of Surgery, 185*(5), 485–491. https://doi.org/10.1016/s0002-9610(03)00044-8

Janssen, N. A. F., Nyga, R., Vanderbeke, L., Jacobs, C., Ergün, M., Buil, J. B., van Dijk, K., Altenburg, J., Bouman, C. S. C., van der Spoel, H. I., Rijnders, B. J. A., Dunbar, A., Schouten, J. A., Lagrou, K., Bourgeois, M., Reynders, M., van Regenmortel, N., Rutsaert, L., Lormans, P., et al. (2021). Multinational observational cohort study of COVID-19-associated pulmonary Aspergillosis(1). *Emerging Infectious Diseases, 27*(11), 2892–2898. https://doi.org/10.3201/eid2711.211174

Jemal, A., Bray, F., Center, M. M., Ferlay, J., Ward, E., & Forman, D. (2011). Global cancer statistics. *CA: a Cancer Journal for Clinicians, 61*(2), 69–90.

Johnson, R. H., Sharma, R., Kuran, R., Fong, I., & Heidari, A. (2021). Coccidioidomycosis: a review. *Journal of Investigative Medicine, 69*(2), 316–323. https://doi.org/10.1136/jim-2020-001655

Kawazu, M., Kanda, Y., Nannya, Y., Aoki, K., Kurokawa, M., Chiba, S., Motokura, T., Hirai, H., & Ogawa, S. (2004). Prospective comparison of the diagnostic potential of real-time PCR, double-sandwich enzyme-linked immunosorbent assay for galactomannan, and a (1DOUBLEHYPHEN>3)-beta-D-glucan test in weekly screening for invasive aspergillosis in patients with hematological disorders. Journal of Clinical Microbiology, 42(6), 2733–2741. https://doi.org/10.1128/JCM.42.6.2733-2741.2004

Lange, B. J., Gerbing, R. B., Feusner, J., Skolnik, J., Sacks, N., Smith, F. O., & Alonzo, T. A. (2005). Mortality in overweight and underweight children with acute myeloid leukemia. *JAMA, 293*(2), 203–211. https://doi.org/10.1001/jama.293.2.203

Leventakos, K., Lewis, R. E., & Kontoyiannis, D. P. (2010). Fungal infections in leukemia patients: How do we prevent and treat them? *Clinical Infectious Diseases : An Official Publication of the Infectious Diseases Society of America, 50*(3), 405–415. https://doi.org/10.1086/649879

Lockhart, S. R., & Guarner, J. (2019). Emerging and reemerging fungal infections. *Seminars in Diagnostic Pathology, 36*(3), 177–181. https://doi.org/10.1053/j.semdp.2019.04.010

Loeffler, J., Hebart, H., Brauchle, U., Schumacher, U., & Einsele, H. (2000). Comparison between plasma and whole blood specimens for detection of Aspergillus DNA by PCR. *Journal of Clinical Microbiology, 38*(10), 3830–3833. https://doi.org/10.1128/JCM.38.10.3830-3833.2000

Maginnis, M. S., & Atwood, W. J. (2009). JC virus: An oncogenic virus in animals and humans? *Seminars in Cancer Biology, 19*(4), 261–269.

Marques, S. A. (2012). Paracoccidioidomycosis. *Clinics in Dermatology, 30*(6), 610–615. https://doi.org/10.1016/j.clindermatol.2012.01.006

Marr, K. A., Carter, R. A., Crippa, F., Wald, A., & Corey, L. (2002). Epidemiology and outcome of mould infections in hematopoietic stem cell transplant recipients. *Clinical Infectious Diseases : An Official Publication of the Infectious Diseases Society of America, 34*(7), 909–917. https://doi.org/10.1086/339202

Marr, K. A., Seidel, K., Slavin, M. A., Bowden, R. A., Schoch, H. G., Flowers, M. E., Corey, L., & Boeckh, M. (2000). Prolonged fluconazole prophylaxis is associated with persistent protection against candidiasis-related death in allogeneic marrow transplant recipients: Long-term follow-up of a randomized, placebo-controlled trial. *Blood, 96*(6), 2055–2061.

Marr, K. A., Boeckh, M., Carter, R. A., Kim, H. W., & Corey, L. (2004). Combination antifungal therapy for invasive aspergillosis. *Clinical Infectious Diseases, 39*(6), 797–802. https://doi.org/10.1086/423380

Mazi, P. B., Rauseo, A. M., & Spec, A. (2021). Blastomycosis. *Infectious Disease Clinics of North America, 35*(2), 515–530. https://doi.org/10.1016/j.idc.2021.03.013

McCarty, T. P., White, C. M., & Pappas, P. G. (2021). Candidemia and invasive candidiasis. *Infectious Disease Clinics of North America, 35*(2), 389–413. https://doi.org/10.1016/j.idc.2021.03.007

Mor, M., Gilad, G., Kornreich, L., Fisher, S., Yaniv, I., & Levy, I. (2011). Invasive fungal infections in pediatric oncology. *Pediatric Blood & Cancer, 56*(7), 1092–1097. https://doi.org/10.1002/pbc.23005

Neves, M., Marinho-Dias, J., Ribeiro, J., & Sousa, H. (2017). Epstein-Barr virus strains and variations: Geographic or disease-specific variants? *Journal of Medical Virology, 89*(3), 373–387.

Pappagianis, D., & Zimmer, B. L. (1990). Serology of coccidioidomycosis. *Clinical Microbiology Reviews, 3*(3), 247–268. https://doi.org/10.1128/CMR.3.3.247

Pfaller, M. A., Castanheira, M., Messer, S. A., Moet, G. J., & Jones, R. N. (2010). Variation in Candida spp. distribution and antifungal resistance rates among bloodstream infection isolates by patient age: Report from the SENTRY Antimicrobial Surveillance Program (2008-2009). *Diagnostic Microbiology and Infectious Disease, 68*(3), 278–283. https://doi.org/10.1016/j.diagmicrobio.2010.06.015

Pfaller, M. A., & Diekema, D. J. (2007). Epidemiology of invasive candidiasis: A persistent public health problem. *Clinical Microbiology Reviews, 20*(1), 133–163. https://doi.org/10.1128/cmr.00029-06

Pham, A. S., Tarrand, J. J., May, G. S., Lee, M. S., Kontoyiannis, D. P., & Han, X. Y. (2003). Diagnosis of invasive mold infection by real-time quantitative PCR. *American Journal of Clinical Pathology, 119*(1), 38–44. https://doi.org/10.1309/RQ05-PP9N-EG6D-ADXR

Plummer, M., de Martel, C., Vignat, J., Ferlay, J., Bray, F., & Franceschi, S. (2016). Global burden of cancers attributable to infections in 2012: A synthetic analysis. *The Lancet Global Health, 4*(9), e609–e616. https://doi.org/10.1016/S2214-109X(16)30143-7

Queiroz-Telles, F., Buccheri, R., & Benard, G. (2019). Sporotrichosis in immunocompromised hosts. *J Fungi (Basel), 5*(1), 8. https://doi.org/10.3390/jof5010008

Roden, M. M., Zaoutis, T. E., Buchanan, W. L., Knudsen, T. A., Sarkisova, T. A., Schaufele, R. L., Sein, M., Sein, T., Chiou, C. C., Chu, J. H., Kontoyiannis, D. P., & Walsh, T. J. (2005). Epidemiology and outcome of Zygomycosis: A review of 929 reported cases. *Clinical Infectious Diseases, 41*(5), 634–653. https://doi.org/10.1086/432579

Singh, A. K., Singh, R., Joshi, S. R., & Misra, A. (2021). Mucormycosis in COVID-19: A systematic review of cases reported worldwide and in India. *Diabetes and Metabolic Syndrome: Clinical Research and Reviews, 15*(4), 102146. https://doi.org/10.1016/j.dsx.2021.05.019

Sipsas, N. V., Lewis, R. E., Tarrand, J., Hachem, R., Rolston, K. V., Raad, I. I., & Kontoyiannis, D. P. (2009). Candidemia in patients with hematologic malignancies in the era of new antifungal agents (2001-2007): Stable incidence but changing epidemiology of a still frequently lethal infection. *Cancer, 115*(20), 4745–4752. https://doi.org/10.1002/cncr.24507

Stockamp, N. W., & Thompson, G. R., 3rd. (2016). Coccidioidomycosis. *Infectious Disease Clinics of North America, 30*(1), 229–246. https://doi.org/10.1016/j.idc.2015.10.008

Sung, L. (2010). Invasive fungal infections in children with cancer. *The Journal of Pediatrics, 156*, A1–S86. https://doi.org/10.1016/j.jpeds.2009.11.076

Sung, L., Lange, B. J., Gerbing, R. B., Alonzo, T. A., & Feusner, J. (2007). Microbiologically documented infections and infection-related mortality in children with acute myeloid leukemia. *Blood, 110*(10), 3532–3539. https://doi.org/10.1182/blood-2007-05-091942

Sutcu, M., & Salman, N. (2015). *Mantar enfeksiyonları* (pp. 480–503). Akademi yayınevi.

Theelen, B., Cafarchia, C., Gaitanis, G., Bassukas, I. D., Boekhout, T., & Dawson, T. L., Jr. (2018). Malassezia ecology, pathophysiology, and treatment. *Medical Mycology, 56*(suppl_1), S10–s25. https://doi.org/10.1093/mmy/myx134

Walsh, T. J., Anaissie, E. J., Denning, D. W., Herbrecht, R., Kontoyiannis, D. P., Marr, K. A., Morrison, V. A., Segal, B. H., Steinbach, W. J., Stevens, D. A., van Burik, J. A., Wingard, J. R., & Patterson, T. F. (2008). Treatment of aspergillosis: Clinical practice guidelines of the Infectious Diseases Society of America. *Clinical Infectious Diseases, 46*(3), 327–360. https://doi.org/10.1086/525258

Wauters, J., Baar, I., Meersseman, P., Meersseman, W., Dams, K., De Paep, R., Lagrou, K., Wilmer, A., Jorens, P., & Hermans, G. (2012). Invasive pulmonary aspergillosis is a frequent complication of critically ill H1N1 patients: A retrospective study. *Intensive Care Medicine, 38*(11), 1761–1768. https://doi.org/10.1007/s00134-012-2673-2

White, P. L., Linton, C. J., Perry, M. D., Johnson, E. M., & Barnes, R. A. (2006). The evolution and evaluation of a whole blood polymerase chain reaction assay for the detection of invasive aspergillosis in hematology patients in a routine clinical setting. *Clinical Infectious Diseases : An Official Publication of the Infectious Diseases Society of America, 42*(4), 479–486. https://doi.org/10.1086/499949

Wingard, J. R., Kubilis, P., Lee, L., Yee, G., White, M., Walshe, L., Bowden, R., Anaissie, E., Hiemenz, J., & Lister, J. (1999). Clinical significance of nephrotoxicity in patients treated with amphotericin B for suspected or proven aspergillosis. *Clinical Infectious Diseases, 29*(6), 1402–1407. https://doi.org/10.1086/313498

Wingard, J. R., Carter, S. L., Walsh, T. J., Kurtzberg, J., Small, T. N., Baden, L. R., Gersten, I. D., Mendizabal, A. M., Leather, H. L., Confer, D. L., Maziarz, R. T., Stadtmauer, E. A., Bolaños-Meade, J., Brown, J., Dipersio, J. F., Boeckh, M., & Marr, K. A. (2010). Randomized, double-blind trial of fluconazole versus voriconazole for prevention of invasive fungal infection after allogeneic hematopoietic cell transplantation. *Blood, 116*(24), 5111–5118. https://doi.org/10.1182/blood-2010-02-268151

Yang, Y., Wang, X. Y., Duan, C., Wang, Z. J., Sheng, H. Y., Xu, X. L., Wang, W. J., & Yang, J. H. (2023). Clinicopathological characteristics and its association with digestive system tumors of 1111 patients with Schistosomiasis japonica. *Scientific Reports, 13*(1), 15115. https://doi.org/10.1038/s41598-023-42456-9

Zavala, S., & Baddley, J. W. (2020). Cryptococcosis. *Seminars in Respiratory and Critical Care Medicine, 41*(1), 69–79. https://doi.org/10.1055/s-0039-3400280

Microorganisms and Other Cancers 10

Hira Pervaiz and Parvez Azam Malik

Abstract

Cancer is among the most aggressive diseases worldwide representing leading cause of mortality. Cancer is a multistep tumorigenesis and transform from normal to cancerous one. The role of various microorganisms has extended to include their capability in modulating different cancer hallmarks. Infection persuaded by viruses, bacteria, fungi, and parasites can lead to tumorigenesis and progress to multiple types of cancers in human. Dysbiosis in microbiome can trigger different inflammatory responses and immune deregulation that can lead to cancer development. In this chapter, microorganisms and their interaction with other cancers like colorectal cancer, breast cancer, and gastric cancer; different factors affecting these cancers; and role of microorganism as a biomarker and as therapeutic potential are discussed.

Keywords

Microbiome · Cancer development · Cancer hallmarks · Dysbiosis · Inflammatory response · Immune deregulation · Tumorigenesis · Biomarker · Therapeutic potential

H. Pervaiz (✉)
Microbiology and Biotechnology Research Lab, Fatima Jinnah Women University, Rawalpindi, Pakistan

P. A. Malik
Department of Pathology, Rai Medical College, Sargodha, Pakistan

N. Masood, A. Yasmin (eds.), *The Microbiome and Cancer*,
https://doi.org/10.1007/978-981-97-6948-3_10

10.1 Introduction

Cancer is one of the most prevailing diseases in the world with the second leading cause of death, related to the social and economic problems (Azevedo et al., 2020). According to International Agency for Research on Cancer, incidence for all cancer cases is 19,965,054 and mortality is 9,736,520 worldwide. The incidence for breast cancer cases is 2,308,931, colorectal cancer 1,926,136, and gastric cancer 968,365 while mortality rate for breast cancer cases is 669,418, colorectal cancer 903,853, and gastric cancer 659,805 worldwide (GLOBOCAN, 2022). In the United States, American Cancer Society evaluates new cancer cases and deaths and assembles the latest information on population-based cancer results taken from the cancer registries and mortality data from National Center for Health Statistics. In the year of 2023 in the United States, 1,958,310 new cancer cases and 609,820 cancer mortalities occurred. Survival rate for cancers thyroid (98%), prostate (97%), testis (95%), and melanoma (94%) is highest, and for pancreas (12%), liver, and esophagus (21%), survival rate is lowest. Progress can be determined better with the mortality rate rather than survival rate because they are not that much affected from detection practice changes. Due to tobacco epidemic in the twentieth century, death rate rose due to elevated lung carcinoma mortalities among men. Among children (1–14 years), cancer is the second common cause of mortality and fourth among adolescents (15–19 years). Leukemia is the most widespread among children consisting of 28% cases and then nervous system tumors consisting of 26% of cases either benign or malignant.

Since 1991, cancer mortality has reduced up to 33% and around 3.8 million cancer mortalities averted. This development is because of decrease in smoking, better screening programs for breast and colorectal cancer, and improved treatment options like adjuvant chemotherapies for colorectal carcinoma and breast carcinoma. Nowadays, improvement in targeted therapy and immune therapy in lung carcinoma contributed in the reducing mortality rate. As well as carcinomas like leukemia, melanoma, and kidney death rate are also reduced whose incidence levels are stable. Better treatment breakthrough helps in managing difficult-to-treat carcinomas (Siegel et al., 2023).

10.1.1 Hallmarks of Cancer

Cancer grows in a process when normal cells cannot govern their proliferation and keep dividing instead of dying, and finally, a mass of tissue is formed which is tumor. Although oncological diagnostics and therapeutics become advance with time, still there is an increase in the cancer cases worldwide. According to research, around 20–25% of malignancies worldwide are due to infections caused by microbes, and their role as carcinogens is because of potential to form and control cancer hallmarks like tumor-promoting inflammation, tumor growth, immune response alterations, angiogenesis, and pro-carcinogenic metabolite production. Microbes has the ability to cause genome instability by damaging host DNA (Yahya et al., 2021).

Tumor-promoting inflammation occurs when heathy cells necrosis results in the release of cell content in the surroundings, activating pro-inflammatory mediators discharge. There are two types of tumor-promoting inflammation factors: intrinsic and extrinsic. Intrinsic factors are genetically linked while extrinsic factors are due to inflammatory cells or mediators (Khusnurrokhman & Wati, 2022). Immune system has established various approaches to deal with bacteria so microorganisms usually does not develop pro-inflammatory response, but disturbing these mechanisms or new pathogenic bacteria enter the body, then immune system reacts to microorganism and activates tumor growth in intestine (Russo et al., 2016). Angiogenesis and angiogenic factors are basics for tumor development in process of invasion and metastasis. Tumor development occurs through the "angiogenic switch" from an avascular stage to a vascular stage in growth, and in some tumors, oncogenes are liable for overexpressing positive regulator of angiogenesis. The survival and development of an organism depends on cell proliferation and elements that regulate cell cycle checkpoints. In tumorigenesis, cell growth and cell cycle are important for embryonic growth, tissue repair, and cell division regulation. Specific cell types proliferate at specific times having specific rates during lifetime of an organism depending on cell type and definite stimuli to tissue injury. In case of gene mutations, if they are not repaired, it can result in uncontrolled cell proliferation (Spoelhof & Ray, 2014). Some microbial metabolites like TLR4 agonists support its associated signaling cascades. In that way, they play an important part in the control of tumor cell growth progression in colon, liver, pancreatic, and breast cancer. In colorectal cancer, TLR4 overexpression enhances proliferation of malignant cells, their invasion, and metastasis, develops tumor-beneficial cellular microenvironment, and suppresses apoptosis (Shatova et al., 2023). High frequent mutations in genome, whether they are point mutation, insertion mutation, deletion mutation, or genetic rearrangement, can lead to genome instability. DNA viruses having large genome are more stable than RNA viruses, but DNA viruses also display sudden sequence change and some have capability to promote genomic instability (Sanjuán et al., 2016). Transient infection of eukaryotic cells with microorganism *Escherichia coli* blocks mitosis and induce megalocytosis. Various oncogenic mechanisms have been suggested for viruses, bacteria, protozoa, and fungi.

10.1.2 Microbes Having Carcinogenic Properties

Microbes exhibit an important part in human health and disease together with influencing the risk of cancer, coevolve inside our body ecosystem and depending upon incoming resources for their existence and replication. So whatever we eat, having excess energy or nutrients, it can affect both microbial and cancerous cells. Microbes can be indigenous or exogenous. There are different factors that suggest the influence and interaction between microbial cells and cancer cells and hence are important in cancer progression.

Globally, microbes are responsible for around 20% of all cancers in humans. The human gut has an assorted ecosystem having viruses, fungi, and bacteria, and archaea and phyla Bacteroidetes, Actinobacteria, Firmicutes, and Proteobacteria of bacteria are the most prevalent. They are beneficial in maintaining intestinal barrier and does not allow harmful microbes to reach epithelium where they develop illness (Whisner & Athena, 2019). The International Agency for Cancer Research has labeled ten microbes as carcinogens, and despite causing disease, they exist among human population in huge proportion, but not necessarily cause cancer, or coinfection with other bacteria enhances cancer development risk (Yasunaga & Matsuoka, 2018). These ten microbes include hepatitis B virus, hepatitis C virus, human papillomavirus, *Helicobacter pylori*, *Schistosoma haematobium*, *Opisthorchis viverrini*, *Clonorchis sinensis*, human immunodeficiency virus (HIV), Epstein-Barr virus (EBV), human T-cell lymphotropic virus, and human herpesvirus. Viruses have direct effect on human genetics. That is why more focus is on them and on bacteria because they cause chronic inflammation that leads to cancer development. Earlier fungi and parasitic protozoa have been neglected, but now they are explored as a significant entity in microbial carcinogenesis (Mukherjee, 2020).

Bacteria responsible for different cancer development are as follows: *Fusobacterium nucleatum* enhances gene expression by changing gut microbes in colorectal cancer patients. If *Helicobacter pylori* increased, then it is an indication of stomach carcinoma, which is one of most common carcinomas worldwide. Ovarian carcinoma is associated with *Chlamydia trachomatis* infection. Fungi has also been responsible for certain carcinoma, for example, long-term candida infection associated with oral cancer due to different metabolite secretions. Some commensal fungi can also develop cancer with the change in lifestyle, for example, excessive alcohol intake. *Candida albicans* can produce enzymes that are linked with cancer progression. Multicellular parasites are also responsible for cancer development, for example, carcinogenic parasites like liver and blood flukes. *Toxoplasma gondii* and *Clonorchis sinensis* altogether cause liver, colorectal, bladder, and brain carcinoma (Fig. 10.1) (Baid, 2023).

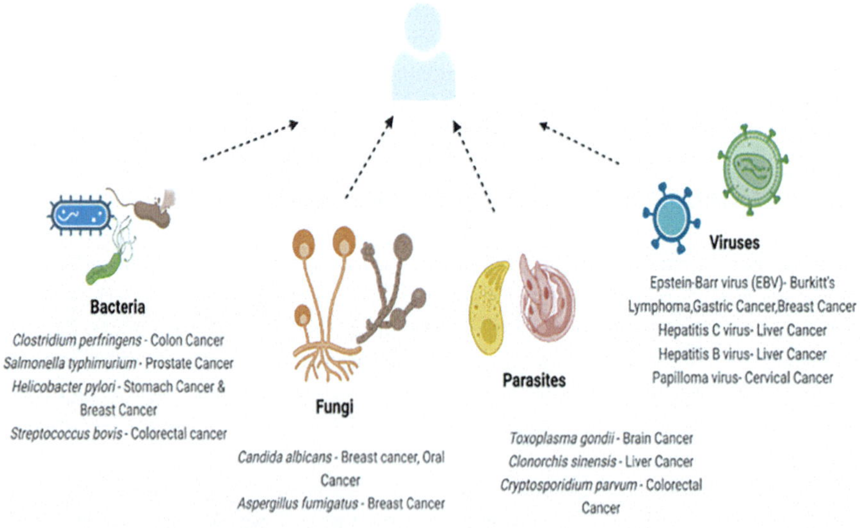

Fig. 10.1 Microbes having carcinogenic properties (Baid, 2023)

10.2 Microorganisms Associated with Colorectal Carcinoma

Colorectal cancer (CRC) is third in incidence and second most leading cause of death worldwide. The lifetime risk of CRC development is around 4.49% in men and 4.15% in women. The incidence is about 32% higher in men and death rate about 42% higher in men than women. Long-term decrease in death rate of CRC is due to altered pattern in risk factors and screening uptake and improved treatment. Screening methodologies including colonoscopy can decrease CRC incidence by detecting premalignant polyps that can then be removed (Siegel et al., 2024).

Despite wide investigation, the accurate etiology of CRC remains unclear. CRC development is a multistep process including accumulation of a series of genetic and epigenetic modifications that affect fitness and clonal expansion of changed cells in the alteration of the normal colonic epithelium to adenomas and cancer (Sears & Garrett, 2014). There is a rising indication that gut microbiome has a vital part in CRC progression (Fig. 10.2) (Ahmed & Umar, 2018).

10.2.1 Factors Influencing Colorectal Carcinoma

10.2.1.1 Dysbiosis

Dysbiosis is differences in composition of gut microbiota that subsequently leads to development of pathological microbes resulting in neurological, metabolic, and gastrointestinal diseases. The genome-wide association (GWAS) studies tells that there exists heritable component to gut microbiome and mutations in the genes

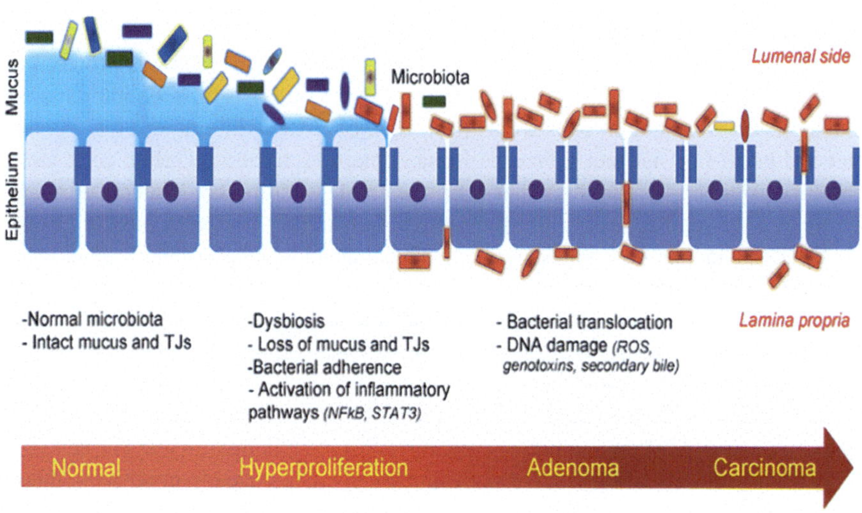

Fig. 10.2 Microbiota dysbiosis, DNA damage, and loss of epithelial barrier leads to colorectal carcinoma (Ahmed & Umar, 2018)

linked with inflammatory response, microbial metabolites, immune regulation, or virulence factors resulting from microbiota composition changes. Dysbiosis is also influenced by diet and usage of antibiotics. Diet that is fat rich and low fiber leads to accumulation of metabolites and increased toxicity. Obesity is linked with deoxycholic levels that encourage inflammation and cancer, lacking rich microbiota in gut of obese persons. The Western dietary pattern described by intake of red and processed meats and excess of refined grains, salt, certain oils, alcohol, high-fat dairy products, and corn-derived fructose reflected to induce inflammatory processes, affect immunity, change the gut microbiome, and upsurge the risk of developing colorectal cancer (CRC). Few drugs bring out significant disruption of the microbial ecosystem and shift the microbiome composition in favor of pathobionts, resulting in dysbiosis (Ahmed & Umar, 2018).

10.2.1.2 Immune Modulation

Gut microbiota keep homeostasis with immune system of host and help sustain T regulatory cells that lessen inflammation through IL-10. Disturbance in this homeostasis by disease-causing microbes causes chronic inflammation that leads to CRC development (Kelly & Mulder, 2012). Around 30% of cases are linked with inflammation represented with major risk for IBD patients, while IBD according to few studies is related with bacterial infections. *Enterobacteriaceae*, *Escherichia coli,* and *Mycobacterium avium* are abundant in IBD patients (Bonnet et al., 2014). Other events responsible for CRC development are oxidative stress due to reactive oxygen or nitrogen species generation, abnormal proliferation, and DNA damage. *Enterococcus faecalis* and *Bacteroides vulgatus* colonization trigger NF-κB signaling and induce chemokine expression. Once initiation occurs, disrupted immune modulation affects tumor growth (Ruiz et al., 2006).

10.2.1.3 Loss of Epithelial Barrier Function

Intestinal epithelial cells (IECs) which is connected by tight junctions protect the luminal content by forming physical barrier. On the outer surface, double-layered mucus secreted by goblet cells forms protective covering while inner layer is contact and germ-free and outer layer is loose containing bacteria. Goblet cells yield numerous mucin glycoproteins that form the chief structural constituent of mucus acting as a physical defense in limiting interaction among microbes and the epithelia. Epithelial barrier function is maintained by intestinal epithelial cells and secreted molecules like mucins (Yang et al., 2017). In the course of epithelial damage, the gut microbiome can move in the colonic mucosa, leading to chronic inflammation. It is assumed that during CRC progression, altered colorectal epithelial cells fail barrier integrity, permitting bacteria and their toxins to enter tumor stroma and involve with toll-like receptors (TLRs) on myeloid cells and trigger downstream adaptor protein MyD88-mediated assembly of pro-inflammatory cytokines like IL-23 (Grivennikov et al., 2012). Notch signaling is recognized to control the intestinal secretion across various species like zebrafish and mammals. Numerous researches have revealed an association among Notch signaling and TLRs through a conserved adaptor protein, MyD88 (Fre et al., 2011).

10.2.2 Microbial Signatures Associated with Colorectal Carcinoma

The formation of colorectal cancer is usually initiated by chronic inflammation, and almost 20% of malignancies in colorectal is headed by chronic inflammation. Tumorous cells produce cytokines and chemokines that attract pro-inflammatory helper T cells, characterized by angiogenic factors, suppression of antitumor T-cell responses, and tissue remodeling enzymes that favor cancer development. In early 1970s, first information was gathered gut microbiota affecting CRC development. Different intestinal microbes have been identified that cause CRC by using high-throughput microbiome sequencing (Francescone et al., 2014).

Streptococcus bovis (*S. bovis*) has been stated as one of the risk factors for colorectal cancer, normally occupied in the gastrointestinal tract. The early sign of *S. bovis involved in colorectal cancer was its induction with* endocarditis or bacteremia (Gupta et al., 2010). The confirmation of linkage in inflammation and colorectal cancer was done after observing relationship among pro-inflammatory potential of *S. bovis* proteins and their cancer-causing properties (Biarc et al., 2004).

F. nucleatum is one of the most extensively used bacterium connected to colorectal cancer progression. According to metagenomics analysis, *Fusobacterium* species is linked with colorectal cancer but it is not known whether it is casual or indirect. After binding of bacteria to mucosal surface, this strain encourages early carcinogenesis. One of the studies shows that strain transcript 400 times elevated in cancer tissues (Castellarin et al., 2012). Fusobacterium adhesin A (FadA) protein is produced by *F. nucleatum* and triggers β-catenin signaling pathway after E-cadherin binding as it is a potential carcinogenic stimulator.

Enterococcus faecalis (*E. faecalis*) is a bacterium present in gut ecosystem where they release superoxide from demethylmenaquinone autoxidation. DNA damage to intestinal epithelial was induced after infecting with *E. faecalis* by forming superoxide. Therefore, in colorectal cancer patients, this bacterium is present in large quantity as compared to healthy persons (Wang et al., 2012).

Enterotoxigenic *Bacteroides fragilis* (*ETBF*) releases *B. fragilis* toxin (BFT) and is responsible for causing diarrhea and inflammatory bowel disease (IBD). This bacterium promotes tumor by increasing signal transducer and activator of transcription 3 (STAT3) and the Th17 response all through colorectal tumorigenesis. Although STAT3 initiation is essential for colorectal tumorigenesis, this is not enough to activate tumorigenesis by *ETBF.* IL-17-dependent nuclear factor kappa B (NF-κB) activation leads to development of a proximal to distal mucosal gradient of CXC chemokines that helps the recruitment of CXCR2-expressing polymorphonuclear immature myeloid cells to bring in side by side *ETBF*-mediated distal colon tumorigenesis (Chung et al., 2018).

Peptostreptococcus anaerobius (*P. anaerobius*) encourages a pro-inflammatory immune microenvironment and stimulates cancer development in the intestine. This bacterium has a part in cancer development by enhancing the pro-inflammatory cytokines expression in an animal model and recruiting tumor-infiltrating immune cells like myeloid-derived immunosuppressive cells. *This strain* rises the reactive oxidative species levels and acts together with toll-like receptor (TLR) 2 and TLR4 in colon cells to stimulate cell proliferation and cholesterol production, and finally, dysplasia of colon cells occurs (Long et al., 2019).

Salmonella infections and colonization can be long-lasting and upsurge the chronic cholecystitis risk and other gastrointestinal ailments, together with cancers. Salmonella stimulates colon cancer by counting on AvrA protein, trigger Wnt/b--catenin, and STAT3 signaling pathways in colon cancer cells (Wang et al., 2018). Salmonella releases typhoid toxin that damages DNA *through* the phosphoinositide 3-kinase (PI3K) pathway in colonic epithelial cells. The increased genomic instability in APC-deficient cells that are exposed to genotoxin is linked with reduced DNA repair capacity to start proper checkpoint responses. *Campylobacter jejuni* releases genetic toxin cytolethal distending toxin with DNAse activity that causes DNA double-strand breaks and stimulates colorectal tumorigenesis. mTOR signaling in mammals has been shown to reduce *C. jejuni*-induced colitis and carcinogenesis (He et al., 2018).

Sulfate-reducing bacteria (SRB) are a microbiota that is important regarding colitis. With the use of methionine and cysteine, these microbes can produce hydrogen sulfide (H_2S). Researches have presented that SRB quantity is elevated in the stool of colorectal cancer patients (Yachida et al., 2019). SRB release H_2S and trigger CRC development by preventing butyrate oxidation and terminating the gut barrier, in addition to inducing DNA damage via reactive oxygen species (ROS) (Louis et al., 2014). Research is in process to apprehend the association between other intestinal microorganisms with CRC development; hence, the bacterial strains mentioned above are not the only ones causing the colorectal carcinoma.

10.2.3 Gut Microbial Markers for the Diagnosis of CRC

Colorectal progression is associated with changes in gut microbiota, representing it as the main etiologic factor in disease progression. Microbiome analysis from fecal and mucosal samples have shown that gut microbiome alteration is related to different stages of CRC, signifying diagnostic prospective in colorectal cancer detection (Fig. 10.3).

The bacteria that can function as CRC biomarkers are combined detection of fecal *Fusobacterium nucleatum* and *Clostridium symbiosum* with FIT for advanced adenoma and colorectal cancer. FIT sensitivity for advanced adenoma can be elevated by detecting this genera group. The other two biomarkers used to differentiate CRC from non-CRC are butyryl-CoA dehydrogenase gene of *Fusobacterium nucleatum* and the RNA polymerase β subunit gene of *P. micra*. Some other biomarkers are *Peptostreptococcus, Prevotella, Parvimonas, Gemella, Porphyromonas*, and *Bacteroides. The specificity of CRC diagnosis and colon adenoma can be increased by using combination of Bacteroides clarus, Clostridium hathewayi, F. nucleatum, and Roseburia intestinalis.*

Fungi and viruses also present in gut and can be used as a CRC biomarker. According to some analysis, Basidiomycota: Ascomycota ratio is elevated in CRC

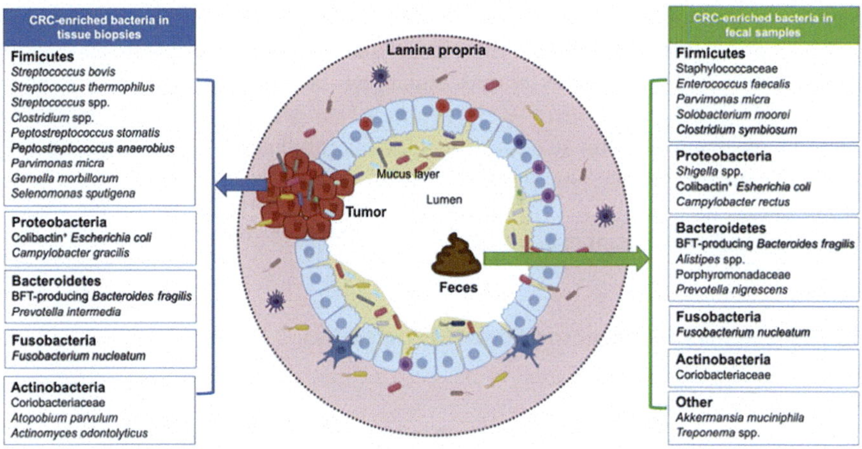

Fig. 10.3 Enriched bacterial strains in colorectal cancer (Liu et al., 2023)

persons in comparison to non-CRC persons, with increase in Malasseziomycetes and reduction in Saccharomycetes and Pneumocystidomycetes (Liu et al., 2023).

10.2.4 Therapeutic Effect

Gut microbiota has been closely related to colorectal cancer so a lot of research are made on its treatment. Therapies that increase the effectiveness of cancer treatment are those that modify gut microbiota together with probiotic administration and fecal microbiota transplantation. Antibiotic administration can have lessened the efficiency of oxaliplatin and CpG oligodeoxynucleotides chemotherapeutic mediators. The antibiotic use enhances disease-causing bacteria *Escherichia shigella* and *Enterobacter* and decreases the anticancer outcome of 5FU. Radiation of pelvic area causes dysbiosis and hindrance to CRC treatment, and radiation-induced gut epithelial damage badly affects the prognosis of CRC. So these aftereffects of radiation can be eliminated with probiotic administration. The mucosal immunity in colon and rectum region is greatly affected by gut microbiome and helps in improving immunotherapy by elevating CD8 and T-cell immune response (Kim & Lee, 2022).

10.3 Microorganisms Associated with Breast Carcinoma

Breast carcinoma is the prominent cancer among females and is the leading cause of cancer-related mortalities in females globally, with more than 600,000 mortalities, marking 6.6% of cancer-related mortalities. In the last decades, rising evidence has established that the gut microbiota has a vital part in carcinogenesis and reaction to anticancer therapy, like in colorectal carcinoma, gastric carcinoma, and breast carcinoma. Researchers have progressively exposed that a great array of microbiome occupy normal breast and breast cancer tissue. The risk factors of breast cancer comprise breast cancer gene (BRCA) mutations, hereditary, radiation, hormone change, obesity, and estrogen levels connected to breast cancer development. Potential mechanisms in the relationship of the microbiome and breast cancer comprised of estrogen metabolism, immune regulation, and obesity. There is also indication that in the breast tissue itself, microbes have a role in managing the tumor microenvironment (Eslami et al., 2020).

10.3.1 Factors Influencing Breast Carcinoma

10.3.1.1 Dysbiosis

The subtle steadiness is disturbed among a host and the resident microbiome and leads to pathological procedures, containing carcinogenesis, and its part in averting tumorigenesis through its numerous biological activities. Few studies represent the gastrointestinal microbiome of BC patients different from that of healthy persons. The fecal microbiome of postmenopausal women with pretreated BC is different from normal persons with normal mammography. BC patients had altered levels of specific bacterial phyla in comparison to control persons; the cancer patients presented higher numbers of *Ruminococcaceae, Clostridiaceae,* and *Faecalibacterium* and lower numbers of *Dorea* and *Lachnospiraceae.* The overall microbial diversity also reported to be lower in cancer patients (Chen et al., 2019a, b). Breast microbiota is lacking in diversity and elevated in BC patients with relative abundance in *Staphylococcus, Enterobacteriaceae,* and *Bacillus* species. Few species can cause double-stranded DNA breaks in cells that initiate breast cancers. There is a significant difference among the microbiota of malignant and benign breast tissue samples, malignancy being linked with an increase in specific bacteria that are generally low in abundance.

The use of antibiotics can affect the target pathogen and occupant of the host, and the influence on nontargeted microbiota depends on specific antibiotic usage, its mode of action, and to what extent it is resistant in the community. The regular use of antibiotics enhances chances of dysbiosis and lower microbial diversity. In contrast to other organs, microbial burden is elevated in digestive system specifically in large intestine. The intestinal microbiota has an important role in the body with the production of metabolites, immunological cytokines, and hormonal intermediates (Eslami et al., 2020).

10.3.1.2 Estrobolome in Breast Carcinoma Risk

Estrobolome is gut bacterial community whose genes can adapt and metabolize estrogen and its derivatives. The estrogen metabolism is administered by gut microbiota especially *Edwardsiella, Bacteroides, Collinsella, Alistipes, Bifidobacterium, Tannerella, Escherichia, Citrobacter, Dermabacter, Clostridium, Faecalibacterium, Propionibacterium, Lactobacillus, Marvinbryantia, and Roseburia.* They secret β-glucuronidases that catalyze deconjugation and estrogen reactivation, finally unbound production of estrogen, permit via enterohepatic circulation, and spread to various organs plus breast tissue. The restarted estrogens bind to estrogen receptors and stimulate the ER + BC development. *Akkermansia muciniphila* release propionate from mucin, and *Ruminococcus obeum* and *Roseburia inulinivorans* produce propionate by sugars metabolization. The butyrates and lactates by *Faecalibacterium prausnitzii, Eubacterium hallii*, Eubacterium rectale, and *Roseburia faecis* modulate oxidative phosphorylation, apoptosis, proliferation, histone deacetylase inhibition, and invasion of breast cancer (Malla et al., 2022).

Estrogen production happens in the ovaries, utilizing endogenous or dietary lipids, and circulate in the blood either bound to carrier proteins or as free unconjugated particles. The liver brings out conjugation and following estrogen inactivation, then once conjugated, the inactive estrogen is conveyed in the intestinal lumen. In the intestinal lumen, the microbiome has a role in the fate of this inactive estrogen. If the person has a microbiota capable of deconjugating estrogen, then freed estrogen is reabsorbed and increases free levels of estrogen. The individuals whose microbiome is less efficient in estrogen deconjugation show greater estrogen excretion in stool, hence lesser circulating estrogen levels (Shapira et al., 2013).

10.3.1.3 Inflammation and Immune Response

The expansion of gut dysbiosis and microbe products causes chronic inflammation that harmfully affects immune system and promotes tumor growth. Alteration in intestinal microbe ecosystem is linked with growth of both intra- and extraintestinal tumors after the beginning of chronic inflammation and alteration in microenvironment and metabolism and this state maybe unalterable.

The gut microbiome ecosystem is associated with incidence of breast cancer by affecting T cells, neutrophils, and few other inflammatory factors. One study showed that in BC patients, there is relation among symbiotic bacteria, neutrophils, and IL-6. So downregulation of inflammatory factors can lessen the breast cancer risk. Some other research shows that oral administration of *Lactobacillus reuteri* taken from human milk to mice can decrease inflammation factors expression (Song et al., 2022).

10.3.2 Breast Tissue and Skin Microbiome

Breast tissue and milk are considered to be sterile but are now known to have varied and unique microbe community. Research associating the microbe composition of nipple aspirate fluid in women having breast cancer history against normal controls established a comparatively higher occurrence of the genus *Alistipes* and lower occurrence of a genus from *Sphingomonadaceae* family. Some other researches validate the microbiota of breast skin swabs. In addition, breast tissue of breast cancer patients relative to health controls is enriched in specific microbes, counting *Fusobacterium, Bacteroidetes, Hydrogenophaga, Atopobium, Staphylococcus, Gluconacetobacter, Enterobacteriaceae, Bacillus,* and *Comamonadaceae* (Chen et al., 2019a, b).

10.3.3 Microbial Signatures Associated with Molecular Subtypes of Breast Cancer

Breast carcinoma is a heterogeneous disease. The clinical categorizing sorts breast cancer into four chief groups: Luminal A (ER+, [PR]-positive, human epidermal growth factor receptor 2 [HER2]-negative, low ki-67), Luminal B (ER+ and PR+/−, HER2+/−, or high ki-67), HER2-positive (HER2+ and ER-), and triple-negative (ER-/PR-/HER2-) based on immunohistochemical. These four subtypes signify biologically discrete disease units, representing that each BC molecular type has a specific response to chemotherapy, target therapy, and immunotherapy. Present investigation has engrossed on the microbes of BC, along with fungi, viruses, and parasites. *Mobiluncus*, *Actinomyces, and Brevundimonas* were enhanced in all subtypes. Fungal genera were also commonly noticed in all categories, counting *Trichosporon*, *Candida*, *Rhodotorula*, *Epidermophyton*, *Geotrichum*, and *Trichophyton*, though more complex fungal diversity exists in ER+ breast cancer compared to the triple-negative samples.

"PathoChip" is used to regulate the microbial signature for various BC subtypes. The PathoChip is a microbial pathogen array comprising of probes, able to sense all publicly available virus sequences and hundreds of parasites, pathogenic bacteria, and fungi. The microbial signatures for every BC molecular subtype are present in Table 10.1). The PathoChip detection results and case reports specified that *Polyomaviridae*, *Ascaris*, *Plasmodium*, and *Hepadnaviridae* were enhanced in ER+/PR+ BC. It is recognized that exclusive microbial signatures are associated with triple-negative breast cancer. *Polyoma viruses*, *Prevotella nigrescens*, *herpesviruses*, *papillomaviruses*, *poxviruses*, *Leishmania*, *Piedraia hortae*, *Arcanobacterium haemolyticum*, *Pleistophora mulleri*, and *Trichuris trichiura* in triple-negative breast tissues are abundant than in normal tissues (Banerjee et al., 2015). Plasmodium-derived molecules have been linked with antitumor properties in adult mice in vivo and in vitro and are useful as targets for tumor immunotherapy in triple-negative breast cancer (Wang et al., 2021).

10.3.4 The Breast Microbiota as a Biomarker for Cancer

The breast microbiome can be useful as biomarker in many ways: Pretreatment, resistance to the therapy can be predicted after determining molecular features of the carcinoma; secondly, during treatment, allowing adjustment in therapy if it is not that much responsive; thirdly, during tumor development, to explain the accomplished resistance to therapeutics. If microbiota is used as a pretreatment biomarker, it will be helpful for oncologists because this would provide more information on cancer aggressiveness and tumor sensitivity to anticancer therapeutics; afterward, treatment procedure can be adjusted.

Although breast microbiota shows excessive potential as a prognostic and predictive biomarker, still more research is required to reflect breast microbiota as a prognostic biomarker. More participants needed to be involved in future studies.

Table 10.1 Breast microbial signatures in molecular subtypes of breast carcinoma (Wang et al., 2021)

Cancer types	Bacterial signatures	Fungal signatures	Viral signatures	Parasitic signatures
ER+/ PR+/HER2-	Arcanobacterium, Bifidobacterium, Cardiobacterium, Citrobacter, Escherichia	Filobasidiella, Mucor, Trichophyton	Hepadnaviridae	Brugia, Paragonimus, Ascaris, plasmodium
ER+/ PR+/ HER2+	Bordetella, campylobacter, chlamydia, Chlamydophila, legionella, Pasteurella	Penicillium	Birnaviridae, Hepeviridae Polyomaviridae, Hepadnaviridae	Ancylostoma, plasmodium, Echinococcus, Schistosoma, trichomonas, Trichostrongylus
ER-/ PR-/ HER2+	Streptococcus	Epidermophyton, Fonsecaea, Pseudallescheria	Nodaviridae	Balamuthia, Ascaris, plasmodium
ER-/ PR-/HER2-	Aerococcus, Arcobacter, Geobacillus, Orientia, Rothia	Alternaria, Malassezia, Piedraia, Rhizomucor		Leishmania, Contracaecum, Schistosoma, Necator, Trichuris, Onchocerca, Toxocara, Trichinella

ER estrogen receptor, *PR* progesterone receptor, *HER2* human epidermal growth factor receptor-2

Considering clinicopathological variability and by doing prospective studies, the prognostic rate of pretreatment breast microbiome composition can be determined. Other factors should be taken into account like age, BMI, menopausal status, ethnicity, and lifestyle-related factors like alcohol intake and dietary habits. Microbiome composition should be recognized that can predict efficacy and resistance to anti-cancer therapeutics after comparison of microbe composition of patients with changes in cancer progression and among responders and nonresponders (Dieleman et al., 2021).

10.3.5 Therapeutic Effect

By changing the microbiome ecosystem, tumor growth can be affected and avert its recurrence. Earlier researches have revealed that after injecting *Lactobacillus acidophilus* into mice with tumor changes the cytokine production and growth of tumor. According to some research in colorectal tumor, oral probiotic supplements having Lactobacillus helveticus lessened the IL-7 production of T cells, changed the gut microbiota, and reduced the cancer cell proliferation (Rong et al., 2019). The epidemiological resemblance occurs among colorectal and breast cancer so it is considered after meddling gut microbiome community whether it has an effect on breast cancer treatment or not. Additionally, microbiota regulation may be useful as an aide to standard cancer treatment. For example, cisplatin, which is a chemotherapeutic drug, can destruct epithelial barrier and translocate intestinal microbes (Perales-Puchalt et al., 2018). Careful use of precise groups of gut microbes in chemotherapy can be helpful in antitumor immunity. For example, fragile bacteriocin is important in targeting antibodies antitumor effect contrary to cytotoxic T lymphocyte-associated protein 4 (Vétizou et al., 2015). Likewise, association among *Bifidobacterium* and PD-L1 therapy has been established (Strouse et al., 2019). Dysbiosis present in nonresponders to anti PD-1 treatment may bring inflammation and seizure of T-cell variation into CD8+ effector cells, linked with remarkable reduction in Sphingomonas proportion. The intestinal injury and changed microbiome ecosystem created by cisplatin may have a role in its antitumor effect (Gui et al., 2015). *P. aeruginosa* with elevated concentrations stop the development of the breast cancer cell lines and low concentration reveals the opposite results. Researches have revealed that *P. aeruginosa*-secreted factors improve the doxorubicin capability to stop cancer proliferation (Chiba et al., 2020). Selective estrogen receptor modulator (SERM) is chiefly endocrine therapy for BC together with tamoxifen and raloxifene having a significant role in averting BC. Raloxifene may be an appropriate therapeutic drug for additional study of *P. aeruginosa* infection (Sui et al., 2012).

Besides, microbe engineering may open novel ways in prevention, diagnosis, and therapy of cancer. Changing the gut microbiome ecosystem may elevate the cancer risk; hence, designing novel antibiotics target specific spectrum of microbiota may be helpful in regulating gastrointestinal microbe as a possible way to lessen breast cancer risk. By altering the signaling pathways and innate and acquired immunity responses, it can be achieved. Engineered probiotics may be helpful in aiming these signaling pathways. But the diversity of microbial community makes it challenging to develop antibiotics and recognize the cancer (Yang et al., 2017).

10.4 Microorganisms Associated with Gastric Carcinoma

Gastric carcinoma is an important health burden globally, ranking fifth in incidence and third leading cause of cancer mortality after lung and colorectal carcinoma and accounting for 782,685 mortalities worldwide. According to International Agency for Research on Cancer (WHO-IARC), by 2040, the yearly burden of gastric cancer will rise to about 1.8 million new cases and 1.3 million mortalities in comparison with those in 2020, and the new cases and mortalities number will increase by around 66%.

The main gastric cancers (GC) are gastric adenocarcinomas (GAC). According to Lauren's classification, GC is histologically categorized into mixed, diffused, intestinal, and nonclassifiable, and according to WHO, GAC is categorized into tubular, papillary, signet ring cell, and mucinous. The Cancer Genome Atlas (TCGA) categorized GC into four main subtypes like genomically stable, Epstein-Barr virus-positive, chromosomal instability, and microsatellite instability. Consecutive procedures like superficial gastritis, dysplasia, intestinal metaplasia, and chronic atrophic gastritis are responsible for the progression of GAC. This is a multifactorial inflammatory condition and the main risk factors are tobacco smoking, less intake of vegetables and fruits, iron deficiency, alcohol consumption, high salt and smoked meat intake, host genetic factors, obesity, and microbial infections. *H. pylori* is the main cause of gastric adenocarcinoma in early stages but other gastric microorganisms are also involved in the progression of GAC (Chattopadhyay et al., 2023).

10.4.1 *H. pylori*-Associated Dysbiosis

Dysbiosis is the compositional and functional variations of the microbiota. Gastric cancer is the main example of the relationship between gut microbe dysbiosis and epithelial cells of the host gut. There is not any confirm arrangement of gastric microbiota in gastric cancer. Some studies suggest microbial diversity is reduced in cancer (Gong et al., 2016), but some studies suggest that gastric microbiome are abundant in GC tissues as compared to normal tissues (Chen et al., 2019a, b). Microbe diversity according to some studies was found to be abundant in advanced-stage GC in comparison to early stage, which did not vary expressively from that in chronic gastritis. *Novosphingobium, Ochrobactrum, Pseudoxanthomonas, Ralstonia,* and *Anoxybacillus were abundant in* early GC, while *Tsukamurella, Salinivibrio, Burkholderia,* and *Uruburuella were enriched in advanced stage GC* (Wang et al., 2020). In some studies, there is not any prominent difference in microbe composition between early and late stage of gastric carcinoma and microbe richness lessened from normal to tumor tissues. *Bacteroides uniformis* and *Prevotella copri* were reduced, whereas *Streptococcus anginosus, Prevotella melaninogenica,* and *Propionibacterium acnes* were abundant in tumor tissue as compared to normal tissues (Liu et al., 2019). *Patescibacteria, Fusobacteria,* and *Bacteroidetes* were enhanced in signet ring cell carcinoma, whereas *Acidobacteria* and *Proteobacteria* were enriched in adenocarcinoma (Yang et al., 2021).

10.4.2 *H. pylori* in Gastric Carcinoma

H. pylori is a gram-negative bacteria and is a main gastric cancer-causing pathogen present in any part of the stomach like cardia and non-cardia. The dysbiosis of gut microbe enhances the chances of non-cardia gastric cancer, especially in presence of *H. pylori* and factors like age and sex. In order to avert gastric cancer, it is recommended to eradicate *H. pylori* from the patients suffering from its infection. *H. pylori* has 40-kb region of chromosomal DNA that encodes protein CagA and type IV secretion system component that forms virulence factor, leading *H. pylori* infection to gastritis by breaking down urea in the gastric mucosa and releasing ammonia in an acidic atmosphere. The virulence factor and cytokines trigger RAS, MEK, and ERK signaling pathway, and inflammation occurs that destroys gastric epithelium and progresses to gastric cancer. This pathogen has an effect on many hormones like ghrelin, gastrin, and leptin and causes further ailments like obesity. Many other studies have shown that *H. pylori* is present in GC tissues and other bacteria.

Proteobacteria is present in HP-positive GC tissues. Some other flora Fusobacteria, Firmicutes, Bacteroidetes, and Actinobacteria also altered with *H. pylori* infection (Yu et al., 2017).

The mechanism by which *H. pylori* cause cancer is due to inflammation response, alteration of gene expression, and activation of signaling pathways (Fig. 10.4). Bacteria in the lumen can be killed with the release of reactive oxygen species and reactive nitrogen species from the stimulation of NADPH oxidase and inducible nitric oxide synthase (Salvatori et al., 2023). Due to the release of Ca^{2+} influx-dependent mitochondrial reactive oxygen intermediates, oxidative stress can be created with the involvement of Cag virulence factor and vacuolating cytotoxin of *H. pylori* (Kim et al., 2007). DNA damage resulted from the activity of NF-κB signaling and generate H_2O_2. It has been stated that DNA methyltransferase can bring out CpG island hypermethylation in carcinogenic inhibitor genes to enhance cancer epigenetically. These epigenetic alterations and oxidative stress have important link with cancer induced by *H. pylori* (Hattori & Ushijima, 2016).

10.4.3 Gastric Microbiota in Gastric Carcinogenesis

GC tissue-specific glands loss and acid secretion reduction cause *H. pylori* depletion and abundance of intestinal commensal bacteria like *Enterococci, Parvimonas, Clostridium, Rhodococcus, Lactobacillus, Carnobacterium,*

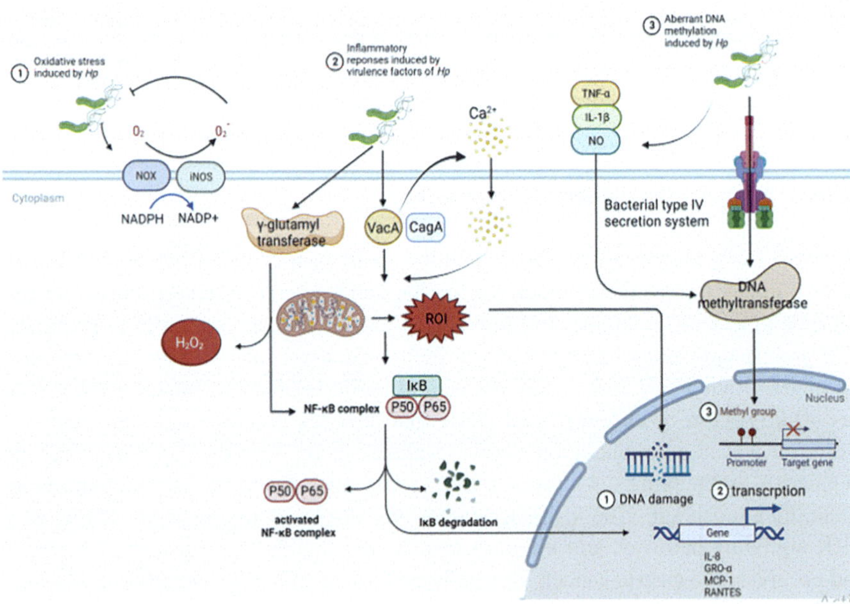

Fig. 10.4 The mechanism of *H. pylori* carcinogenesis (Wang et al., 2023)

Achromobacter, and *Citrobacter* (Gantuya et al., 2020). Some lactic acid bacterial strains used as probiotics have interaction with *H. pylori, preventing its attachment to epithelial cells, bacteriocins, and discharge inflammatory factors* (Sakarya & Gunay, 2014).

Many researches presented a link between specific microbes and gastric carcinoma. In GC patients, gastric microbiota is enriched with microbial flora from oral or intestinal origin. Oral microbiota like *Veillonella, Streptococcus, Parvimonas, Fusobacterium, Peptostreptococcus, Slackia,* and *Haemophilus* present in GC patients. One study explored the composition, diversity, and bacterial symbiotic associates and gastric microbial functional profile in three microenvironments, counting 230 normal, 247 perineural, and 229 tumor tissues. They stated that *S. anginosus, P. melaninogenica,* and *P. acnes* are abundant in tumor microhabitats and *B. uniformis* and *P. copri* are considerably reduced, and *B. fragilis* and *A. muciniphila* having same variation pattern among peritumor and tumor tissues. Bacterial enrichment was lessened in tumor microhabitats, and network linked with gastric bacteria was simplified (Liu et al., 2019). In the same gastric microbiome habitat, numerous bacterial phylotypes varied among intestinal and diffuse gastric cancer types; however, gastric microbiota arrangement did not differ considerably. This shows that the gastric habitat of gastric cancer governs complete structure of the gastric microbiota rather than its stage, type, or cellular differentiation.

Current clinical studies have shown dysregulation of fungal flora among tumorigenic gastric mucosa and normal gastric mucosa in gastric cancer having lesser Sordariales, Rozellomycota, Pezizomycetes, and Chaetomiaceae in GC group according to taxonomic classification. Solicoccozyma was more abundant and differentially developed. *Candida albicans* abundance helps GC development by lessening the diversity of microbes in the stomach, reported in gastric cancer. These two researches showed that changed GC-associated fungal composition and *Candida albicans* and Solicoccozyma are recognized as fungal biomarkers for gastric cancer (Huang et al., 2023).

10.4.4 *H. pylori* and Immune Regulatory Effect

H. pylori infections can accelerate innate and adaptive immune responses. The host cell's pattern recognition receptors (PRRs) identify *HF. pylori* pathogen-associated molecular patterns (PAMPs) and activate the first stage of the innate immune response (Uno et al., 2007). The main constituent of PRRs is toll-like receptors (TLRs) that can bind to double-stranded RNA, CpG repeats, lipoproteins, lipoteichoic acid, lipopolysaccharide, flagellin, and unmethylated nucleic acids of *H. pylori* (Satoh & Akira, 2016). The expression of interferon can be stimulated after TLRs activate NF-Kb, interferon regulatory factor, and activator protein-1 after identifying PAMPs. In adaptive immunity, chief intermediates of host immune response to *H. pylori* infection are cluster of differentiation CD4 + T cells (Karkhah et al., 2019). CD4 + T cells were more enriched in gastric cancer but CD8 + T cells were abundant in normal tissues samples (Huang et al., 2014). Inflammation is responsible for the acid-secreting parietal cells loss that leads to an upsurge in stomach pH, resulting in reduced *H. pylori* enrichment and bigger colonization by other bacteria (Pereira-Marques et al., 2019). The inflammation caused by *H. pylori* produces reactive oxygen species, and reactive nitrogen species results in DNA damage like point mutations, dysregulate signal transduction pathways, double-strand DNA breaks, and brings apoptosis of gastric epithelial cells (Shimizu et al., 2017). In *H. pylori*-positive gastric epithelial cells, it can cause cancer due to genetic instability as the DNA repair system was impaired (Han et al., 2020). *H. pylori* also triggers immunosuppressive responses. Forkhead box protein and regulatory T cells are important immunosuppress regulators. Foxp3 mutation is linked with autoimmune disease (Deng et al., 2010).

10.4.5 Microbiome as Biomarker of Gastric Carcinoma

Early GC diagnosis can be done by sequencing methods for microbiome of gut. *Escherichia, Oscillospira*, *Desulfovibrio*, and *Faecalibacterium* in feces have chief variation in gastric cancer that has the potential for diagnosis of gastric cancer. Another two genera *Lactobacillus* and *Megasphaera are also the* diagnostic biomarker as they are abundant in GC patients. In a widespread observational study, *Streptococcus constellatus* and *Streptococcus anginosus* were used as GC biomarker. In case control studies, longitudinal study has proved that *Moryella* and *Vibro* are definite microbes in early gastric neoplasia that shows that alterations in the gut microbiome can be used as progressive biomarkers (Wang et al., 2023).

10.5 Conclusion

Cancer is one of the most multifaceted, distressing, and poorly understood human pathologies associated with discrete microbial alterations and global changes in microbial community structure. It has been known for decades that microorganisms have role in carcinogenesis but exact molecular mechanism remains unclear. The bacteria, viruses, fungi, and protozoa trigger tumorigenesis and they interact with host cell to produce hallmarks of the cancer. Alteration in human microbiota causes chronic inflammation, gut dysbiosis, and immune modulation that lead to cancer development. Microbiota can act as a biomarker for the diagnosis of cancer and in the therapy as well. Microbial contributions in cancer biology will take central stage in the future for cancer research and further investigation is required to understand their molecular mechanism in tumorigenesis and impact in diagnosis and therapeutics.

References

Ahmed, I., & Umar, S. (2018). Microbiome and colorectal cancer. *Current Colorectal Cancer Reports, 14*, 217–225.

Azevedo, M. M., Pina-Vaz, C., & Baltazar, F. (2020). Microbes and cancer: Friends or faux? *International Journal of Molecular Sciences, 21*(9), 3115.

Baid, S. (2023). *Cancer microbiology: Connecting the dots*. American Society for Microbiology.

Banerjee, S., Wei, Z., Tan, F., Peck, K. N., Shih, N., Feldman, M., Rebbeck, T. R., Alwine, J. C., & Robertson, E. S. (2015). Distinct microbiological signatures associated with triple negative breast cancer. *Scientific Reports, 5*(1), 15162.

Biarc, J., Nguyen, I. S., Pini, A., Gosse, F., Richert, S., Thierse, D., Van Dorsselaer, A., Leize-Wagner, E., Raul, F., Klein, J. P., & Schöller-Guinard, M. (2004). Carcinogenic properties of proteins with pro-inflammatory activity from streptococcus infantarius (formerly S. Bovis). *Carcinogenesis, 25*(8), 1477–1484.

Bonnet, M., Buc, E., Sauvanet, P., Darcha, C., Dubois, D., Pereira, B., Déchelotte, P., Bonnet, R., Pezet, D., & Darfeuille-Michaud, A. (2014). Colonization of the human gut by E. Coli and colorectal cancer risk. *Clinical Cancer Research, 20*(4), 859–867.

Castellarin, M., Warren, R. L., Freeman, J. D., Dreolini, L., Krzywinski, M., Strauss, J., Barnes, R., Watson, P., Allen-Vercoe, E., Moore, R. A., & Holt, R. A. (2012). Fusobacterium nucleatum infection is prevalent in human colorectal carcinoma. *Genome Research, 22*(2), 299–306.

Chattopadhyay, I., Gundamaraju, R., & Rajeev, A. (2023). Diversification and deleterious role of microbiome in gastric cancer. *Cancer Reports, 6*(11), e1878.

Chen, J., Douglass, J., Prasath, V., Neace, M., Atrchian, S., Manjili, M. H., Shokouhi, S., & Habibi, M. (2019a). The microbiome and breast cancer: A review. *Breast Cancer Research and Treatment, 178*, 493–496.

Chen, X. H., Wang, A., Chu, A. N., Gong, Y. H., & Yuan, Y. (2019b). Mucosa-associated microbiota in gastric cancer tissues compared with non-cancer tissues. *Frontiers in Microbiology, 10*, 1261.

Chiba, A., Bawaneh, A., Velazquez, C., Clear, K. Y., Wilson, A. S., Howard-McNatt, M., Levine, E. A., Levi-Polyachenko, N., Yates-Alston, S. A., Diggle, S. P., & Soto-Pantoja, D. R. (2020). Neoadjuvant chemotherapy shifts breast tumor microbiota populations to regulate drug responsiveness and the development of metastasis. *Molecular Cancer Research, 18*(1), 130–139.

Chung, L., Orberg, E. T., Geis, A. L., Chan, J. L., Fu, K., Shields, C. E., Dejea, C. M., Fathi, P., Chen, J., Finard, B. B., & Tam, A. J. (2018). Bacteroides fragilis toxin coordinates a pro-carcinogenic inflammatory cascade via targeting of colonic epithelial cells. *Cell Host & Microbe, 23*(2), 203–214.

Deng, L., Zhang, H., Luan, Y., Zhang, J., Xing, Q., Dong, S., Wu, X., Liu, M., & Wang, S. (2010). Accumulation of foxp3+ T regulatory cells in draining lymph nodes correlates with disease progression and immune suppression in colorectal cancer patients. *Clinical Cancer Research., 16*(16), 4105–4112.

Dieleman, S., Aarnoutse, R., Ziemons, J., Kooreman, L., Boleij, A., & Smidt, M. (2021). Exploring the potential of breast microbiota as biomarker for breast cancer and therapeutic response. *The American Journal of Pathology, 191*(6), 968–982.

Eslami, S. Z., Majidzadeh-A, K., Halvaei, S., Babapirali, F., & Esmaeili, R. (2020). Microbiome and breast cancer: New role for an ancient population. *Frontiers in Oncology, 10*, 120.

Francescone, R., Hou, V., & Grivennikov, S. I. (2014). Microbiome, inflammation and cancer. *Cancer Journal (Sudbury, Mass.), 20*(3), 181.

Fre, S., Bardin, A., Robine, S., & Louvard, D. (2011). Notch signaling in intestinal homeostasis across species: The cases of drosophila, zebrafish and the mouse. *Experimental Cell Research, 317*(19), 2740–2747.

Gantuya, B., El Serag, H. B., Matsumoto, T., Ajami, N. J., Uchida, T., Oyuntsetseg, K., Bolor, D., & Yamaoka, Y. (2020). Gastric mucosal microbiota in a Mongolian population with gastric cancer and precursor conditions. *Alimentary Pharmacology & therapeutics, 51*(8), 770–780.

GLOBOCAN. (2022). https://gco.iarc.fr/

Gong, D., Gong, X., Wang, L., Yu, X., & Dong, Q. (2016). Involvement of reduced microbial diversity in inflammatory bowel disease. *Gastroenterology Research and Practice, 2016*, 1.

Grivennikov, S. I., Wang, K., Mucida, D., Stewart, C. A., Schnabl, B., Jauch, D., Taniguchi, K., Yu, G. Y., Österreicher, C. H., Hung, K. E., & Datz, C. (2012). Adenoma-linked barrier defects and microbial products drive IL-23/IL-17-mediated tumour growth. *Nature, 491*(7423), 254–258.

Gui, Q. F., Lu, H. F., Zhang, C. X., Xu, Z. R., & Yang, Y. H. (2015). Well-balanced commensal microbiota contributes to anti-cancer response in a lung cancer mouse model. *Genetics and Molecular Research, 14*(2), 5642–5651.

Gupta, A., Madani, R., & Mukhtar, H. (2010). Streptococcus bovis endocarditis, a silent sign for colonic tumour. *Colorectal Disease, 12*(3), 164–171.

Han, T., Jing, X., Bao, J., Zhao, L., Zhang, A., Miao, R., Guo, H., Zhou, B., Zhang, S., Sun, J., & Shi, J. H. (2020). Pylori infection alters repair of DNA double-strand breaks via SNHG17. *The Journal of Clinical Investigation, 130*(7), 3901–3918.

Hattori, N., & Ushijima, T. (2016). Epigenetic impact of infection on carcinogenesis: Mechanisms and applications. *Genome Medicine, 8*, 1–3.

He, Z., Gharaibeh, R. Z., Newsome, R. C., Pope, J. L., Dougherty, M. W., Tomkovich, S., Pons, B., Mirey, G., Vignard, J., Hendrixson, D. R., & Jobin, C. (2018). Campylobacter jejuni promotes colorectal tumorigenesis through the action of cytolethal distending toxin. *Gut, 21*, gutjnl-2018.

Huang, X. M., Liu, X. S., Lin, X. K., Yu, H., Sun, J. Y., Liu, X. K., Chen, C., Jin, H. L., Zhang, G. E., Shi, X. X., & Zhang, Q. (2014). Role of plasmacytoid dendritic cells and inducible costimulator-positive regulatory T cells in the immunosuppression microenvironment of gastric cancer. *Cancer Science, 105*(2), 150–158.

Huang, H., Zhong, W., Wang, X., Yang, Y., Wu, T., Chen, R., Liu, Y., He, F., & Li, J. (2023). The role of gastric microecological dysbiosis in gastric carcinogenesis. *Frontiers in Microbiology, 14*, 1218395.

Karkhah, A., Ebrahimpour, S., Rostamtabar, M., Koppolu, V., Darvish, S., Vasigala, V. K., Validi, M., & Nouri, H. R. (2019). Helicobacter pylori evasion strategies of the host innate and adaptive immune responses to survive and develop gastrointestinal diseases. *Microbiological Research, 218*, 49–57.

Kelly, D., & Mulder, I. E. (2012). Microbiome and immunological interactions. *Nutrition reviews., 70*(suppl_1), S18–S30.

Khusnurrokhman, G., & Wati, F. F. (2022). Tumor-promoting inflammation in lung cancer: A literature review. *Annals of Medicine and Surgery, 79*, 104022.

Kim, J., & Lee, H. K. (2022). Potential role of the gut microbiome in colorectal cancer progression. *Frontiers in Immunology, 12*, 807648.

Kim, J. M., Kim, J. S., Lee, J. Y., Kim, Y. J., Youn, H. J., Kim, I. Y., Chee, Y. J., Oh, Y. K., Kim, N., Jung, H. C., & Song, I. S. (2007). Vacuolating cytotoxin in helicobacter pylori water-soluble proteins upregulates chemokine expression in human eosinophils via Ca2+ influx, mitochondrial reactive oxygen intermediates, and NF-κB activation. *Infection and Immunity, 75*(7), 3373–3381.

Liu, X., Shao, L., Liu, X., Ji, F., Mei, Y., Cheng, Y., Liu, F., Yan, C., Li, L., & Ling, Z. (2019). Alterations of gastric mucosal microbiota across different stomach microhabitats in a cohort of 276 patients with gastric cancer. *E Bio Medicine, 40*, 336–348.

Liu, Y., Lau, H. C., Cheng, W. Y., & Yu, J. (2023). Gut microbiome in colorectal cancer: Clinical diagnosis and treatment. *Genomics, Proteomics & Bioinformatics, 21*(1), 84–96.

Long, X., Wong, C. C., Tong, L., Chu, E. S., Ho Szeto, C., Go, M. Y., Coker, O. O., Chan, A. W., Chan, F. K., Sung, J. J., & Yu, J. (2019). Peptostreptococcus anaerobius promotes colorectal carcinogenesis and modulates tumour immunity. *Nature Microbiology, 4*(12), 2319–2330.

Louis, P., Hold, G. L., & Flint, H. J. (2014). The gut microbiota, bacterial metabolites and colorectal cancer. *Nature reviews Microbiology, 12*(10), 661–672.

Malla, R. R., Marni, R., Kumari, S., Chakraborty, A., & Lalitha, P. (2022). Microbiome assisted tumor microenvironment: Emerging target of breast cancer. *Clinical Breast Cancer, 22*(3), 200–211.

Mukherjee, A. (2020). A systematic review on parasite induced carcinogenesis. *International Journal of Innovative Science and Research Technology, 5*(6), 1095–1099.

Perales-Puchalt, A., Perez-Sanz, J., Payne, K. K., Svoronos, N., Allegrezza, M. J., Chaurio, R. A., Anadon, C., Calmette, J., Biswas, S., Mine, J. A., & Costich, T. L. (2018). Frontline science: Microbiota reconstitution restores intestinal integrity after cisplatin therapy. *Journal of Leukocyte Biology, 103*(5), 799–805.

Pereira-Marques, J., Ferreira, R. M., Pinto-Ribeiro, I., & Figueiredo, C. (2019). Helicobacter pylori infection, the gastric microbiome and gastric cancer. *Helicobacter pylori in Human Diseases: Advances in Microbiology, Infectious Diseases and Public Health, 11*, 195–210.

Rong, J., Liu, S., Hu, C., & Liu, C. (2019). Single probiotic supplement suppresses colitis-associated colorectal tumorigenesis by modulating inflammatory development and microbial homeostasis. *Journal of Gastroenterology and Hepatology., 34*(7), 1182–1192.

Ruiz, P. A., Shkoda, A., Kim, S. C., Sartor, R. B., & Haller, D. (2006). IL-10 gene-deficient mice lack TGF-beta/Smad-mediated TLR2 degradation and fail to inhibit proinflammatory gene expression in intestinal epithelial cells under conditions of chronic inflammation. *Annals of the New York Academy of Sciences, 1072*(1), 389–394.

Russo, E., Taddei, A., Ringressi, M. N., Ricci, F., & Amedei, A. (2016). The interplay between the microbiome and the adaptive immune response in cancer development. *Therapeutic Advances in Gastroenterology., 9*(4), 594–605.

Sakarya, S., & Gunay, N. (2014). Saccharomyces boulardii expresses neuraminidase activity selective for α2, 3-linked sialic acid that decreases Helicobacter pylori adhesion to host cells. *APMIS, 122*(10), 941–950.

Salvatori, S., Marafini, I., Laudisi, F., Monteleone, G., & Stolfi, C. (2023). Helicobacter pylori and gastric cancer: Pathogenetic mechanisms. *International Journal of Molecular Sciences., 24*(3), 2895.

Sanjuán, R., Pereira-Gómez, M., & Risso, J. (2016). Genome instability in DNA viruses. In *Genome Stability* (pp. 37–47). Academic Press.

Satoh, T., & Akira, S. (2016). Toll-like receptor signaling and its inducible proteins. *Microbiology Spectrum, 4*(6), 10–1128.

Sears, C. L., & Garrett, W. S. (2014). Microbes, microbiota, and colon cancer. *Cell Host & Microbe, 15*(3), 317–328.

Shapira, I., Sultan, K., Lee, A., & Taioli, E. (2013). Evolving concepts: How diet and the intestinal microbiome act as modulators of breast malignancy. *International Scholarly Research Notices, 2013*, 1.

Shatova, O. P., Zabolotneva, A. A., & Shestopalov, A. V. (2023). Molecular ensembles of microbiotic metabolites in carcinogenesis. *Biochemistry (Moscow), 88*(7), 867–879.

Shimizu, T., Chiba, T., & Marusawa, H. (2017). Helicobacter pylori-mediated genetic instability and gastric carcinogenesis. *Molecular Pathogenesis and Signal Transduction by Helicobacter pylori, 400*, 305–323.

Siegel, R. L., Miller, K. D., Wagle, N. S., & Jemal, A. (2023). Cancer statistics, 2023. *CA: A Cancer Journal for Clinicians, 73*(1), 17–48.

Siegel, R. L., Giaquinto, A. N., & Jemal, A. (2024). Cancer statistics, 2024. *CA: A Cancer Journal for Clinicians, 74*, 12.

Song, X., Wei, C., & Li, X. (2022). The relationship between microbial community and breast cancer. *Frontiers in Cellular and Infection Microbiology, 12*, 831.

Spoelhof, B., & Ray, S. D. (2014). *Encyclopedia of toxicology* (pp. 761–765). Academic Press.

Strouse, C., Mangalam, A., & Zhang, J. (2019). Bugs in the system: Bringing the human microbiome to bear in cancer immunotherapy. *Gut Microbes, 10*(2), 109–112.

Sui, S. J., Lo, R., Fernandes, A. R., Caulfield, M. D., Lerman, J. A., Xie, L., Bourne, P. E., Baillie, D. L., & Brinkman, F. S. (2012). Raloxifene attenuates Pseudomonas aeruginosa pyocyanin production and virulence. *International Journal of Antimicrobial Agents, 40*(3), 246–251.

Uno, K., Kato, K., Atsumi, T., Suzuki, T., Yoshitake, J., Morita, H., Ohara, S., Kotake, Y., Shimosegawa, T., & Yoshimura, T. (2007). Toll-like receptor (TLR) 2 induced through TLR4 signaling initiated by helicobacter pylori cooperatively amplifies iNOS induction in gastric epithelial cells. *American Journal of Physiology-Gastrointestinal and Liver Physiology, 293*(5), G1004–G1012.

Vétizou, M., Pitt, J. M., Daillère, R., Lepage, P., Waldschmitt, N., Flament, C., Rusakiewicz, S., Routy, B., Roberti, M. P., Duong, C. P., & Poirier-Colame, V. (2015). Anticancer immunotherapy by CTLA-4 blockade relies on the gut microbiota. *Science, 350*(6264), 1079–1084.

Wang, T., Cai, G., Qiu, Y., Fei, N., Zhang, M., Pang, X., Jia, W., Cai, S., & Zhao, L. (2012). Structural segregation of gut microbiota between colorectal cancer patients and healthy volunteers. *The ISME Journal, 6*(2), 320–329.

Wang, J., Lu, R., Fu, X., Dan, Z., Zhang, Y. G., Chang, X., Liu, Q., Xia, Y., Liu, X., & Sun, J. (2018). Novel regulatory roles of Wnt1 in infection-associated colorectal cancer. *Neoplasia, 20*(5), 499–509.

Wang, L., Xin, Y., Zhou, J., Tian, Z., Liu, C., Yu, X., Meng, X., Jiang, W., Zhao, S., & Dong, Q. (2020). Gastric mucosa-associated microbial signatures of early gastric cancer. *Frontiers in Microbiology, 11*, 1548.

Wang, N., Sun, T., & Xu, J. (2021). Tumor-related microbiome in the breast microenvironment and breast cancer. *Journal of Cancer, 12*(16), 4841.

Wang, M., Yang, G., Tian, Y., Zhang, Q., Liu, Z., & Xin, Y. (2023). The role of the gut microbiota in gastric cancer: The immunoregulation and immunotherapy. *Frontiers in Immunology, 14*, 14.

Whisner, C. M., & Athena, A. C. (2019). The role of the microbiome in cancer initiation and progression: How microbes and cancer cells utilize excess energy and promote one another's growth. *Current Nutrition Reports, 8*, 42–51.

Yachida, S., Mizutani, S., Shiroma, H., Shiba, S., Nakajima, T., Sakamoto, T., Watanabe, H., Masuda, K., Nishimoto, Y., Kubo, M., & Hosoda, F. (2019). Metagenomic and metabolomic analyses reveal distinct stage-specific phenotypes of the gut microbiota in colorectal cancer. *Nature Medicine, 25*(6), 968–976.

Yahya, E. B., Abdulsamad, M. A., & Allaq, A. A. (2021). Recent advances in the role of microorganisms in cancer incidence: Mechanisms and health precautions. *Biomedical Research and Therapy, 8*(9), 4525–4539.

Yang, J., Tan, Q., Fu, Q., Zhou, Y., Hu, Y., Tang, S., Zhou, Y., Zhang, J., Qiu, J., & Lv, Q. (2017). Gastrointestinal microbiome and breast cancer: Correlations, mechanisms and potential clinical implications. *Breast Cancer, 24*, 220–228.

Yang, J., Zhou, X., Liu, X., Ling, Z., & Ji, F. (2021). Role of the gastric microbiome in gastric cancer: From carcinogenesis to treatment. *Frontiers in Microbiology, 12*, 641322.

Yasunaga, J. I., & Matsuoka, M. (2018). Oncogenic spiral by infectious pathogens: Cooperation of multiple factors in cancer development. *Cancer Science, 109*(1), 24–32.

Yu, G., Torres, J., Hu, N., Medrano-Guzman, R., Herrera-Goepfert, R., Humphrys, M. S., Wang, L., Wang, C., Ding, T., Ravel, J., & Taylor, P. R. (2017). Molecular characterization of the human stomach microbiota in gastric cancer patients. *Frontiers in Cellular and Infection Microbiology, 7*, 302.

Protists and Cancer

11

Aiman Sajjad, Farhat Jabeen, and Saba Zafar

Abstract

Protists, a diverse group of eukaryotic unicellular microorganisms that are neither animals, plants, nor fungi, protozoa have historically been investigated in relation to parasitic illnesses and the effects they have on human health. But new and surprising research has revealed a link between protists and cancer that was previously unknown. This chapter highlights the new area of study that looks at the complex interactions between protists and many aspects of cancer treatment, progression, and development. The term "cancer" refers to over 100 malignant disorders. All malignancies are thought to begin with a single aberrant cell that spreads uncontrollably. The conversation includes current research that shows protists can affect the tumor microenvironment, change immune responses, and possibly even be involved in carcinogenic processes. It also looks into the various ways that protists could affect cancer cells, such as direct contacts, molecular signaling cascades, and host cellular machinery modification. Protist integration into the larger framework of cancer biology may open the door to novel strategies for cancer detection, treatment, and prevention, ultimately strengthening our capacity to fight this difficult and complicated disease.

Keywords

Protists · Cancer · Tumor microenvironment · Carcinogenic processes · Cancer biology

A. Sajjad · F. Jabeen (✉) · S. Zafar
Department of Zoology, Government College University Faisalabad, Faisalabad, Pakistan
e-mail: farhatjabeen@gcuf.edu.pk

N. Masood, A. Yasmin (eds.), *The Microbiome and Cancer*,
https://doi.org/10.1007/978-981-97-6948-3_11

11.1 Introduction

The term "protists" refers to a very diverse group that was once considered to be a single kingdom, but more recent genetic research has shown that it actually belongs to several kingdoms. Protists are an extremely diverse category of organisms that can be unicellular, multicellular, heterotrophic, or autotrophic (Carr et al., 2023; Romano et al., 2023). Historically, any eukaryote that wasn't an animal, plant, or fungus was referred to as a "protist." Protists include, among other things, microalgae like diatoms and dinoflagellates, macroalgae like kelps and seaweeds, and significant disease-causing microbes like Plasmodium, the parasite that causes malaria, which kills hundreds of thousands of people annually in spite of efforts to prevent and treat the illness (Vanrell & Romano, 2023; Rodriguez et al., 2024).

Although protists have long been associated with pathogenicity, recent studies have revealed a surprising and new connection between protists and cancer. This discovery provides fresh insights into the intricate relationships that protists have with many parts of cancer biology, potentially impacting aspects of cancer formation, progression, and treatment (Zheng et al., 2024).

The word "cancer" refers to a broad range of more than 100 malignant conditions, all of which are thought to arise from a single defective cell that proliferates uncontrollably. Cancer is a serious hazard to public health worldwide, with a projected 19.3 million new cases and nearly ten million deaths from the disease in 2020 (Bray et al., 2021). Numerous environmental factors, including infections and poor lifestyles, can have a major role in the development of cancer. Globally, 15.4% of cancer cases are caused by infectious illnesses (Plummer et al., 2016), and this percentage rises to 20% in developing nations. In fact, by 2050, infections are expected to be the primary cause of cancer in humans (Sung et al., 2021; Zur Hausen, 2007). The International Agency for Research on Cancer (IARC) has classified 11 species of pathogenic microbes as "carcinogenic to humans" and group 1 carcinogen (Plummer et al., 2016). Even though research on parasite-related carcinogenesis has been focused on humans and animals thus far, knowledge of these parasites' function in humans has the potential to significantly advance both preventive and novel therapeutic options for all living species (Rolph & Cavanaugh, 2022). Furthermore, there are numerous ways in which a full understanding of this phenomenon would advance human medicine.

The purpose of this chapter is to examine the possible consequences of protists in relation to cancer and to examine their potential as new targets for therapeutic approaches. A comprehensive literature search has been carried out by searching reliable sources including Science Direct, Elsevier, and PubMed using important search terms like "protists and cancer," "parasitic protists and cancer," and "protist-host interactions in cancer."

The investigation aims to reveal the complex relationships that exist between protists and diverse facets of cancer formation, progression, and treatment across a range of cancer types. The chapter seeks to discover possible targets for therapeutic treatments and medication development by exploring the molecular mechanisms driving protist-cancer interactions.

11.2 Types of Protists

Protists are a varied group of mostly unicellular creatures found in the major eukaryotic lineages. Some of them, like waste decomposition (slime and water molds), some animal-like protists, are medically important because they cause a variety of human diseases, but most of them are free-living organisms that are vital to the environment because they perform functions like photosynthesis (plantlike algae) (Hardianto et al., 2024).

11.3 Types of Cancer and Protist Interactions

Cancer is a broad category that includes many different diseases, each with its own origin, behavior, and treatment issues. Breast cancer is one of the most common types of cancer. It is caused by cells found in the breast tissue, mainly in the ducts or lobules (Muhammad et al., 2024). In contrast, lung cancer mostly affects the lungs, and small cell lung cancer (SCLC) and non-small cell lung cancer (NSCLC) are the two primary subtypes of the illness (Yatabe, 2024). In the colon or rectum, colorectal cancer first appears as benign polyps that eventually turn into malignant tumors (Liu et al., 2024). Men's prostate glands are affected by prostate cancer, which frequently begins inside the gland and advances slowly (Kotrikadze et al., 2024). Abnormal cell proliferation in the tissues of the skin is the cause of skin cancer, including melanoma, squamous cell carcinoma, and basal cell carcinoma (Linares et al., 2015). Blood in the urine and frequent urination are signs of bladder cancer, which develops in the lining of the bladder. Abnormal white blood cell production is triggered by leukemia, a malignancy of the blood-forming tissues (Porserud et al., 2024). Pancreatic cancer starts in the pancreas and is infamous for its aggressive nature and late-stage diagnosis, whereas lymphomas affect the lymphatic system, particularly the lymph nodes and spleen (Daamen et al., 2024). Whereas brain cancer includes tumors that originate in the brain or spinal cord, with primary tumors produced from brain cells and metastatic tumors spreading from other places, ovarian cancer starts in the ovaries and is frequently discovered in advanced stages (Biricioiu et al., 2024; Young Han et al., 2024). The fact that these main cancer forms only make up a small portion of the disease's wide range highlights how complex cancer is and how urgently more research, prevention initiatives, and efficient treatment plans are needed.

Globally, freshwater and moist soils are home to the heterotrophic protist Naegleria gruberi, which is mostly found in its amoeba form. The most notable characteristic of Naegleria gruberi is its exceptionally quick transition from amoebae to flagellates, which have two anterior 9 + 2 flagella. A change, it entails creating a brand-new cytoplasmic microtubule cytoskeleton in its entirety.

The highly controlled mechanism of cells crawling like amoeboids is essential for both the migration of cells during development and the metastasis of human tumor cells (Cantoni et al., 2022). According to Cramer et al. (1997), there are three fundamental components to this amoeboid movement that are linked to the dynamics of the actin cytoskeleton: adhesion to the substrate, retraction of the cell body, and protrusion of the leading edge. Localized actin polymerization at the leading-edge membrane causes protrusion, which in turn causes the extension of a wide structure resembling a sheet called a lamellipod. Molecular grips to the substrate are created via adhesions as the lamellipod moves forward. The process of retraction involves the cell body being drawn forward and the adhesions being broken. Of these processes, protrusion is the well understood. Gaining more insight into the molecular mechanisms of cell motility is essential to identifying key processes influencing tumor cells' spreading behavior. Without a doubt, changes in cell shape and motility play a major role in the invasion, migration, and metastatic behavior of cancer cells (Yilmaz & Christofori, 2010).

11.4 Protist's Impact on the Tumor Microenvironment

Extracellular matrix, secreted substances, and both resident and infiltrating host cells make up a heterogeneous mixture that is a tumor. It is more than just a group of cancerous cells. Tumor cells cause significant changes in the chemistry, physiology, and molecular composition of the host tissues to aid in the growth and dissemination of their tumors. A growing tumor's microenvironment is a complex, dynamic system. Tumor microenvironment composition varies with tumor type; nevertheless, common features include extracellular matrix, stromal cells, immune cells, and blood vessels. According to Truffi et al. (2020), the "tumor microenvironment is not just a silent bystander, but rather an active promoter of cancer progression." Early in tumor growth, a dynamic and reciprocal relationship is formed between cancer cells and components of the tumor microenvironment, which supports cancer cell survival, local invasion, and metastatic spread (Anderson & Simon, 2020).

Numerous regulatory genes and signaling modules are involved in the evolution of tumors, many of which are yet unknown (Ibrahim et al., 2011; Tacutu et al., 2011). The shape and surface alterations of the cells, their interactions with the surrounding cells in their microenvironment, and the mobility of the malignant tumor cells are all significant during these processes (Marques et al., 2009).

This significance is further highlighted by the fact that over 90% of cancer-related deaths are attributable to metastasis formation and the resulting secondary tumors (Jemal et al., 2007). Primary tumor cells infiltrate surrounding tissues, enter the bloodstream through the systemic circulation, move through the vasculature and lymphatics, stop in far-off capillaries and lymphatic vessels, extravasate into the perivascular tissue, and eventually grow from micrometastases into macroscopic secondary tumors (Marques et al., 2009). This is a complex process.

11.5 Protists in Carcinogenic Processes

There are two primary theories about the evolutionary history of cancer: The first suggests that atavism is the cause of cancer, while the second postulates somatic evolution through mutations (SMT). Since atavistic development could be effectively resisted on its own, evidence supporting and extending the atavistic theory (AT) would be greatly appreciated (Niculescu, 2018).

11.6 The Somatic Mutations Theory

According to the somatic mutation theory (SMT), cancer develops from a single somatic cell that has accumulated several DNA mutations, as established by Soto and Sonnenschein (2004). This somatic cell and its progeny undergo a transition from a quiescent state, which is assumed to be the default state of most metazoan cell proliferation, to an unregulated state of proliferation due to mutations in the genes controlling proliferation and the progression of the cell cycle. There hasn't been much testing done on the SMT, and there are a lot of lines of evidence that show problems this theory can't solve. With the increasing identification of oncogenes and tumor-suppressor genes over time, further ad hoc hypotheses were introduced. One is the tissue organization field theory (TOFT), which holds that cancer is a disorder of tissue organization, similar to organogenesis, and that proliferation is the default state of all cells (Soto & Sonnenschein, 2011).

11.7 The Atavistic Theory

The atavistic theory of cancer, proposed in 2011 by physicians Davis and Lineweaver, postulated that the biological origin of cancer may be discovered during this phase of unicellularity to multicellularity transition, characterized by instability in the early regulatory networks that regulated the expression of unicellular (UG) and multicellular (MG) genes. PZ Meyers (P-Zombie Meyers) severely mocked this notion in 2012 in a way that was deemed unscientific (http://scienceblogs.com/pharyngula/2012/11/20/aaargh-physicists-again). The atavist idea states that highly conserved cell survival mechanisms that are locked in the eukaryotic genome are released in cancer (Thomas et al., 2017). It illustrates how our earliest single-celled ancestors were able to efficiently respond through reproduction to stress and adverse life circumstances.

UG genes are highly upregulated during the transition into the atavistic life-form, whereas MG genes are downregulated. Modern eukaryotes encode portions of the DNA of their common ancestor because of their shared ancestry. Humans have a "dark genome" made up of between 104 and 2 x104 protein-noncoding genes that have become inactive over time. It makes up about 98.8% of the DNA in a cell. Additionally, 1% of the human genome consists of "cancer genes" that are transcriptionally up- or downregulated in cancer cells (Pedraza-Fariña, 2006; Walen,

2017). Stressors that can cause tumors have the ability to awaken these deeply dormant atavistic genes. There is a significant relationship between the age of the genes and their level of expression, according to recent research by Trigos et al. (2017) on UGs and MGs expressed in malignancies.

Old genes express themselves most strongly when cancer moves from a low to a high grade (Garber et al., 2001; Perou et al., 2000; Bussey et al., 2017). The populations of cancer cells in tumors are heterogeneous and hierarchical. The sub-pool of cancer-initiating cells and non-cancer-initiating cells are distinguished by their unique molecular signatures, differentiation, and functional capabilities, the cancer-initiating cells often exhibit self-renewal properties and the ability to drive tumor growth and recurrence, while non-cancer-initiating cells may be more differentiated and lack such tumor-propagating potential (Morii, 2017). Stem cells having the ability to self-renew and multidifferentiate are known as CICs and non-CICs. Precursor and stem cell lineage (SPCL) of invasive pathogen protists such as Entamoeba invadens has a conserved dual capability for primitive terminal differentiation (PTD) and self-renewal. How human cells under environmental stress activate silent atavist dark genome gene networks that imitate intricate protist life cycles is not entirely understood. Either way, it appears that highly harmful atavistic/invasive programs derived from ancestral networks are still encoded in the human DNA (Niculescu, 2017).

11.8 Mechanisms of Interaction between Protists and Cancer Cells

Here, taking into consideration several current theories of carcinogenesis (the description and in-depth investigation of the development of these ideas is beyond our purview), we discuss two methods for carrying out the intermediate phase of carcinogenesis. Individual cancer cells are fused with healthy and other cancerous cells in the first approach. The second strategy includes multiple mechanisms by which cell-in-cell structures are formed during the tumorigenic process (Demin et al., 2022).

Since both paths and even directions are convoluted and could cross, we'll address each one separately. All cell populations in humans and other animals are polymorphic, but malignant cell populations are especially so. They frequently appear to have merely random patterns and to be chaotic (Bignold, 2004). This piece offers a "perspective from the outside." Its premise is that during their cell (life) cycles, cancer cells employ survival strategies that are inherent to invertebrates and unicellular eukaryotes (protists). These strategies are implemented through the mechanisms that cause multinucleated cells and cell-in-cell complexes to disintegrate into mononuclear cells. We focused particularly on the multinuclearity and cell fusion "polymorphism" of free-living protists. These events are routinely and independently pragmatic in these creatures belonging to different macrotaxons that are phylogenetically distant.

11.9 Cell-Cell Fusion

Adhesion distance (about 10 nm) closeness always precedes cell fusion, which is accomplished via the application of particular cell-cell adhesion proteins rather than at random. In multicellular organisms, homologous and heterologous cells can fuse together normally (Chen et al., 2007; Hernández & Podbilewicz, 2017).

It is well known that cancer cells can merge (Miller et al., 1988). Tumors frequently go through this process (Shabo et al., 2020). The emergence of theories of carcinogenesis and/or metastasis based on the phenomenon of hybrid cancer cells—binucleates, synkaryons, or syncytial multinucleated cancer cells—production has resulted from a cautious approach to cancer cell fusions, considering them like a secret adversary (Weiler & Dittmar, 2019). More recently, the Orthotopic Tumor Model (https://www.creative-biolabs.com/drug-discovery/therapeutics/orthotopic-tumor-models.htm), one of the most well-known tumor models, has been used to directly demonstrate that when cloned in vivo, a single multinucleated cancer cell can produce orthotopic, metastatic tumors (Weihua et al., 2011). It is true that in the latter instance, the authors observe that "Because cytokinesis was absent, multinucleation may occur in a single mononuclear cell." However, the capacity to create "normal-cancer cell" hybrids is the most significant aspect of such a fusion. Because of a reduction in the proportion of cell surface tumor markers and their eventual total elimination, these hybrids are immune system evading. As demonstrated, these hybrid "invisible cells" can appear at any point during the course of carcinogenesis. Immunotherapy is not accessible to them (Weiler & Dittmar, 2019).

Furthermore, the analysis of these fusions shows that following bone marrow cell transplantation, these hybrid cells cannot be identified using the current techniques that involve using DNA probes (on the Y chromosome or short tandem DNA repeats) particular to transplants in the affected organ, lymph, tissue fluid, or blood. Based on this, the exciting "Dark Matter Hypothesis" has recently been published, proposing that "normal-cancer cell" hybrids—"hybrids which cannot as yet be detected and quantified, but which contribute to tumor growth and progression"—are evolving continuously within the human body.

Recent data supporting this notion confirms a considerable impact from these fusions. In fact, it has recently been demonstrated that heterologous cell hybridization is a major factor in the development of brain tumor malignancy (Gutmann & Kettenmann, 2019).

11.10 Cell-in-Cell Phenomena

Based only on the number of reviews published in the previous 5 years (Davies et al., 2020), cell cannibalism, entosis, and emperipolesis—non-phagocyte cell-in-cell phenomena—are considered the "premium brand of the modern oncology fashion." As a result, the possibility of distinguishing between two quite similar processes of encircling the living cells on histopathological markers and the related molecular pathways will be briefly discussed here.

Only a small portion of the triggers of emperipolesis has been investigated. Specifically, it was discovered that diseases and histiocyte neoplasia are linked to both CD68 overexpression and CD1a (Leu-6) underexpression. A cell's release from the extracellular matrix initiates entosis, which is further accelerated by differences in the actomyosin contraction rates of nearby cells. E- and P-cadherins mediate these processes, which cause RhoA and contractile actomyosin in these cells to polarize. Consequently, the recipient (host) cell is dedicated to engulfing by RhoAN19 inhibition, whereas the invading cell typically engulfs following RhoAN19 overexpression.

Entosis usually results in the trapped cell dying lysosomally (Gupta et al., 2017). This process is generally recognized as cellular cannibalism and non-cell autonomous type IV programmed death (Martins et al., 2017). The production of binucleates and the postmitotic blockage of cytokinesis in the host cell can result from the death of a captured cell during entosis. Nevertheless, in the event that the host cell exhibits behaviors resembling transcytosis, the cells absorbed during entosis might not perish. These cells are able to separate from the recipient cell and, in certain situations, even go through mitosis and divide to produce daughter interiorized cells (Durgan & Florey, 2018).

Cannibalism of cancer cells results in the creation of cell-in-cell formations that include living cells and fragments of necrotic or apoptotic cells (Fais & Overholtzer, 2018). As a result, there are similarities between these phenomena and phagocytosis (also known as efferocytosis) and emperipolesis. The mechanics behind this process have only partially been studied. It is known to be mediated by the CD68 surface marker, which is suggestive of emperipolesis, and the evolutionarily conserved transmembrane protein TM9, which is involved in phagocytosis.

Actin and the caveolar network's vesicles, as well as the cortical protein ezrin, which is necessary for metastatic cells to feed, are additional particular players in this process. Cannibalism of tumor cells "may represent a sign of regression to a simpler, ancestral or primeval life style, similar to that of unicellular organisms, such as amoebas," claims Lozupone and Fais (2015). Indeed, we found that amoebas and metastatic melanoma cells both possess the transmembrane protein TM9SF4, which is connected to the cannibalistic tendencies of both cells.

Cancer cell cannibalism comes in two flavors: homotypic and heterotypic. Nonsibling cancer cells are consumed in the first instance. The second category involves stromal cancer cells or cancer cells with a distinct histotype. It is thought that cancer cell cannibalism causes temporary polyploidy, which helps these cancer cells survive while receiving anticancer treatment (Krajcovic & Overholtzer, 2012).

11.11 Protists' Somatogamy: The Merging of Vegetative Cells

In the life cycles of protists, the fusing of vegetative (somatic) cells can occur completely or partially. Researchers first described these asexual (agamic) fusions as early as the late nineteenth and early twentieth centuries, and they called this phenomenon plasmogamy (plastogamy), cytogamy, or somatogamy.

11.12 Cell-in-Cell Structures in Protists

Protists are the first organisms that exhibit cell cannibalism, one of the events leading to the formation of cell-in-cell structures. It is believed to be the consumption of living things by cells that act as "superpredators" inside their own species. It has long been known and researched that certain types of amoeboid creatures, such as flagellates and ciliates, are cannibals.

Particularly among the latter, species belonging to broad genera like Blepharisma, Stylonychia, Stentor, Frontonia, and others (Adl & Berger, 2000) inherently exhibit this feature. It is interesting that, contrary to what early research has showed experimentally, hunger is not the primary driver of cannibalism. The majority of the time, its appearance can be attributed to a disruption in the biological mechanisms involved in identifying individuals within one's own species (Demin et al., 2022).

It appears that in unicellular eukaryotes, this occurrence has no "genetic consequences" for the cannibal cell. With the exception of those outlined by Weyer (1930), the findings of research on this subject are consistent. This study reports that during cannibalism, the ciliate Gastrostyla steini only digested the ingested victim's cytoplasm; the macro- and micronuclei of the prey were maintained for an extended period of time. Furthermore, the author took into account the findings of a study on the inheritance of recently acquired traits in these cells' progeny, which supports the idea that the cannibal cell's nuclear apparatus may have incorporated the nuclei of its prey, changing its genotype. Unfortunately, no one has yet to validate the study's findings.

We can state with confidence that the author has not been able to provide conclusive evidence for the preservation of absorbed individual nuclei and their fusion with predator nuclei, given the methodological level of investigations of the nuclear machinery.

11.13 Immune Responses and Protists

In the past few decades, various species of protozoa and their components have demonstrated antitumor potential via immunological and nonimmune mechanisms; hence, the quest for novel techniques for tumor intervention may contribute to improving the present effectiveness of tumor treatment. Cancer still affects a large number of people worldwide, even with improvements in diagnosis and care. Although the prognosis for some individuals has been improved by current tumor treatment procedures, which include surgery, chemotherapy, radiation, targeted therapy, and immunotherapy, these treatments are not helpful for all patients. As a result, several cancers still have comparatively poor 5-year survival rates (Zheng et al., 2024).

Protozoa such as Trypanosoma cruzi and Toxoplasma gondii have demonstrated antitumor effects in vitro and in studies with animals that harbor tumors. These investigations have uncovered a variety of mechanisms by which protozoa fight cancer, indicating promising avenues for their use. In this study, we discuss the potential for clinical development and application of several protozoa, as well as their antitumor capacity and antitumor pathways (Zheng et al., 2024; Sardinha-Silva et al., 2022).

11.14 Toxoplasma Gondii

The obligatory intracellular protozoan parasite T. gondii infects around one-third of all people on the planet. Generally, this parasite is spread by eating raw meat, drinking water tainted with oocysts, or coming into contact with the excrement of pets, especially cats (Montoya & Liesenfeld, 2004). People who have T. gondii infections typically show little to no symptoms, as the infections are commonly dormant. However, T. gondii can cause severe symptoms that affect the muscles, brain, and eyes when the immune system is weakened or damaged (Dupont et al., 2012).

Remarkably, studies have demonstrated that T. gondii infection can markedly improve the prognosis in tumor-bearing animal models and dramatically slow down tumor progression (Seyedeh et al., 2015; Wang et al., 2019). This is predicated on the idea that an infection with T. gondii can elicit a number of powerful immunological responses from the body (Dupont et al., 2012). Furthermore, a number of research have demonstrated the strong link between T. gondii infection and malignancy. In its tachyzoite stage, for instance, T. gondii displays glucose-independent growth; this is similar to the metabolic characteristics of tumor development (Nitzsche et al., 2017). Moreover, it has been shown that in mice-harboring tumors, anti-T. gondii antibodies bind exclusively to tumor cells (Mohamadi et al., 2019). These findings suggest that toxoplasma can significantly enhance the organism's defenses against cancer. The precise relationship between T. gondii and cancer is still unknown (El Skhawy & Eissa, 2023).

T. gondii is one of the protozoa with the most research and application potential among those showing antitumor characteristics. Unfortunately, the amount of study that has been done on its tumor-targeting properties has limited its usefulness for treating multiple solid tumors or nonsolid tumors. Moreover, the antitumor potential of other significant T. gondii components is still mostly unknown, despite the fact that the ROP and GRA proteins have drawn a lot of interest. It is significant to remember that prior technologies have severely limited the antitumor research that is currently being conducted on T. gondii. However, recent advancements in single-cell technologies, machine learning, and multi-omics approaches may help further the study of T. gondii in antitumor research.

11.15 Trypanosoma Cruzi

Cardiomyopathy and arrhythmia are two of the most prevalent symptoms of Chagas disease, which is mostly found in 21 Latin American nations and is caused by the bacterium T. cruzi (Pérez-Molina & Molina, 2018). Since its discovery a century ago, T. cruzi's possible antitumor activities have been thoroughly investigated and progress has been achieved. In the middle of the twentieth century, two Soviet scientists identified and recorded the potential antitumor properties of T. cruzi extracts, providing a novel approach to treating cancers. There is some competition between tumors and T. cruzi because both have been shown to have immune evasion mechanisms and rapid division within the human body. Additionally, there is ample evidence of T. cruzi's preferential invasion of tumor cells (Kallinikova et al., 2001).

Moreover, several antineoplastic medicines and anti-T. cruzi therapies have shown therapeutic effects against both cancers and T. cruzi, much like the protozoa previously stated (Chennamaneni et al., 2009). In light of this data, T. cruzi's antitumor potential merits more investigation.

Despite being the first protozoa to be studied for their ability to fight cancer, Trypanosoma's progress in this area has been sluggish recently because of its lesser prevalence than that of T. gondii or Plasmodium. Trypanosoma is highly pathogenic, and long-term infection can produce persistent inflammation that can accelerate the cell death plus regeneration, increase DNA damage, and raise the likelihood of oncogene expression—all of which can contribute to the development of cancer. Partially confirmed is its association with gynecological, gastrointestinal, and hematological system malignancies (Ribeiro Franco et al., 2023).

Consequently, there is presently not enough study on T. cruzi toxicity attenuation, and the host's harm or side effects from its strains and components cannot be well regulated. Trypanosoma toxic attenuation treatment is essential for its application in antitumor therapy. Additionally, similar to other protozoa, its antitumor activity cannot be shown in humans due to constrained settings that restrict present study to animal tests like mice. Therefore, more research is still needed to confirm the usefulness of many research findings.

11.16 Conclusions and Future Perspectives

In conclusion, the investigation of the connection between protists and cancer is an emerging topic of study that will have a big impact on our knowledge of cancer biology and possible treatments. Based on available data, protists—typically investigated in relation to parasitic diseases—may also be involved in the intricate processes involved in the onset, spread, and management of cancer.

The complex nature of protist-cancer relationships is highlighted by the interaction between protists and cancer cells within the tumor microenvironment, the modulation of immune responses, and possible involvement in carcinogenic processes. Through the clarification of the molecular mechanisms that underpin these interactions, scientists might pinpoint new targets for the creation of drugs and other therapeutic interventions aimed at combating different forms of cancer.

Protists and cancer research is expected to go in a number of the important ways in the future. First and principal, a detailed characterization of the protist species and their distinct functions in various cancer types and stages is required. Investigating the variety of protists, their modes of action, and their effects on tumor biology is necessary to achieve this.

Furthermore, it is crucial to comprehend the intricate host-protist interactions and how they affect the development of cancer. This entails figuring out the complex immunological reactions, signaling pathways, and biological mechanisms that protists use to influence cancer cells and the surrounding tumor microenvironment.

Additionally, the advancement of cancer therapy paradigms may be possible through the creation of novel therapeutic approaches that target protists. Future study could explore many avenues to exploit the potential of protists as therapeutic targets, such as repurposing existing medications, designing novel therapeutic molecules, or investigating immunotherapeutic techniques.

To sum up, the incorporation of protists into the wider context of cancer biology presents fresh perspectives and potential treatment avenues for addressing this intricate illness. We can ultimately improve our capacity to diagnose, treat, and prevent cancer in a variety of patient populations by advancing our understanding of the complex relationships that exist between protists and cancer cells.

References

Adl, S. M., & Berger, J. D. (2000). Timing of life cycle morphogenesis in synchronous samples of Sterkiella histriomuscorum. II. The sexual pathway. *Journal of Eukaryotic Microbiology, 47*(5), 443–449.

Anderson, N. M., & Simon, M. C. (2020). The tumor microenvironment. *Current Biology, 30*(16), R921–R925.

Bignold, L. P. (2004). Chaotic genomes make chaotic cells: The mutator phenotype theory of carcinogenesis applied to clinicopathological relationships of solid tumors. *Cancer Investigation, 22*(3), 338–343.

Biricioiu, M. R., Sarbu, M., Ica, R., Vukelić, Ž., Kalanj-Bognar, S., & Zamfir, A. D. (2024). Advances in mass spectrometry of gangliosides expressed in brain cancers. *International Journal of Molecular Sciences, 25*(2), 1335.

Bray, F., Laversanne, M., Weiderpass, E., & Soerjomataram, I. (2021). The ever-increasing importance of cancer as a leading cause of premature death worldwide. *Cancer, 127*(16), 3029–3030.

Bussey, K. J., Cisneros, L. H., Lineweaver, C. H., & Davies, P. C. (2017). Ancestral gene regulatory networks drive cancer. *Proceedings of the National Academy of Sciences, 114*(24), 6160–6162.

Cantoni, D., Osborne, A., Taib, N., Thompson, G., Martín-Escolano, R., Kazana, E., & Tsaousis, A. D. (2022). Localization and functional characterization of the alternative oxidase in Naegleria. *Journal of Eukaryotic Microbiology, 69*(4), e12908.

Carr, M., Hopkins, K., & Ginger, M. L. (2023). The protistan origins of animals and fungi. In *Evolution of fungi and fungal-like organisms* (pp. 3–38). Springer International Publishing.

Chen, E. H., Grote, E., Mohler, W., & Vignery, A. (2007). Cell–cell fusion. *FEBS Letters, 581*(11), 2181–2193.

Chennamaneni, N. K., Arif, J., Buckner, F. S., & Gelb, M. H. (2009). Isoquinoline-based analogs of the cancer drug clinical candidate tipifarnib as anti-Trypanosoma cruzi agents. *Bioorganic & Medicinal Chemistry Letters, 19*(23), 6582–6584.

Cramer, L. P., Siebert, M., & Mitchison, T. J. (1997). Identification of novel graded polarity actin filament bundles in locomoting heart fibroblasts: Implications for the generation of motile force. *The Journal of Cell Biology, 136*, 1287–1305.

Daamen, L. A., Parikh, P. J., & Hall, W. A. (2024). The use of MR-guided radiation therapy for pancreatic cancer. *Seminars in Radiation Oncology, 34*(1), 23–35. WB Saunders.

Davies, P. C., & Lineweaver, C. H. (2011). Cancer tumors as Metazoa 1.0: Tapping genes of ancient ancestors. *Physical Biology, 8*(1), 015001.

Davies, S. P., Terry, L. V., Wilkinson, A. L., & Stamataki, Z. (2020). Cell-in-cell structures in the liver: A tale of four E's. *Frontiers in Immunology, 11*, 650.

Demin, S., Berdieva, M., & Goodkov, A. (2022). Cell-cell fusions and cell-in-cell phenomena in healthy cells and cancer: Lessons from protists and invertebrates. *Seminars in Cancer Biology, 81*, 96–105. Academic Press.

Dupont, C. D., Christian, D. A., & Hunter, C. A. (2012). Immune response and immunopathology during toxoplasmosis. *Seminars in Immunopathology, 34*, 793–813. Springer-Verlag.

Durgan, J., & Florey, O. (2018). Cancer cell cannibalism: Multiple triggers emerge for entosis. *Biochimica et Biophysica Acta (BBA)-Molecular Cell Research, 1865*(6), 831–841.

El Skhawy, N., & Eissa, M. M. (2023). Shedding light on a mysterious link between toxoplasma Gondii and cancer: A review. *Experimental Parasitology, 108544.*

Fais, S., & Overholtzer, M. (2018). Cell-in-cell phenomena in cancer. *Nature Reviews Cancer, 18*(12), 758–766.

Franco, P. I. R., do Carmo Neto, J. R., Miguel, M. P., Machado, J. R., & Celes, M. R. N. (2023). Cancer and Trypanosoma cruzi: Tumor induction or protection? *Biochimie, 207*, 113–121.

Garber, M. E., Troyanskaya, O. G., Schluens, K., Petersen, S., Thaesler, Z., Pacyna-Gengelbach, M., & Petersen, I. (2001). Diversity of gene expression in adenocarcinoma of the lung. *Proceedings of the National Academy of Sciences, 98*(24), 13784–13789.

Gupta, N., Jadhav, K., & Shah, V. (2017). Emperipolesis, entosis and cell cannibalism: Demystifying the cloud. *Journal of Oral and Maxillofacial Pathology: JOMFP, 21*(1), 92.

Gutmann, D. H., & Kettenmann, H. (2019). Microglia/brain macrophages as central drivers of brain tumor pathobiology. *Neuron, 104*(3), 442–449.

Hardianto, H., Mahanal, S., Susanto, H., & Prabaningtyas, S. (2024). Protist literacy: A novel concept of protist learning in higher education. *Eurasia Journal of Mathematics, Science and Technology Education, 20*(2), em2399.

Hernández, J. M., & Podbilewicz, B. (2017). The hallmarks of cell-cell fusion. *Development, 144*(24), 4481–4495.

Ibrahim, A. F., Weirauch, U., Thomas, M., Grünweller, A., Hartmann, R. K., & Aigner, A. (2011). MicroRNA replacement therapy for miR-145 and miR-33a is efficacious in a model of colon carcinoma. *Cancer Research, 71*(15), 5214–5224.

Jemal, A., Siegel, R., & Ward, E. (2007). Cervical cancer. *CA: Cancer Journal for Clinicians, 57*, 43–66.

Kallinikova, V. D., Matekin, P. V., Ogloblina, T. A., Leikina, M. I., Kononenko, A. F., Sokolova, N. M., & Pogodina, L. S. (2001). Anticancer properties of flagellate protozoan Trypanosoma cruzi Chagas, 1909. *Biology Bulletin of the Russian Academy of Sciences, 28*, 244–255.

Kotrikadze, N., Khutsishvili, E., Alibegashvili, M., Ramishvili, L., Chigogidze, T., Gabunia, N., & Ahmad, S. (2024). Plastic orchiectomy and some indicators of improvement of vital parameters and general condition of the organism in blood of the men with prostate cancer. *Indian Journal of Biochemistry and Biophysics (IJBB), 61*(1), 39–47.

Krajcovic, M., & Overholtzer, M. (2012). Mechanisms of ploidy increase in human cancers: A new role for cell cannibalism. *Cancer Research, 72*(7), 1596–1601.

Linares, M. A., Zakaria, A., & Nizran, P. (2015). Skin cancer. *Primary Care: Clinics in Office Practice, 42*(4), 645–659.

Liu, F. H., Huang, J. Y., & Chang, Y. C. (2024). Dental visits and colon-rectum cancer: A nationwide population-based nested case-control study in Taiwan. *Journal of Dental Sciences, 19*(1), 601–606.

Lozupone, F., & Fais, S. (2015). Cancer cell cannibalism: A primeval option to survive. *Current Molecular Medicine, 15*(9), 836–841.

Marques, J. C., Fuziwara, C. S., Yamashita, A. S., Rezende, E., Da-Silva, M. J., & Kimura, E. T. (2009). Effects of let-7 microRNA on cell growth and differentiation of papillary thyroid cancer. *Translational Oncology, 2*(4), 236–241.

Martins, I., Raza, S. Q., Voisin, L., Dakhli, H., Law, F., De Jong, D., & Perfettini, J. L. (2017). Entosis: The emerging face of non-cell-autonomous type IV programmed death. *Biomedical Journal, 40*(3), 133–140.

Miller, F. R., McInerney, D., Rogers, C., & Miller, B. E. (1988). Spontaneous fusion between metastatic mammary tumor subpopulations. *Journal of Cellular Biochemistry, 36*(2), 129–136.

Mohamadi, F., Shakibapour, M., Sharafi, S. M., Andalib, A. R., Tolouei, S., & Yousofi Darani, H. (2019). Anti-toxoplasma gondii antibodies attach to mouse cancer cell lines but not normal mouse lymphocytes. *Biomedical Reports, 10*(3), 183–188.

Montoya, J., & Liesenfeld, O. (2004). Toxoplasmosis lancet 363, 1965–1976. *CrossRef| PubMed| CAS| Web of Science Times Cited, 807.*

Morii, E. (2017). Heterogeneity of tumor cells in terms of cancer-initiating cells. *Journal of Toxicologic Pathology, 30*(1), 1–6.

Muhammad, N., Hanif, M., & Yang, P. (2024). Beyond cisplatin: New frontiers in metallodrugs for hard-to-treat triple negative breast cancer. *Coordination Chemistry Reviews, 499*, 215507.

Niculescu, V. F. (2017). Development, cell line differentiation and virulence in the primitive stem–/progenitor cell lineage of Entamoeba. *Journal of Stem Cell Research and Medicine, 2*(1), 1–8.

Niculescu, V. F. (2018). Carcinogenesis: Recent insights in protist stem cell biology lead to a better understanding of atavistic mechanisms implied in cancer development. *MOJ Tumor Research, 1*(1), 18–29.

Nitzsche, R., Günay-Esiyok, Ö., Tischer, M., Zagoriy, V., & Gupta, N. (2017). A plant/fungal-type phosphoenolpyruvate carboxykinase located in the parasite mitochondrion ensures glucose-independent survival of toxoplasma gondii. *Journal of Biological Chemistry, 292*(37), 15225–15239.

Pedraza-Fariña, L. G. (2006). Cancer issue: Mechanisms of oncogenic cooperation in cancer initiation and metastasis. *The Yale Journal of Biology and Medicine, 79*(3–4), 95.

Pérez-Molina, J. A., & Molina, I. (2018). Chagas disease. *The Lancet, 391*(10115), 82–94.

Perou, C. M., Sørlie, T., Eisen, M. B., Van De Rijn, M., Jeffrey, S. S., Rees, C. A., & Botstein, D. (2000). Molecular portraits of human breast tumours. *Nature, 406*(6797), 747–752.

Plummer, M., de Martel, C., Vignat, J., Ferlay, J., Bray, F., & Franceschi, S. (2016). Global burden of cancers attributable to infections in 2012: A synthetic analysis. *The Lancet Global Health, 4*(9), e609–e616.

Porserud, A., Karlsson, P., Nygren-Bonnier, M., Aly, M., & Hagströmer, M. (2024). The feasibility of an exercise intervention after robotic-assisted radical cystectomy for urinary bladder cancer, prior to the CanMoRe trial. *Pilot and Feasibility Studies, 10*(1), 12.

Rodriguez, A., Suo, X., & Liu, D. (2024). Classification of medically important parasites. In *Molecular medical microbiology* (pp. 2907–2919). Academic Press.

Rolph, K. E., & Cavanaugh, R. P. (2022). Infectious causes of neoplasia in the domestic cat. *Veterinary Sciences, 9*(9), 467.

Romano, P. S., Akematsu, T., Besteiro, S., Bindschedler, A., Carruthers, V. B., Chahine, Z., & Vanrell, M. C. (2023). Autophagy in protists and their hosts: When, how and why? *Autophagy Reports, 2*(1), 2149211.

Sardinha-Silva, A., Alves-Ferreira, E. V., & Grigg, M. E. (2022). Intestinal immune responses to commensal and pathogenic protozoa. *Frontiers in Immunology, 13*, 963723.

Seyedeh, M. S., Nahid, S., Nahid, M., Shima, D. P., Morteza, Y., & Hossein, Y. D. (2015). Low titer of antibody against toxoplasma gondii may be related to resistant to cancer. *Journal of Cancer Research and Therapeutics, 11*(2), 305–307.

Shabo, I., Svanvik, J., Lindström, A., Lechertier, T., Trabulo, S., Hulit, J., & Pawelek, J. (2020). Roles of cell fusion, hybridization and polyploid cell formation in cancer metastasis. *World Journal of Clinical Oncology, 11*(3), 121.

Soto, A. M., & Sonnenschein, C. (2004). The somatic mutation theory of cancer: Growing problems with the paradigm? *BioEssays, 26*(10), 1097–1107.

Soto, A. M., & Sonnenschein, C. (2011). The tissue organization field theory of cancer: A testable replacement for the somatic mutation theory. *BioEssays, 33*(5), 332–340.

Sung, H., Ferlay, J., Siegel, R. L., Laversanne, M., Soerjomataram, I., Jemal, A., & Bray, F. (2021). Global cancer statistics 2020: GLOBOCAN estimates of incidence and mortality worldwide for 36 cancers in 185 countries. *CA: a Cancer Journal for Clinicians, 71*, 209–249.

Tacutu, R., Budovsky, A., Yanai, H., Wolfson, M., & Fraifeld, V. (2011). Immunoregulatory network and cancer-associated genes: Molecular links and relevance to aging. *Network Biology, 1*(2), 112.

Thomas, F., Ujvari, B., Renaud, F., & Vincent, M. (2017). Cancer adaptations: Atavism, de novo selection, or something in between? *BioEssays, 39*(8), 1700039.

Trigos, A. S., Pearson, R. B., Papenfuss, A. T., & Goode, D. L. (2017). Altered interactions between unicellular and multicellular genes drive hallmarks of transformation in a diverse range of solid tumors. *Proceedings of the National Academy of Sciences, 114*(24), 6406–6411.

Truffi, M., Sorrentino, L., & Corsi, F. (2020). Fibroblasts in the tumor microenvironment. *Tumor Microenvironment: Non-Hematopoietic Cells*, 15–29.

Vanrell, M. C., & Romano, P. S. (2023). Close encounters: pathogenic protists-host cell interactions. In: *Phagocytosis-Main Key of Immune System*. IntechOpen.

Walen, K. H. (2017). Mitotic slippage process concealed cancer-sought chromosome instability mechanism (S-CIN). *Journal of Cancer Therapy, 8*(6), 608–623.

Wang, P., Li, S., Zhao, Y., Zhang, B., Li, Y., Liu, S., & You, F. (2019). The GRA15 protein from toxoplasma gondii enhances host defense responses by activating the interferon stimulator STING. *Journal of Biological Chemistry, 294*(45), 16494–16508.

Weihua, Z., Lin, Q., Ramoth, A. J., Fan, D., & Fidler, I. J. (2011). Formation of solid tumors by a single multinucleated cancer cell. *Cancer, 117*(17), 4092–4099.

Weiler, J., & Dittmar, T. (2019). Cell fusion in human cancer: The dark matter hypothesis. *Cells, 8*(2), 132.

Weyer, G. (1930). Untersuchungen über die Morphologie und Physiologie des Formwechsels der Gastrostyla steinii Engelmann. *Arch. f. Protistenk, 71*, 139–228.

Yatabe, Y. (2024). Molecular pathology of non-small cell carcinoma. *Histopathology, 84*(1), 50–66.

Yilmaz, M., & Christofori, G. (2010). Mechanisms of motility in metastasizing cells. *Molecular Cancer Research, 8*(5), 629–642.

Young Han, C., Bedia, J. S., Yang, W. L., Hawley, S. J., Bergan, L., Hopper, M., & Bast, R. C., Jr. (2024). Autoantibodies, antigen-autoantibody complexes and antigens complement CA125 for early detection of ovarian cancer. *British Journal of Cancer*, 1–8.

Zheng, Z., Lu, X., Zhou, D., Deng, X. F., Liu, Q. X., Liu, X. B., & Dai, J. G. (2024). A novel enemy of cancer: Recent investigations into protozoan anti-tumor properties. *Frontiers in Cellular and Infection Microbiology, 13*, 1325144.

Zur Hausen, H. (2007). *Infections causing human cancer*. John Wiley & Sons.

Cancer Management and Therapies Related to Microorganisms

12

Hira Gul and Nosheen Masood

Abstract

This chapter presents an overview of the intricate relationship between microorganisms and cancer management and therapies. While traditionally studied in the context of parasitic illnesses, microorganisms, including bacteria, fungi, and protists, have garnered attention for their potential roles in cancer progression, treatment response, and even prevention. The discussion encompasses various aspects, from the impact of microbial infections on cancer risk to the emerging field of using microorganisms as therapeutic agents against malignancies. Key topics include the modulation of the tumor microenvironment by microorganisms, their effects on immune responses, and their potential involvement in carcinogenesis processes. Moreover, the chapter explores the mechanisms through which microorganisms interact with cancer cells, such as direct interactions, molecular signaling pathways, and manipulation of host cellular machinery. Additionally, it highlights recent advances in utilizing microorganisms for cancer detection, treatment, and prevention strategies, underscoring their potential to revolutionize cancer management paradigms. By shedding light on this evolving area of research, this chapter aims to stimulate further investigations and foster the development of innovative approaches to combat cancer using microbial-based interventions.

H. Gul (✉)
Microbiology and Biotechnology Research Lab, Department of Biotechnology, Fatima Jinnah Women University, Rawalpindi, Pakistan

N. Masood
Department of Biotechnology, Fatima Jinnah Women University, Rawalpindi, Pakistan

N. Masood, A. Yasmin (eds.), *The Microbiome and Cancer*,
https://doi.org/10.1007/978-981-97-6948-3_12

Keywords

Cancer management · Microorganisms · Tumor microenvironment · Carcinogenesis · Therapeutic agents · Cancer detection

12.1 What Makes Treating Cancer So Challenging?

We have spent billions of dollars on performing experiments, from the eradication of smallpox to the sequencing of the human genome, but we haven't found a solution for such a disease (cancer) that affects many people and their families at any given time. Cancer development occurs when normal body cells acquire mutations. Cellular defense systems often identify mutations and damage to DNA, resultantly either repair the damage or cause cell death. Certain types of mutations enable cancerous cells to propagate uncontrolled and thus infect surrounding tissues and can possibly metastasize to different organs (Nair et al., 2014). Once cancer spreads, it becomes nearly impossible to combat it. The onset of metastasis makes cancer extremely fatal. Over 100 distinct forms of cancer exist, and no miracle drug has yet been discovered for management of all of them. Cancer treatment options commonly consist of surgery for tumor removal, along with combination of radiation and chemotherapy to eradicate any remaining malignant cells. Sometimes, immunotherapy, hormone therapies, and targeted pharmaceuticals have been used to treat certain types of cancer. Healthy bodily tissues are disrupted with these therapies, which directly affect the entire body. Finding an alternate treatment for a fatal illness is desperately needed to save millions of lives. Therefore, complementary therapies such as biotherapy, gene therapy, and photodynamic therapy (PDT) have been suggested to enhance the efficacy of cancer therapy. Recently, the use of microorganisms (bacteria, virus, and fungi) in cancer prevention and treatment has significantly evolved (Rommasi, 2022).

12.2 Why Do Microorganisms Serve a Purpose in Cancer Management?

Bacteria account for more than 75% of Earth's population. Emerging research indicates that these microbes could potentially be able to aid in the battle against cancer, even though we generally think of them as something to be avoided (Shanmugaraj et al., 2020). Numerous characteristics of bacteria make them valuable for the treatment of cancer: They can readily penetrate tumors due to their unique nature of motility; they thrive in the immune-deficient and hypoxic conditions found in tumors; and they are capable to deplete nutrients from the tumor microenvironment that are essential for the survival of cancer cells (Laliani et al., 2020). In addition to these properties, easily modifiable genome and great ability to induce an immune response provide opportunity to scientists in exploring their role in cancer management. Genetically modified bacteria have been shown in various recent studies to enhance antitumor immunity or boost the efficacy of alternate cancer therapies, highlighting the likelihood that these tiny beings could play a significant role as big allies in the fight against cancer (Laliani et al., 2020).

Since viruses are widely recognized to function as effective vaccine agent, recently, some viruses are engineered for the treatment of cancer. Oncolytic viruses have been researched as novel virotherapy for cancer with the capacity to induce anticancerous immune response concomitant with lysis-dependent cytoreductive action. To further support the clinical efficacy of oncolytic virotherapy, there is mounting evidence that a tumor infection by an oncolytic virus opens a wide range of options for immunomodulating interventions, which includes dendritic cell-based vaccines, generic immunostimulant strategies, systemic chemotherapy, and antigenic libraries (Patyar et al., 2010).

Fungi can synthesize an array of enzymes and antibacterial compounds, thus rendering them valuable microbes in medical applications (Patyar et al., 2010). The promising feature of marine-derived fungi to synthesize a wide range of chemical compounds and enzymes that effectively inhibit the growth of cancerous cells is the reason that makes them a prospective source of anticancer drugs. Furthermore, the molecules released during the secondary metabolic process induces DNA fragmentation and change of cell morphology which leads to the apoptosis of tumor cells (Liang et al., 2022).

12.3 Historical Perspective of Bacteria as an Anticancer Agent

"If you don't like bacteria, you are on the wrong planet" quote by Stewart brand explains well the role of bacteria in all matters of life. About 150 years ago, the first reports were made regarding the link between bacteria and cancer. At that time, two German physicians, W. Busch and F. Fehleisen, independently noted that the symptoms of cancer patients had improved following an accidental erysipelas infection. Being the first medical physician to evaluate and inquire about these observations, Bush injected the cilia infection (erysipelas) in a cancer patient and observed him for a period (Liang et al., 2022). Almost a century ago, the therapeutic potential of bacteria as an anticancer agent came to light when an American physician William Coley observed that one of his patients with neck cancer started to recover following an erysipelas (bacterial infection of skin's dermis layer). Intentionally, he used mixture of bacteria *Streptococcus pyogenes* and *Bacillus prodigiosus*, also called *Serratia marcescens* (heat-killed Streptococcus and Serratia marcescens), to treat patients with soft tissue and bone sarcomas. That combination has been known as Coley's toxin in history. In the late twentieth century, bladder cancer has been effectively managed with a weakened strain of bacillus Calmette-Guérin (BCG) also known as *Mycobacterium bovis.* BCG via intramuscular (IM) injections significantly decrease the likelihood of cancer recurrence in bladder cancer patients. The only medication for the treatment of high-risk non-muscle invasive bladder cancer (NMIBC) that has received FDA authorization is BCG (Sedighi et al., 2019).

12.4 Role of Therapeutic Microorganisms in Cancer Prevention

12.4.1 Bacteriotherapy for Targeting Tumor

Advances in molecular biology and bacteriology have expanded the scope of bacteria-mediated cancer prevention and therapy. Bacterial endotoxins, which are lipopolysaccharides, have been tested as potential therapeutic products for cancer treatments (Sedighi et al., 2019). Tumors can be killed with bacterial toxins, and bacterial immunotoxins can serve as the basis for cancer vaccinations. Currently, researchers have been using bacteria as an agent to eradicate brain tumors with differing degrees of success (Din et al., 2023).

Microorganisms utilized for anticancer bacteriotherapy can either be living, attenuated, or even genetically modified bacteria. The exploitation of these forms of microbes in different cases has been associated with positive outcomes in cancer management (Duong et al., 2019).

Anticancer medicines can be synthesized and produced via anaerobic bacterial spores. Additionally, these bacteria play a significant role in drug delivery and as carriers for genes. Once spores reach an oxygen-deprived organ, they begin to germinate and proliferate and rapidly begin to exert their anticancer effect (Perez-Chanona & Jobin, 2014).

E. coli-mediated bacterial infections have shown the potential to inhibit tumor tissues. These infections cause differentiation of killer T cells+CD8 cells that are considered responsible for IFN-γ production and elevating the expression level of MHC class I on cancerous cells, thus boosting CD8 + T cell's uptake into tumor recognized sites. Improving the immune system of the body is the main task performed by bacteria in cancer management (Shen et al., 2023). There are multiple strategies to combat cancer cells by using bacteria, but the following are the most significant ones: (1) reducing the availability of nutrients required for the metabolism of cancer cells, (2) the capacity to proliferate in necrotic areas and induce hypoxic circumstances for the tumor, (3) using it as a vehicle for anticancer medication delivery, (4) release various substance, and (5) the formation of biofilms (Stern et al., 2015). Different bacterial strains that can target different tumor sites for cancer management are shown in Table 12.1.

Table 12.1 Bacterial strains in cancer treatment

Bacterial strain	Specialized function	Carcinoma category
Escherichia coli	Working as carrier	Nasopharyngeal
Salmonella typhimurium	Forming biofilms	Colon cancer
Pseudomonas aeruginosa	Immunity booster	Lung cancer non-small cell
Salmonella typhimurium	Colonizing and invading solid tumors	Breast carcinoma
Klebsiella pneumoniae	Releasing substances	Cervix cancer

12.4.2 Bacteriobots (Bacteria-Mediated Microrobots)

The notion of "bacterial robots" is a cutting-edge, entirely novel approach to cure cancer using bacteria. To boost bacterial efficacy as bacteriobots, they must be encapsulated in biocompatible and biodegradable microbeads that pose no harm to human cells. Utilizing suitable microbeads is essential to synthesize effective bacterial robots for targeting tumor, as these beads encapsulate the bacteria, thus shielding them from immune system opsonization and intrusions (Bo & Wang, 2023). The rate and process that trapped bacteria in the microbeads are released should be taken into consideration in the next step. To make the bacterial robot move with greater efficiency and to simply enable it for the agents to reach the target tissue, flagellum from flagellate bacteria should eventually be added to the microbial granules. Salmonella typhimurium was researched as one of the potent bacterial-based microrobot (Zong et al., 2023). In this approach, bacteria were encapsulated using biocompatible and biodegradable alginate granules and then the flagella of S. typhimurium were attached to the microbeads as a motility segment. Finaly, robotic bacteria will be able to effectively target the tumor sites in the body (Suri et al., 2023).

Another study demonstrates live nanobots containing nonpathogenic E. coli bacteria were used to treat cancer. This live robot was equipped with bio-nano sensor technology that was synthesized naturally and directly bonded to VEGF (vascular endothelial growth factor) (Guduru et al., 2023). According to the accessible data, robotic bacteria in cancerous cells have stronger cancerous tissue targeting, enhanced chemotactic motility, and swifter migration rates than healthy cells. Additionally, this novel generation of bacterial robots also serves as microsensors and micro-stimuli. Thus, they have the potential of delivering drugs and therapeutic nanoparticles to treat solid carcinomas (Shah et al., 2023). Figure 12.1 shows the binding of bacteriobots to target sites and their effect on immune cells.

12.4.3 Microbiota in Cancer Therapy

Numerous studies show evidence that the gastrointestinal tract's microbial population is essential for the suppression of tumors. The microbiota can suppress tumor cells by modifying the host immune system. Due to the release of anticancer substances, gastrointestinal bacteria exert their inhibitory function. Butyrate is one of the chemical compounds produced by bacteria (Elkrief et al., 2019). It inhibits the proliferation of colorectal cancer cells (CRC) via deactivating the histone deacetylase-dependent Akt/ERK signal pathway, both in vivo and in vitro, hence imparting its anticancer actions. Additionally, its ability to effectively combat lymphoma cancer cell lines has been documented. On the other hand, butyrate plays an effective role in reduction of intestinal inflammation by stimulating T-cell regulatory differentiation and reducing STAT3 and NF-κB pathway activity. Furthermore, butyrate regulates the signaling of carcinogenic molecules and reduces tumors by modulating methylation and microRNAs (Elkrief et al., 2019).

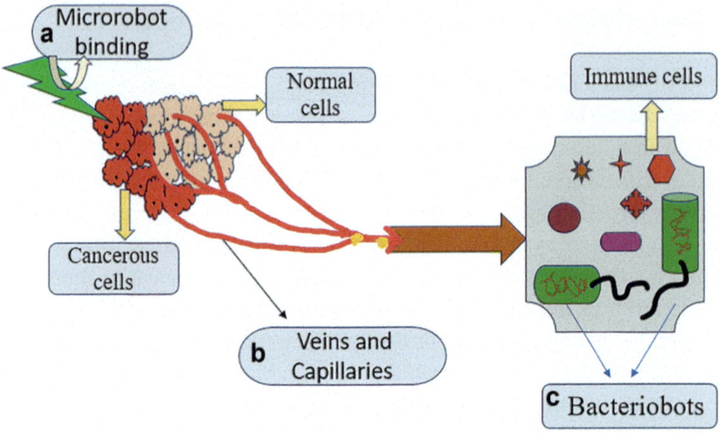

Fig. 12.1 Graphic in vivo depiction of bacteria used in bacteriotherapy: Bacteria precisely bind to tumor cells while remain unbound to normal cells (**a**), bacterial motility through veins and capillaries (**b**), and stimulation of immune cells to respond against cancerous cells (**c**)

The C-jun N-terminal kinase (JNK) pathway can be activated by ferric generated by Lactobacillus casei, which can cause tumor cells to undergo apoptosis. JNK-induced DNA damage show tendency to kill cancer cells (Inamura, 2020). Numerous studies revealed that lactobacilli may trigger the immune system against tumors invasion by affecting the Th1 response, dendritic cells, and NK cells, directly leading to the elimination of precancerous or cancerous cells. In this way, they can combat carcinomas and activate host immune cells (Inamura, 2020).

Bacterial lipopolysaccharide (LPS), a substance mostly found in gram-negative bacteria's outer cell membrane, has potential to act on host cell receptors' surfaces to activate them (Matson et al., 2021). Like the TLR4 (toll-like receptor 4), a member of the PRRs (pattern recognition receptor) family, these receptors are stimulated by bacterial structures, thereby inducing the T-cell-mediated response to carcinogenic cells (Gharaibeh & Jobin, 2018).

12.4.4 Gene Therapy for Cancer: Bacteria as Carriers

Bacteria can aid in the process of gene therapy by carrying a particular gene of interest and the induction of target protein synthesis in the cancer-ridden environment. Cytotoxic peptides, anticancer drugs, therapeutic proteins, prodrug-to-drug-converting enzymes in solid tumors, and even coding and noncoding delivery of gene can all be administered by bacteria as a compatible transmitter or vector (Cecchin et al., 2023). Balancing the toxicity of bacteria with the dose required to achieve their effectiveness has proven to be one of the most significant challenges in using them as anticancer agents as these bacteria can be quite hazardous to the host at large dosages (Chakrabarty, 2016). However, the selective targeting of solid tumors for direct treatment poses another significant challenge in cancer gene therapy. The use of genetically engineered bacteria to express proteins as a vector for gene therapy in cancer has been shown as one approach in overcoming such limitations (Ebrahimzadeh et al., 2021).

Certain bacteria species have demonstrated an extraordinary capacity to recognize, target, and eliminate malignancies, including *Escherichia coli, Clostridium, Lactococcus, Shigella, Bifidobacteria, Listeria, Vibrio*, and *Salmonella.* Their tumoricidal effects seemed to be robust and potent (Quillien et al., 2023).

12.4.5 Bacterial Application in Vaccines

Of all the cancer immunotherapy approaches, oral DNA vaccination is one of the most promising methods. However, the primary factor limiting the efficacious immune response included low transfer levels of DNA. A novel oral DNA vaccination technique for efficient cancer immunotherapy that is mediated by bacteria coated in cationic nanoparticles is discovered (Quillien et al., 2023). Through the incorporation of cationic CP nanoparticles to attenuated Salmonella, a protective layer of nanoparticles is formed around the bacteria that not only substantially improves the bacteria's ability to cope with acid and enzymes in the stomach and intestines yet also safeguards Salmonella from phagosomes (Mohan et al., 2022).

Oral administration of the DNA vaccine encoding the autologous VEGFR2 via the hybrid vector depicts significant T-cell activation and cytokine synthesis and dissemination. The most important aspect of this approach is that it efficiently inhibits tumor proliferation, which suppresses tumor necrosis and angiogenesis in tumor vessels (Li et al., 2023a).

As evidenced by this action, encapsulating live attenuated bacteria with polymer nanoparticles is a viable approach to comprehensive, reliable, and effective engineering of DNA vaccine vector. A multitude of microbiological choices enable the delivery, customization, and utilization of nanoparticles in a wide range of medical applications (Bo & Wang, 2023).

12.4.6 Bacterial Peptides and Enzymes in Cancer Therapy

See Table 12.2.

12.5 Role of Viruses in Cancer Management

Virotherapy is the application of viruses that can specifically locate and kill tumor cells through various ways without impacting healthy cells as a means of treating cancer. This technique applies biotechnology to reprogram viruses so they can be used as therapeutic agents to cure cancer (Alemany, 2013). Three primary categories of virotherapy are available: (a) virus vectors for gene therapy, (b) anticancer oncolytic viruses, and (c) viral immunotherapy. All these strategies are based on

Table 12.2 Numerous varieties of antibacterial and anticancer bacterial peptides and enzymes (Chinnadurai et al., 2023)

Bacterial peptides	Bacterial species	Cytotoxic function
Halitoralins	*Halobacillus litoralis YS3106* (isolated from sea)	Antitumor action against human gastric cancer cells
Arenamides	*Salinispora arenicola* (cyclohexadepe peptide)	Poses cytotoxic impact on different cancer cells including human colorectal carcinoma HCT-116 by inhibition of activation of TNF cleavage pathways and directly affects NF-kappa B (NF-κB) activity (Chinnadurai et al., 2023)
Lucentamycins	*Nocardiopsis lucentensis CNR-712* (isolated from marine actinomycete)	HCT-116-human colon cancer cells
Urukthapelstatin	*Mechercharimyces asporophorigenens YM11-542* (from a marine bacterium)	Urukthapelstatin is a widely recognized cyclic thiopeptide antibiotic that works against human colorectal carcinoma cell line, HCT-116 (Ghaly et al., 2023)
Idoglobomides A and B	*Bacillus licheniformis* (produced by marine bacteria)	Inhibit growth of human gastric cell lines
Mixirins	*Bacillus* species (marine bacterium)	Cyclic acetate peptide has three significant members, maxirin A, B, and C, that shows potential in inhibiting the growth of human colon HCT 116 cancer
Pep27anal2	*Streptococcus pneumoniae*	Pep27ana is composed of 27 amino acids. It has antimicrobial as well as anticancer properties. It can slow down the growth in SNU-601-gastric cancer cell lines (Karpiński & Adamczak, 2018)

(continued)

Table 12.2 (continued)

Bacterial peptides	Bacterial species	Cytotoxic function
Entap	Produced by *Enterococcal* strains	Entap works against cancer by preventing the growth of new cells, terminating the cell cycle at the G1 stage, and triggering autophagic death. It shows cytotoxic effect against human colorectal adenocarcinoma cells-HT-29 and AGS-gastric adenocarcinoma cells (Karpiński & Adamczak, 2018)
Helicobacter pylori (H. pylori) ribosomal protein	HPRP-A1 and HPRP-A2 (enantiomer isomer) is derived from the ribosomal protein L1 (RPL1) N-terminus of *Helicobacter pylori*	Executes an action by (1) activation of the associated caspase-3, caspase-8, and caspase-9 pathway, which results in apoptosis, and (2) ability to stop cellular growth by blocking the cell cycle at G0/G1 and G2/M, interfering with mitochondrial function, and promoting the production and concentration of ROS (reactive oxygen species). Poses anticancer activity against human gastric cancer cells (SGC-7901 & BGC-823) and HepG2 (human liver cancer cell line) (Kordi et al., 2023)
Bacterial enzyme		
ADI (arginine deiminase)	*Mycoplasma arginine or Mycoplasma hominis*	The PEGylated form of the enzyme arginine deaminase is ADI-PEG20. It hinders the growth and proliferation of cancer cells, reduces metabolism and biological activity of tumor cells, induces caspase-independent apoptosis, and stimulates autophagy to exert its anticancer effect. In several types of cancer cell lines, primarily liver cell carcinoma cell lines, this enzyme exhibits exceptional therapeutic potential (Lath et al., 2023)

therapeutic techniques, such as overexpressing the targeted genes, using RNA techniques to silence or reduce the expression of cancerous genes called gene knockout, and utilizing viruses as vectors in delivering the gene that triggers tumor cells to undergo apoptosis and die—a process known as "suicide gene delivery" (Bifulco et al., 2023).

12.5.1 Tumor Targeting-Oncolytic Virus

Oncolytic viruses cause membrane ruptures and destruction in tumorous cells, resultantly releasing antigens that the immune system can quickly recognize and then launch an attack on the cancer cells. This is the mechanism by which these viruses generate their anticancer operation (Breitbach et al., 2011). Currently, oncolytic viruses have gained increasing attention as therapeutic agents regarding the management and handling of the symptoms of cancer (Table 12.3) (Breitbach et al., 2016). As an example, the measles virus shows the tendency to target the CD46 that exhibits overexpression on various types of myeloma cells. In comparison with non-tumorous cells, tumor cells are distinguished by uncontrolled metabolism, rapid growth, and division, which makes it possible for many viruses to preferentially target those cells for replication (Fukuhara et al., 2016).

Furthermore, antiviral responses such type I and II interferons (IFNs) or tumor necrosis factor (TNF) are frequently ineffective against cancer cells. The 80 Newcastle disease virus (NDV), also called RNA oncolytic virus despite being a malignant virus, has been recognized for having the ability to selectively infect tumor cells and displays anticancer impact (Breitbach et al., 2016). Another oncovirus, the hepatitis B virus (HBV), also shows anticancer effectiveness through upregulating TGF-β, downregulating the intracellular level of tumor-suppressor protein p53, and elevating the expression of CCNE1, TERT, and MLL4 (Guse et al., 2011).

Another example is an oncolytic adenovirus, which most likely induces cancer by invading the host cells and using their vital processes to inject its own nucleic acid (Guse et al., 2011). Despite this, this virus is like the NDV in such a way that it is utilized extensively in cancer prevention and management and mostly requires

Table 12.3 List of oncolytic viruses and their phase of clinical trials

Family of virus	Name of virus	Phase of clinical trial	Specificity in cancer management
Paramyxoviridae	MN-NIS	Phase I/II	Ovarian myeloma, fallopian tube transitional cells, and peritoneal carcinoma
Parvoviridae	ParvOryx(H-1PV)	I and II	Pancreatic and glioblastoma carcinoma
Poxviridae	Pexa-Vec (JX-594)	II and III	Melanoma and hepatocellular
Herpesviridae	G207	I and II	Astrocytoma and glioma
	rQNestin 34.5	I	Glioblastoma, glioma, and astrocytoma
	Talimogene	III	Head and neck cancer, squamous cell, pancreatic, and melanoma
	G47	II	Prostate and glioblastoma cancer
Adenoviridae	CG0070	III	Non-muscle invasive bladder cancer
	ICOVIR-5	I	Melanoma
	VCN 01	I	Pancreatic adenocarcinoma
	DNX 2440	I	Breast and colorectal
	TILT 123	I	Melanoma and solid tumors
	LOAd-703	I and II	Ovarian, biliary, and pancreatic
	H101	III	Hepatocellular
	Delta 24 RGD	I	Glioblastoma
Reoviridae	Reolysin	II and III	Lung cancer, ovarian cancer, glioblastoma, and squamous cells of neck and head

engineering and genetical modification. Some of them have been genetically engineered to target specific cancer cell receptors (Javid et al., 2023).

RIGVIR, a nonpathogenic intestinal cytopathic human orphan virus, was the first oncolytic virus authorized by regulatory authorities for cancer therapy. It was approved in Latvia in 2004 for the treatment of melanoma (Lawler et al., 2017). Despite having received approval in China in 2005, the modified adenovirus H101 (Oncorine, recombinant human adenovirus type 5 injection) has not currently been acknowledged globally for its therapeutic effectiveness (Li et al., 2023b).

The US Food and Drug Administration (FDA) licensed T-VEC for melanoma in 2015, indicating it as the first oncolytic viral immunotherapy (Lundstrom, 2023). Approved for subgroups of melanoma patients, T-VEC (Imlygic®) is a modified herpes simplex virus (HSV) that releases the immune-stimulating GM-CSF protein in cancer cells with a reduced risk of harming healthy cells. It encourages the death of tumor cells by infecting them and it is allowed for melanoma patients (Mondal et al., 2020).

12.5.2 Virus Immunotherapy

The concept of viral immunotherapy includes the use of viruses as carriers of immune-stimulating agents, such as antigens specific to tumors, to enable the host immune system and to identify and combat tumor cells (Prestwich et al., 2008). One of the most prevalent viruses utilized as a delivery system for cancer immunotherapy is 78 oncolytic adenoviruses (Shi et al., 2020). Cancer vaccines are utilized in conjunction with adjuvants which are widely recognized as materials or cells that can support immune response. One cancer vaccine used to treat advanced prostate cancer is called Sipuleucel-T (Provenge) (Tur-Planells et al., 2023). Another vaccination approved for the treatment of advanced melanoma skin cancer is talimogene laherparepvec. Chimeric antigen receptor T cells (CART) are also considered as another form of viral immunotherapy (Woller et al., 2014). This approach relies on the use of T cells that have been genetically altered to generate a synthetic T-cell receptor, which kills carcinogenic cells (Xu et al., 2023b). To accomplish this, after purification, T cells will be transduced to the vector (viral) with a gene that encodes the genetically engineered chimeric antigen receptor. To ensure safety, the most frequently utilized viral vector for this purpose is one that is produced by partially deleting the U3 region of lentiviral and incorporating gammaretrovirus into it (Zheng et al., 2019).

The combination of conventional therapies with virotherapy agents is another approach that generates stronger anticancer effects without cross-resistance and enables the usage of low virus dosage, thus reducing the toxicity of virus for normal cells (Kelly & Russell, 2007).

12.6 Fungi as an Anticancer Agent

The potential of marine-derived fungi as a source of anticancer drugs lies in their capacity to synthesize a variety of chemicals and enzymes that efficiently inhibit the growth of cancer cells. L-asparaginase works by breaking down asparagine, the primary component of cancer cells (Chen et al., 2013). The substances synthesized during the secondary metabolic process cause DNA fragmentation and can alter cell shape, which causes cancer cells to die.

Fungi can produce a wide range of enzymes, antibacterial and anticancer compounds, which makes them useful in therapeutics (Fig. 12.2). The fungi that yield the greatest number of anticancer compounds are *Aspergillus niger*, *A. fumigatus*, *A. oryzae*, *A. flavus*, *A. versicolor*, *A. terreus*, *Penicillium citrinum*, *P. chrysogenum*, and *P. polonicum*. These fungi have been used to examine their anticancer activity against cancers of the pancreatic cancer, uterine cervix, colorectal cancer, breast, ovarian, and colon cancer (Deshmukh et al., 2018).

According to reports, fungi produce a higher quantity of antibacterial and anticancer chemicals than bacteria (Gohar et al., 2020). However, the genes that group together in a genetic package known as BGCs (biosynthetic gene clusters) are responsible for these production efficiencies and the functions of the active chemicals. However, marine fungi are a rich source of bioactive substances and have produced a variety of anticancer substances (How et al., 2022).

Asparaginase, an enzyme that has been isolated from several different sources, including *A. niger*, *A. tubingensis*, *A. terreus*, *Fusarium* sp., *Talaromyces*

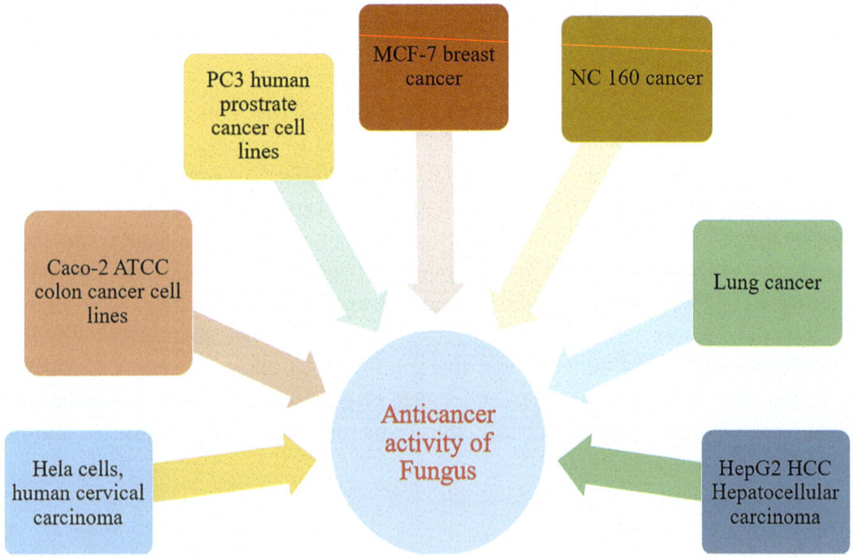

Fig. 12.2 Therapeutic application of fungi

pinophilus, Trichosporon asahii, Trichoderma viride, and *Sarocladium strictum*, has been shown to have the greatest anticancer properties (Kornienko et al., 2015).

Furthermore, from the secondary metabolic production of marine fungi, a variety of novel anticancer products have been identified and isolated (Noman et al., 2021). These include Gliotoxin, which was isolated from *Aspergillus* sp., Versicolactone B from *A. terreus*, Preussin from *A. candidus*, Patulin and Deoxytryptoquivaline from *A. giganteus*, and Octadecanoic from *Aspergillus* sp. (Saftien et al., 2023).

The isolated chemical substance from *P. concentricum's* secondary metabolism showed anticancer efficacy against a line of human Caucasian colon adenocarcinoma cells (Shamsaei et al., 2021). These substances included 3-hydroxy, benzenemethanol, 2-bromogentisyl alcohol, 2-chlorogentisyl alcohol, and epoxydon 6-dehydroxy-6-bromogabosine C. Against the melanoma cell line and the noncancerous keratinocyte cell line, the pentacyclic decalinoylspirotetramic acid and pyrenosetin D from *Pyrenochaetopsis* sp. displayed anticancer behavior (Xu et al., 2023a).

12.7 Conclusion

Although chemotherapy and radiotherapy have been considered as the cornerstone of cancer treatment, their adverse effects are linked to a range of sensitivities in cancer patients. As a result, bacteria-mediated cancer treatment has garnered a lot of attention in context with earlier anticancer therapies that were ineffective and showed limited effects. Bacteriotherapy is a revolutionary method in cancer treatment that can be used independently or in conjunction with standard therapeutic drugs to enhance their benefits. Under the right circumstances, viruses have the potential to kill tumor tissue in cancer patients. It is noteworthy that for some human pathogens, the damage induced to tumors is much greater than the harm inflicted to normal host tissues. Most viruses can now be engineered or adapted to eliminate their pathogenicity without diminishing their oncolytic potency. Attenuated versions of prevalent human pathogens make up most oncolytic viruses that are currently undergoing clinical testing. They have been genetically modified in recent years to mitigate their toxicity, boost their oncolytic effectiveness, or optimize their selectivity for cancerous cells. Considering the remarkable diversity of fungal species and the structures (metabolites) they make, it is astonishing that relatively few of these drugs have progressed to clinical trials, especially considering the critical need for novel cancer chemotherapeutics. Furthermore, this field of natural product research still has a lot to offer the field of anticancer research, estimations that only a small percentage of fungi have been investigated, along with technical challenges associated with the cultivation and biochemical-guided isolation of metabolites, meaning that this area of research still has much to offer.

References

Alemany, R. (2013). Viruses in cancer treatment. *Clinical and Translational Oncology, 15*, 182–188.

Bifulco, M., Di Zazzo, E., Napolitano, F., Malfitano, A. M., & Portella, G. (2023). History of how viruses can fight cancer: From the miraculous healings to the approval of oncolytic viruses. *Biochimie, 206*, 89–92.

Bo, Y., & Wang, H. (2023). Biomaterial-based in situ cancer vaccines. *Advanced Materials*, 2210452.

Breitbach, C. J., De Silva, N. S., Falls, T. J., Aladl, U., Evgin, L., Paterson, J., Sun, Y. Y., Roy, D. G., Rintoul, J. L., & Daneshmand, M. (2011). Targeting tumor vasculature with an oncolytic virus. *Molecular Therapy, 19*(5), 886–894.

Breitbach, C. J., Lichty, B. D., & Bell, J. C. (2016). Oncolytic viruses: Therapeutics with an identity crisis. *eBioMedicine, 9*, 31–36.

Cecchin, R., Troyer, Z., Witwer, K., & Morris, K. V. (2023). Extracellular vesicles: The next generation in gene therapy delivery. *Molecular Therapy, 31*, P1225.

Chakrabarty, A. M. (2016). Bacterial azurin in potential cancer therapy. *Cell Cycle, 15*(13), 1665–1666.

Chen, J., Zhang, X. D., & Jiang, Z. (2013). The application of fungal beta-glucans for the treatment of colon cancer. *Anti-Cancer Agents in Medicinal Chemistry (Formerly Current Medicinal Chemistry-Anti-Cancer Agents), 13*(5), 725–730.

Chinnadurai, R. K., Khan, N., Meghwanshi, G. K., Ponne, S., Althobiti, M., & Kumar, R. (2023). Current research status of anti-cancer peptides: Mechanism of action, production, and clinical applications. *Biomedicine & Pharmacotherapy, 164*, 114996.

Deshmukh, S. K., Prakash, V., & Ranjan, N. (2018). Marine fungi: A source of potential anticancer compounds. *Frontiers in Microbiology, 8*, 2536.

Din, S. R. U., Saeed, S., Khan, S. U., Arbi, F. M., Xuefang, G., & Zhong, M. (2023). Bacteria-driven cancer therapy: Exploring advancements and challenges. *Critical Reviews in Oncology/Hematology*, 104141.

Duong, M. T.-Q., Qin, Y., You, S.-H., & Min, J.-J. (2019). Bacteria-cancer interactions: Bacteria-based cancer therapy. *Experimental & Molecular Medicine, 51*(12), 1–15.

Ebrahimzadeh, S., Ahangari, H., Soleimanian, A., Hosseini, K., Ebrahimi, V., Ghasemnejad, T., Soofiyani, S. R., Tarhriz, V., & Eyvazi, S. (2021). Colorectal cancer treatment using bacteria: Focus on molecular mechanisms. *BMC Microbiology, 21*(1), 1–12.

Elkrief, A., Derosa, L., Zitvogel, L., Kroemer, G., & Routy, B. (2019). The intimate relationship between gut microbiota and cancer immunotherapy. *Gut Microbes, 10*(3), 424–428.

Fukuhara, H., Ino, Y., & Todo, T. (2016). Oncolytic virus therapy: A new era of cancer treatment at dawn. *Cancer Science, 107*(10), 1373–1379.

Ghaly, G., Tallima, H., Dabbish, E., Badr ElDin, N., Abd El-Rahman, M. K., Ibrahim, M. A., & Shoeib, T. (2023). Anti-cancer peptides: Status and future prospects. *Molecules, 28*(3), 1148.

Gharaibeh, R. Z., Jobin, C. (2018). Microbiota and cancer immunotherapy: in search of microbial signals. Gut:gutjnl-2018-317220.

Gohar, U. F., Mukhtar, H., Mushtaq, A., Farooq, A., Saleem, F., Hussain, M. A., & Ghani, M. U. (2020). Fungi: A potential source of biopharmaceuticals. *Agrobiological Records, 2*, 49–62.

Guduru, V. J., Mulaparthi, A. R., Karumuri, L. C., Gariskuti, H., & Cholla, S. S. (2023). DNA Nanorobots in cancer therapy-a review. *Journal of Pharma Insights and Research, 1*(2), 001–006.

Guse, K., Cerullo, V., & Hemminki, A. (2011). Oncolytic vaccinia virus for the treatment of cancer. *Expert Opinion on Biological Therapy, 11*(5), 595–608.

How, C. W., Ong, Y. S., Low, S. S., Pandey, A., Show, P. L., & Foo, J. B. (2022). How far have we explored fungi to fight cancer? *Seminars in Cancer Biology*, 976–989. Elsevier.

Inamura, K. (2020. Elsevier). Roles of microbiota in response to cancer immunotherapy. *Seminars in Cancer Biology*, 164–175.

Javid, H., Sharbaf Mashhad, A., Yazdani, S., Akbari Oryani, M., Akbari, S., Rezagholinejad, N., Tajaldini, M., & Karimi-Shahri, M. (2023). The role of viruses in cancer development versus cancer therapy: An oncological perspective. *Cancer Medicine, 12*, 11127.

Karpiński, T. M., & Adamczak, A. (2018). Anticancer activity of bacterial proteins and peptides. *Pharmaceutics, 10*(2), 54.

Kelly, E., & Russell, S. J. (2007). History of oncolytic viruses: Genesis to genetic engineering. *Molecular Therapy, 15*(4), 651–659.

Kordi, M., Borzouyi, Z., Chitsaz, S., Hadi Asmaei, M., Salami, R., & Tabarzad, M. (2023). Antimicrobial peptides with anticancer activity: Today status, trends and their computational design. *Archives of Biochemistry and Biophysics, 733*, 109484.

Kornienko, A., Evidente, A., Vurro, M., Mathieu, V., Cimmino, A., Evidente, M., van Otterlo, W. A., Dasari, R., Lefranc, F., & Kiss, R. (2015). Toward a cancer drug of fungal origin. *Medicinal Research Reviews, 35*(5), 937–967.

Laliani, G., Sorboni, S. G., Lari, R., Yaghoubi, A., Soleimanpour, S., Khazaei, M., Hasanian, S. M., & Avan, A. (2020). Bacteria and cancer: Different sides of the same coin. *Life Sciences, 246*, 117398.

Lath, A., Santal, A. R., Kaur, N., Kumari, P., & Singh, N. P. (2023). Anti-cancer peptides: Their current trends in the development of peptide-based therapy and anti-tumor drugs. *Biotechnology and Genetic Engineering Reviews, 39*(1), 45–84.

Lawler, S. E., Speranza, M.-C., Cho, C.-F., & Chiocca, E. A. (2017). Oncolytic viruses in cancer treatment: A review. *JAMA Oncology, 3*(6), 841–849.

Li, C.-X., Qi, Y., Chen, Y., Zhang, Y., Li, B., Feng, J., & Zhang, X.-Z. (2023a). Tuning bacterial morphology to enhance anticancer vaccination. *ACS Nano, 17*(9), 8815–8828.

Li, M., Zhang, M., Ye, Q., Liu, Y., & Qian, W. (2023b). Preclinical and clinical trials of oncolytic vaccinia virus in cancer immunotherapy: A comprehensive review. *Cancer Biology & Medicine, 20*(9), 646.

Liang, S., Wang, C., Shao, Y., Wang, Y., Xing, D., & Geng, Z. (2022). Recent advances in bacteria-mediated cancer therapy. *Frontiers in Bioengineering and Biotechnology, 10*, 1880.

Lundstrom, K. (2023). Viral vectors in gene therapy: Where do we stand in 2023? *Viruses, 15*(3), 698.

Matson, V., Chervin, C. S., & Gajewski, T. F. (2021). Cancer and the microbiome—Influence of the commensal microbiota on cancer, immune responses, and immunotherapy. *Gastroenterology, 160*(2), 600–613.

Mohan, C. D., Rangappa, S., Nayak, S. C., Jadimurthy, R., Wang, L., Sethi, G., Garg, M., & Rangappa, K. S. (2022). Bacteria as a treasure house of secondary metabolites with anticancer potential. *Seminars in Cancer Biology*, 998–1013. Elsevier.

Mondal, M., Guo, J., He, P., & Zhou, D. (2020). Recent advances of oncolytic virus in cancer therapy. *Human Vaccines & Immunotherapeutics, 16*(10), 2389–2402.

Nair, N., Kasai, T., & Seno, M. (2014). Bacteria: Prospective savior in battle against cancer. *Anticancer Research, 34*(11), 6289–6296.

Noman, E., Al-Shaibani, M. M., Bakhrebah, M. A., Almoheer, R., Al-Sahari, M., Al-Gheethi, A., Radin Mohamed, R. M. S., Almulaiky, Y. Q., & Abdulaal, W. H. (2021). Potential of anti-cancer activity of secondary metabolic products from marine fungi. *Journal of Fungi, 7*(6), 436.

Patyar, S., Joshi, R., Byrav, D. P., Prakash, A., Medhi, B., & Das, B. (2010). Bacteria in cancer therapy: A novel experimental strategy. *Journal of Biomedical Science, 17*, 1–9.

Perez-Chanona, E., & Jobin, C. (2014). From promotion to management: The wide impact of bacteria on cancer and its treatment. *BioEssays, 36*(7), 658–664.

Prestwich, R. J., Harrington, K. J., Pandha, H. S., Vile, R. G., Melcher, A. A., & Errington, F. (2008). Oncolytic viruses: A novel form of immunotherapy. *Expert Review of Anticancer Therapy, 8*(10), 1581–1588.

Quillien, L., Buscail, L., & Cordelier, P. (2023). Pancreatic cancer cell and gene biotherapies: Past, present, and future. *Human Gene Therapy, 34*(3-4), 150–161.

Rommasi, F. (2022). Bacterial-based methods for cancer treatment: What we know and where we are. *Oncology and Therapy, 10*(1), 23–54.

Saftien, A., Puschhof, J., & Elinav, E. (2023). Fungi and cancer. *Gut, 72*, 1410.

Sedighi, M., Zahedi Bialvaei, A., Hamblin, M. R., Ohadi, E., Asadi, A., Halajzadeh, M., Lohrasbi, V., Mohammadzadeh, N., Amiriani, T., & Krutova, M. (2019). Therapeutic bacteria to combat cancer; current advances, challenges, and opportunities. *Cancer Medicine, 8*(6), 3167–3181.

Shah, S., Famta, P., Srinivasarao, D. A., Shahrukh, S., Jain, N., Vambhurkar, G., Singh, S. B., & Srivastava, S. (2023). Unravelling the potential of microbots in cancer therapy. *Applied Materials Today, 34*, 101887.

Shamsaei, S., Getso, M., Ahmadikia, K., Yarahmadi, M., Farahani, H., Aslani, R., Mohammadzade, A., Raissi, V., Soleimani, A., & Arghavan, B. (2021). Recent findings on the role of fungal products in the treatment of cancer. *Clinical and Translational Oncology, 23*, 197–204.

Shanmugaraj, B., Priya, L. B., Mahalakshmi, B., Subbiah, S., Hu, R.-M., Velmurugan, B. K., & Baskaran, R. (2020). Bacterial and viral vectors as vaccine delivery vehicles for breast cancer therapy. *Life Sciences, 250*, 117550.

Shen, X., Zhu, C., Liu, X., Zheng, H., Wu, Q., Xie, J., Huang, H., Liao, Z., Shi, J., & Nan, K. (2023). Engineered bacteria for augmented in situ tumor vaccination. *Biomaterials Science, 11*, 1137.

Shi, T., Song, X., Wang, Y., Liu, F., & Wei, J. (2020). Combining oncolytic viruses with cancer immunotherapy: Establishing a new generation of cancer treatment. *Frontiers in Immunology, 11*, 683.

Stern, C., Kasnitz, N., Kocijancic, D., Trittel, S., Riese, P., Guzman, C. A., Leschner, S., & Weiss, S. (2015). Induction of CD 4+ and CD 8+ anti-tumor effector T cell responses by bacteria mediated tumor therapy. *International Journal of Cancer, 137*(8), 2019–2028.

Suri, K., D'Souza, A., Huang, D., Bhavsar, A., & Amiji, M. (2023). Bacterial extracellular vesicle applications in cancer immunotherapy. *Bioactive Materials, 22*, 551–566.

Tur-Planells, V., García-Sastre, A., Cuadrado-Castano, S., & Nistal-Villan, E. (2023). Engineering non-human RNA viruses for cancer therapy. *Vaccine, 11*(10), 1617.

Woller, N., Gürlevik, E., Ureche, C.-I., Schumacher, A., & Kühnel, F. (2014). Oncolytic viruses as anticancer vaccines. *Frontiers in Oncology, 4*, 188.

Xu, J., Zeng, Y., Yu, C., Xu, S., Tang, L., Zeng, X., Huang, Y., Sun, Z., Xu, B., & Yu, T. (2023a). Visualization of the relationship between fungi and cancer from the perspective of bibliometric analysis. *Heliyon, 9*, 8.

Xu, L., Sun, H., Lemoine, N. R., Xuan, Y., & Wang, P. (2023b). Oncolytic vaccinia virus and cancer immunotherapy. *Frontiers in Immunology, 14*, 1324744.

Zheng, M., Huang, J., Tong, A., & Yang, H. (2019). Oncolytic viruses for cancer therapy: Barriers and recent advances. *Molecular Therapy-Oncolytics, 15*, 234–247.

Zong, R., Ruan, H., Liu, C., Fan, S., & Li, J. (2023). Bacteria and bacterial components as natural bio-Nanocarriers for drug and gene delivery Systems in Cancer Therapy. *Pharmaceutics, 15*(10), 2490.

SPRINGER NATURE

GPSR Compliance

The European Union's (EU) General Product Safety Regulation (GPSR) is a set of rules that requires consumer products to be safe and our obligations to ensure this.

If you have any concerns about our products, you can contact us on ProductSafety@springernature.com

In case Publisher is established outside the EU, the EU authorized representative is:

Springer Nature Customer Service Center GmbH
Europaplatz 3
69115 Heidelberg, Germany

The manufacturer's authorised representative in the EU is Springer Nature Customer Service Centre GmbH, Europaplatz 3, 69115 Heidelberg, Germany. If you have any concerns regarding our products, please contact ProductSafety@springernature.com

Printed and bound by CPI Group (UK) Ltd, Croydon, CR0 4YY

30/09/2026

02225253-0009